Home Technology Integration Fundamentals and Certification

Cisco Learning Institute
and
Academic Business Consultants

PEARSON
Prentice
Hall

Upper Saddle River, New Jersey
Columbus, Ohio

Editor in Chief: Stephen Helba
Assistant Vice President and Publisher: Charles E. Stewart, Jr.
Production Editor: Alexandrina Benedicto Wolf
Design Coordinator: Diane Ernsberger
Cover Designer: Tim Hunter
Production Manager: Matt Ottenweller
Marketing Manager: Ben Leonard

This book was set in Palatino by The GTS Companies. It was printed and bound by Courier/Kendallville.
The cover was printed by Phoenix Color Corp.

Pearson Education Ltd.   Pearson Education Australia Pty. Limited
Pearson Education Singapore Pte. Ltd.   Pearson Education North Asia Ltd.
Pearson Education Canada, Ltd.   Pearson Educación de Mexico, S.A. de C.V.
Pearson Education–Japan   Pearson Education Malaysia Pte. Ltd.

10 9 8 7 6 5 4 3 2 1
ISBN 0-13-114826-5

Preface

Dozens of individuals participate in the construction or retrofitting of a home or small commercial building. Electricians, cabling technicians, security installers, low-voltage experts, and home entertainment installers all participate in the development of the modern home's "smart" installations. The convergence of these fields demands the skills of a new type of technician: one that can "integrate" all of these diverse technologies into one systematic deliverable solution. One of the best ways to build technical skills is to participate in industry-based, hands-on training.

The Cisco Learning Institute is a 501 (C) (3) public benefit corporation founded and originally funded by Cisco Systems. The Institute is dedicated to enhancing the way students learn and teachers teach using technology. Using the basics of a curriculum provided by Cisco Systems, the Cisco Learning Institute developed a complete online curriculum designed to direct students' learning. Academic Business Consultants, Inc., helped with the authoring of the original material. Leviton Manufacturing provides the lab configuration that is so critical to the hands-on aspect of the program, and HAI and Premise Systems provide the industry-leading software that helps to control the devices in the lab and home. BlueVolt provides students around the world with online hosting services for the curriculum and the assessment system. All of these organizations have worked very hard to provide the HTI industry and its students with the very best in industry-based training. For more information, visit these websites:

http://hti.ciscolearning.org
http://www.comptia.org
http://www.internethomealliance.com
http://www.leviton.com
http://www.bluevolt.com
http://www.prenhall.com
http://www.homeauto.com
http://www.premisesystems.com

INTENDED AUDIENCE

This book and the companion curriculum and labs are intended to serve the needs of individuals interested in understanding the field of home technology integration. The material presented in these training products is designed to provide students with the background knowledge, hands-on experience, and overall confidence to prepare for the CompTIA HTI+ exams and (more importantly) for a solid career in a growing and exciting field.

All of the HTI+ exam objectives are covered either in this text, the companion lab manual, or the online curriculum. The authors assume that the users of these products have some knowledge of electricity.

SAFETY FIRST

The activities covered in this program deal with both high- and low-voltage electricity applications. The authors and the publishers of this material urge all students and teachers to exercise great caution when dealing with electricity. Although the labs in this curriculum have been carefully tested, and all of the training sites have been given careful directions for constructing the lab walls, each individual should exercise great caution with the lab equipment. Electrical shock can cause injury or death.

THE CISCO LEARNING INSTITUTE HOME TECHNOLOGY INTEGRATION TRAINING PROGRAM

The Cisco Learning Institute Home Technology Integration Training Program is a carefully designed total training solution that includes a textbook, a lab manual, hands-on labs, and an online curriculum. All of these components are engineered to work together to provide the best possible teaching and learning experience.

Textbook

The textbook provides the student and the teacher with a completely portable reference to the entire curriculum. The text includes detailed objectives for each chapter, as well as chapter summaries, key terms, and review questions. The text is a complete reference to the curriculum and provides the basis for independent study and HTI+ exam preparation. Also packaged with the main text is an examGEAR™ CD containing hundreds of questions to help readers prepare for the exam.

Chapters in the main text are organized as follows:

Chapter 1 discusses the overall field of home technology integration, including an overview of the industry as well as the role of the HTI.

Chapter 2 covers standards, codes, and regulations that govern and guide the home technology field, including the guidelines from IEEE, ICEA, ISO/IEC, and NRC-IRC.

Chapter 3 presents the fundamentals of building a computer network, including general concepts, infrastructure, and technologies, as well as computer subsystems and networking products.

Chapter 4 analyzes the area of telecommunications, including equipment location, configuration, and connecting devices.

Chapter 5 discusses audio and video considerations, physical products, equipment location, and configuration, as well as in-house services and the standards and codes that regulate this aspect of the field.

Chapter 6 introduces the field of security and surveillance: design considerations, physical products, and all standards and installation procedures.

Chapter 7 addresses home-automation systems, including lighting, HVAC, water, and access systems. Each of these systems is addressed separately and then as a total unit.

Chapter 8 provides a complete overview of structured cabling systems. Twisted-pair, shielded, coax, and fiber cable are covered, along with the appropriate discussions of codes and standards.

Chapter 9 discusses the high-voltage applications: 120- and 140-volt systems are discussed in detail, along with power backup systems.

Chapter 10 covers the critical area of system integration. Students learn how to integrate the components of the various systems in the wired home.

Chapter 11 leads the technician through the layout, design, and presentation of the residential installation.

Chapter 12 covers the rough-in process, including cable placement and device location, for new and existing homes.

Chapter 13 explains the trim-out process including the Structured Media Center and all cable installations.

Chapter 14 covers the all-important topics of finishing, system testing, and troubleshooting.

Chapter 15 summarizes HTI installations for multiple dwelling units.

Chapter 16 concludes the book with a discussion of customer support and training.

Lab Manual

The accompanying lab manual, which is part of the print bundle, includes documentation for all 51 labs. The manual provides students with a ready reference to use with the labs in class and as a convenient place to record lab activities. The labs are designed to give students a true view of the situations that they will encounter in the field. The equipment used in the labs are manufactured by Leviton Manufacturing and HAI, two of the largest and best-known manufacturers of low-voltage home automation devices. The manual offers an easy-to-follow guide for completing the labs and recording results.

Certification Kit

The Certification Kit on the accompanying CD provides a variety of tools that should prove useful as students prepare for the HTI+ exam. The CD includes a variety of student interactive activities and simulations that provide work-based practice for the technician. In addition, supplemental test questions have been added to provide a rich practice environment for the HTI+ exam.

Hands-on Labs

The goal of this course is to help produce knowledgeable technicians who are capable of installing and servicing complex integrated systems within the home. No amount of reading can replace the value of hands-on experience. The labs use state-of-the-art tools and equipment manufactured by industry leaders. Leviton equipment is used for all the labs dealing with low-voltage structure wiring, and audio and video installations (and for most of the security installations). HAI and Premise software are used for the control and programming labs.

Online Curriculum

The full curriculum is available online from either a local training site or from BlueVolt, our national provider. The curriculum includes text, illustrations, and many exciting simulations and interactive activities designed to foster understanding and build confidence. The curriculum is available 24/7 and can be accessed from any computer connected to the Internet.

ACKNOWLEDGMENTS

The Cisco Learning Institute wants to thank several organizations and individuals for their help and support during the developmental process: Bridget Bisnette, Cisco Systems; Ian Hendler, Leviton Manufacturing; Bruce Luie, Premise Systems; Lisa Bourdeaux, BlueVolt; Jay McLellan, HAI; and David McLean, Fluke Networks.

Many individuals worked on the creation of the program. Academic Business Consultants did an excellent job of managing the authoring and development process. Chuck Stewart provided craft and skill to the manuscript as developmental editor; Nancy Sixsmith did a superb job as our copyeditor. Our editor at Prentice Hall, Charles Stewart, did an excellent job guiding the project. Thanks also to our production editor, Alex Wolf, who provided outstanding production coordination.

We are also indebted to the following technical reviewers: Ron Kovac, Ball State University; Russell Hillpot, Gadsden State Community College; Dennis Quatrine, Henry Ford Community College; and Tom Moffses, North Florida Community College.

Thanks to all of our students and instructors. We hope this program will provide you with the skills to enter this exciting field.

Dave Dusthimer
Publisher, Cisco Learning Institute

Contents

The Residential Environment

OBJECTIVES

Upon completion of this chapter, the HTI will be able to complete the following tasks:

- Define an integrated home network
- Identify the types of subsystems that make up an integrated home network
- Identify the basic components of an integrated home network
- Describe the home network integration process and its key players
- Recognize the needs and expectations of the customer
- Monitor current and future trends in the industry
- Identify tools needed by the HTI to install an integrated home network

INTRODUCTION

For years, the electronics industry has awaited the arrival of home automation. The vision of home automation with a robot in every home is now being replaced with a more practical concept of integrating electronic devices and personal computers to create a single, centrally controlled environment. Homeowners see the benefits of such a system. They want to connect everyone and everything within the home. This is achieved by integrating subsystems. This new concept is referred to as the *integrated home network*, and those who install home networks are called *home technology integrators* **(HTIs)**. Home network integration can be achieved on any scale, based on what the homeowner wants and the technology that is currently available. Some home network systems installed today are designed to provide scalability as technologies become available, eventually allowing fully integrated homes.

This chapter will give an overall explanation of integrated home networks by identifying the types of subsystems that can be integrated, the

Integrated home network

Home technology integrators (HTIs)

1

process of designing and installing an integrated home network, and the key players involved in the process. This chapter also discusses the requirements for home networks, the current market, and the future demand for home networks. Finally, the chapter prepares the HTI by introducing common tools used for installing cabling.

AN INTEGRATED HOME NETWORK

In the past, home networking has been about PCs, their peripherals, and access to the Internet. Traditionally, a home network consisted of multiple PCs and peripherals linked together by networking wires and wireless devices. The network was used for shared Internet access, peripheral sharing, and file and application sharing.

This trend is changing as consumer electronics transition from the analog to the digital world and begin offering integrated networked solutions for the home. The integrated home network is shifting from simply connecting different computer peripherals to connecting different applications and hardware, such as the PC, the television, and a variety of household appliances that serve multiple functions.

Today, an integrated home network connects residential subsystems to each other and to the Internet. In addition to PCs, today's home networks include consumer electronics devices such as televisions, VCRs, and CD players; along with traditional home appliances such as microwave ovens, refrigerators, washers and dryers, heating and air conditioning thermostats, home security systems, and home automation controls. Furthermore, home automation and security functions are implemented by linking consumer electronics, appliances, and system controls. These subsystems are accessed and controlled by an application that resides on a home network control system, typically having a control processor and interface. This is basically a network computer with a user-friendly interface. For example, the television, DVD, VCR, stereo system, and lighting could be linked together and controlled from a PC or from several PCs connected to the network to form a surveillance system that can be linked to an alarm system. Figure 1–1 illustrates the possibilities for networking a home.

Figure 1–1
The Integrated Home Network and its Subsystems

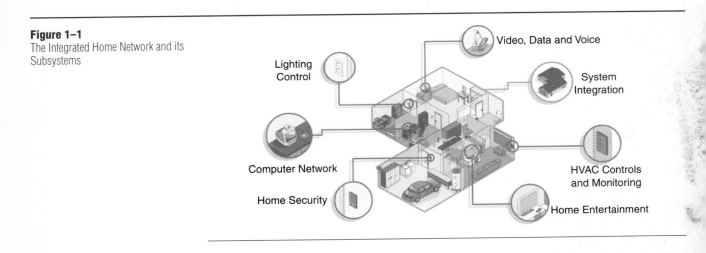

The residential subsystems covered in this course include the following:

- Computer network
- Telecommunications
- Audio/video

- Security and surveillance
- Lighting management
- Heating, ventilation, and air conditioning (HVAC) management
- Water for sprinklers, pools, spas, and fountains
- Access, including entry controls for gates and doors
- Miscellaneous systems such as lift systems, fireplace ignition, and skylights

Components of the Integrated Home Network

An integrated home network is the collection of components that process, manage, transport, and store information that enables the connection and integration of multiple computing, control, monitoring, and communication devices in the home. The rapidly increasing consumer adoption of personal PCs and Internet access, along with advances in telecommunications technology and progress in the development of smart devices and consumer electronics, have increasingly emphasized the need for an integrated home network.

A typical integrated home network includes the following four components:

1. **High-speed Internet access:** *High-speed Internet access* through xDSL or cable technologies is ideal for the integrated home network. Figure 1–2 illustrates the projected growth of high-speed Internet access from 2000 to 2005.

High-speed Internet access

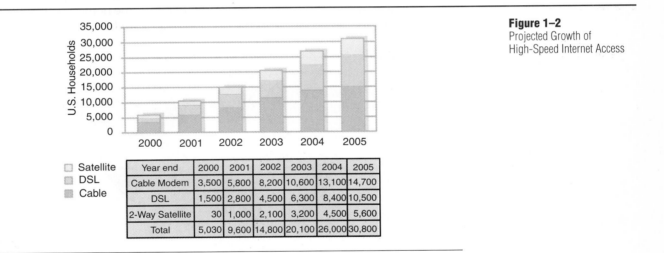

Year end	2000	2001	2002	2003	2004	2005
Cable Modem	3,500	5,800	8,200	10,600	13,100	14,700
DSL	1,500	2,800	4,500	6,300	8,400	10,500
2-Way Satellite	30	1,000	2,100	3,200	4,500	5,600
Total	5,030	9,600	14,800	20,100	26,000	30,800

Figure 1–2
Projected Growth of High-Speed Internet Access

2. **Residential gateways:** The *residential gateway*, as shown in Figure 1–3, is the interface device that serves as the entry/exit point that bridges the Internet and computer network technologies in the home. Residential gateways are detailed in Chapter 3.

Residential gateways

Figure 1–3
Residential Gateway

3. Network wiring infrastructure: Home networking technologies provide interconnectivity for all networking products within the home. A wide variety of technologies exist for interconnecting devices. Although traditional wired Ethernet systems offer a proven solution for businesses, most homeowners do not want to rewire their existing homes. Fortunately, the emergence of "no new wiring" technologies can solve the home networking problem for retrofits. The new technologies include wireless communications, phone line, and power line solutions.

Intelligent appliances

4. Intelligent appliances: *Intelligent appliances* have the capability to communicate and interoperate with other devices within the home network. With access to the Internet, some of these devices can be controlled remotely. Intelligent appliances generally fall into one of the following categories:

Appliances—Use *physical processing* features (mechanical functioning). They include kitchen appliances, lighting, and entry controls.
Electronics—Use both *physical and logical processing* features. They include DVD players, stereos, and security systems.
Computers—Use *logical processing* features (digital intelligence).

Home Network Integration Process
Every home network installation consists of three phases: design and engineering phase, installation phase, and customer support phase. Each phase, as shown in Figure 1–4, is critical to the success of a project.

Figure 1–4
Home Network Integration Process

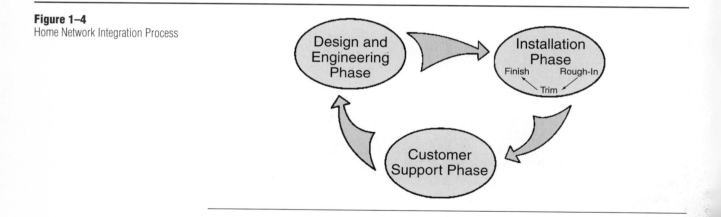

1. **Design and engineering phase**
   Gather network and lifestyle requirements
   Design the home network
   Document how the network connects devices
   Secure customer's approval for project
2. **Installation phase** (consists of three steps)
   Step 1   Rough-in the wiring throughout the home
   Step 2   Trim the wire by terminating it
   Step 3   Connect the components and test the network
3. **Customer support phase**
   Train the homeowner on the proper way to use and care for the network
   Provide customer service and support to create opportunities for additional sales

Key Players
The home network integration process involves many people from the HTI to suppliers of intelligent devices. The two key players in the process are the HTI

and the customer, which in most cases is the homeowner. The HTI is involved in every phase of the process. Homeowners are involved in the first and last phases.

Builders, architects, manufacturers, and suppliers also play a role. Their roles depend on the type of project (new construction versus retrofit), the degree of integration, and the requirements of the homeowner.

- **Home Technology Integrator (HTI):** The HTI provides the customer with the knowledge and expertise to develop the right solutions using the right technology. The HTI possesses an understanding of all subsystems and how to integrate them, as well as a strong connection to the industry to provide cutting-edge knowledge to match customer needs and existing and future technology.
- **Homeowner:** Determines the requirements with the HTI, and makes decisions about the type of features needed and the budget available for home integration.
- **Customer:** In the context of this course, the *customer* refers to anyone who hires the HTI to design and install a home network. The customer can be the homeowner, a home builder, or a developer.
- **Service Providers:** *Service providers* provide services such as Internet, gas, electricity, and telephone.
- **Homebuilders and Developers:** Work with the home integrator to incorporate structured wiring and other components into a new home. This group works closely with their subcontractors to tie the entire project together.
- **Architects and Designers:** Determine whether any special structural design requirements are necessary to accommodate the home network. Architects are especially important in the retrofit market to help integrate the connectivity solutions into the overall design of the home.
- **Manufacturers:** Develop components and appliances that use the Internet to receive and send data about the device. Manufacturers are developing devices that are "plug-and-play."
- **Distributors and Retailers:** Supply the devices and components for the home network. They partner with manufacturers to develop and promote devices and components that are "plug-and-play" to make connection to the home network simple.
- **Media, Analysts, and Trade Associations:** Inform the public about existing and new products. The home technology integrator should be aware of developments in the industry and be prepared for questions or concerns that homeowners may have, based on information presented by the media.

Service providers

Home Integration Industry Organizations
There are several home integration industry organizations with which you should be familiar.

- **Internet Home Alliance (IHA):** The IHA's mission is to advance the home technology industry, particularly products and services that require a broadband or persistent connection to the Internet.

Web link: http://www.internethomealliance.com/

Figure 1–5
IHA Logo

- **Continental Automated Buildings Association (CABA):** CABA promotes advanced technologies for the automation of homes and buildings. Its objectives are to facilitate and encourage industry-wide interoperability of protocols and standards, to encourage research and development in home and building automation, and to be the definitive source for North American integrated systems and automation in homes and buildings.

   Web link: http://www.caba.org/

Figure 1–6
CABA Logo

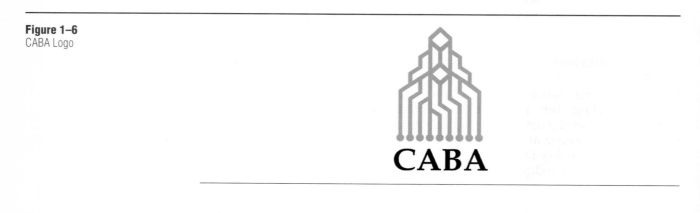

- **Custom Electronic Design & Installation Association (CEDIA):** CEDIA is a global trade association of companies that specialize in planning and installing electronic systems for the home. These systems include home networking; communication systems; multiroom entertainment systems; and integrated whole-house subsystems providing control of lighting, security, and HVAC systems.

   Web link: http://www.cedia.net/

Figure 1–7
CEDIA Logo

- **Consumer Electronics Association (CEA):** CEA is an organization with a belief that market-based competition among all communication channels is the best way to promote rapid deployment of broadband technologies. In January 2002, the Home Automation and Networking Association (HANA) merged with CEA and formed the TechHome Division of CEA. The mission of the TechHome Division is to increase sales and support of consumer electronics products by HTIs.

Web link: http://www.ce.org/

Figure 1–8
CEA Logo

Other Organizations

There are many organizations, groups, and associations that publish standards, codes, regulations, and methods for installation and/or testing of materials affecting the residential environment. Residential system and component manufacturers adopt a lot of these to ensure standardization. Some local enforcement agencies may also adopt or adhere to such standards as evidence of quality in installation. Some of the major agencies and organizations include the following:

Electrical
- National Electrical Contractors Association (NECA)
  http://www.necanet.org/
- National Electrical Code (NEC)
  http://www.nfpa.org/nec/

Electronic
- Video Electronic Standards Association (VESA)
  http://www.vesa.org/
- National Systems Contractors Association (NSCA)
  http://www.nsca.org/
- American Electronics Association (AEA)
  http://www.aeanet.org/

Computer
- Wireless LAN Industry Forum (WLIF)
  http://www.wlif.com/
- International Computer Security Association (ICSA)
  http://www.icsa.net/
- National Institute of Standards and Technology (NIST)
  http://www.nist.gov/

Telecommunications
- Telecommunications Industry Association (TIA)
  http://www.tiaonline.org/
- Building Industry Consulting Services, International (BICSI)
  http://www.bicsi.org
- Canadian Wireless Telecommunications Association (CWTA)
  http://www.cwta.ca/

Security
- International Association of Home Safety and Security Professionals
  http://www.iahssp.org/
- Security Industry Association (SIA)
  http://www.siaonline.org/

- National Burglar and Fire Alarm Association (NBFAA)
  http://www.alarm.org/
- National Fire Protection Association (NFPA)
  http://www.nfpa.org/
- National Institute for Certification in Engineering Technologies (NICET)
  http://www.nicet.org/
- Canadian Alarm and Security Association (CANASA)
  http://www.canasa.org

HVAC

- Plumbing-Heating-Cooling Contractors Association (PHCC)
  http://www.naphcc.org/
- American Society of Heating, Refrigerating and Air-Conditioning Engineers (ASHRAE)
  http://www.ashrae.org/

Home Building

- National Association of Home Builders (NAHB)
  http://www.nahb.com/
- National Conference of States on Building Codes and Standards (NCSBCS)
  http://www.ncsbcs.org/

RESIDENTIAL NETWORKING MARKET

In a market study, In-Stat/MDR (http://www.instat.com) reported that the home networking equipment market nearly quadrupled in size from $150 million in 1999 to $585 million by the end of 2001, as shown in Figure 1–9. That pace is not predicted to slow down. In fact, an article in *Archi-Tech Magazine* (September 2001) entitled "The Networked Home" (http://www.architech mag.com/issuebackup/sep01.html) forecast that the number of networked homes in the U.S. would reach 1.7 million in 2005, up from 650,000 in 2000.

Figure 1–9
Bar Graph Showing the Growth of the Home Network Equipment Market

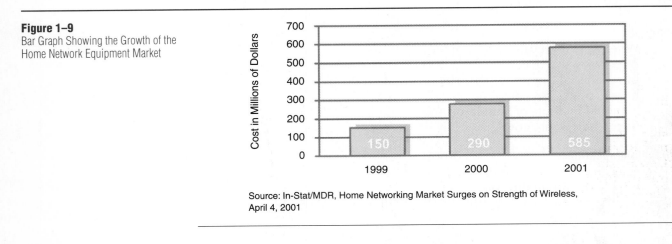

Source: In-Stat/MDR, Home Networking Market Surges on Strength of Wireless, April 4, 2001

Because the number of networked homes is expected to double or even triple within the next three years, this explosive growth presents a unique opportunity for the HTI. Some of the key market drivers for the deployment of residential networking technologies are as follows:

- **Increasing usage of the Internet:** Consumer demand for Internet applications is increasing. These applications include voice and video; and data-in applications such as streaming video, web browsing, e-mail, MP3 files, Internet telephones, digitized photographs, and online shopping and gaming.

- **Increasing need for high-speed access to the Internet:** Internet users are requiring access services that provide much higher speeds than dial-up access.
- **Decreasing prices for broadband services:** The prices for broadband services are decreasing, and the demand for these services is increasing.
- **Growth in telecommuters:** Due to the growth in telecommuting, there are increasing numbers of people who work at home and need high-speed access to the Internet to connect and communicate with their workplace. Telecommuters use technologies such as Virtual Private Networks (VPNs) to access the office network from home.
- **Networking multiple PCs and appliances:** Because of the growth in the number of multiple PC households, demands for faster Internet access have accelerated. The networking of appliances and their Internet access requires a high-speed and always-on connection.
- **Increasing availability of applications requiring high bandwidth:** As the number of multimedia and interactive applications that require high-bandwidth capabilities increases, the demand for home networking technologies will also increase. For example, the number of video-based applications over the Internet has grown substantially in the last several years, and these applications require a very fast method of accessing the Internet.

Understanding the Needs of the Homeowner

The home network is designed and implemented based on the requirements of the homeowner. The devices used and the functions that they provide are a direct result of what the customer wants the network to accomplish and how they intend to live on a daily basis with the technology that the network requires. The HTI can advise on the technologies available for the home network and how they can be implemented to retrofit existing homes. New construction is another market for residential networking. Homes built today incorporate technology into the structural design and provide the capability to interconnect the house. The basis for incorporating technology into a home network is driven by the following factors:

- **Converging media:** Digitization of data, voice, video, and communications is driving consumer product convergence, along with computing power being embedded within everyday consumer appliances.
- **Internet-based lifestyles:** The availability of a wide variety of web-based applications is changing the way people utilize the Internet for everyday living. The Internet enables a new generation of consumer applications, including voice and video communications, e-commerce solutions, personalized news services, home security and automation, utilities resource management, and entertainment title distribution.
- **Deregulation:** There is a great deal of deregulation taking place in global infrastructure industries such as telecommunications, utilities, and cable. These companies are deploying new techniques of improving their business by utilizing the Internet and providing additional services to the home.
- **Digitization of consumer products:** There has been a trend toward the digitization of consumer products, which leads to higher quality, greater accuracy, higher reliability, faster speed, power management, and lower cost. It has also led to the development of a whole new category of products such as Personal Video Recorders, satellite modems are bringing faster Internet access to the home, and MP3 players have changed portable digital music.

In a study conducted at the beginning of 2002 by In-Stat/MDR, a leading high-tech research firm, 42% of homeowners cited sharing Internet access as the most important reason for owning a home network. Sharing files was chosen by 17%, and print sharing was chosen by 11%.

Web link: http://www.instat.com

Figure 1–10 shows the primary reasons for owning a home network.

Figure 1–10
Primary Reasons for Wanting a Home Network

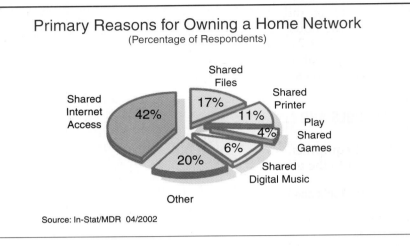

Primary Reasons for Owning a Home Network
(Percentage of Respondents)

Source: In-Stat/MDR 04/2002

As homeowners learn the advanced possibilities of what can be accomplished beyond the home computer network, they will want the ability to integrate a security system; heating/cooling control systems; home entertainment equipment; programmable lighting configurations; and, at some point, "smart" appliances.

The most important characteristics of a home network for the homeowner include the following:

- **Ease of use:** Controls and devices should be easy to use and maintain for people with limited exposure to technology.
- **Reliability:** The network must be reliable. All home networking devices, as well as common household appliances such as microwave ovens and cordless phones, should work in harmony and without interference.
- **Scalability:** A home network needs the capability to expand to accommodate future needs of the homeowner as well as new technologies and devices. This protects the homeowner's initial investment and results in a lower lifetime cost.
- **No new wiring:** Homeowners in older homes do not want the mess and expense of running wires behind existing walls or the unsightliness of them along baseboards. The *"no new wiring" technologies*—such as wireless communication, phone line, and power line—provide a solution for retrofit projects.
- **Security and privacy:** Customers want to own their own data and completely manage access to their data.

"No new wiring" technologies

Future of Home Networking

As the overall cost of technology continues to decrease, home network technology will increase in acceptance and accessibility. Also, standards in connectivity will increase the simplicity of networking the home subsystems into one integrated system.

As the home networking market has expanded at an ever-increasing rate, large companies have been encouraged to collaborate in projects that promote the convergence between personal computers, digital television, and audio/video (A/V) platforms. This collaboration is important to ensure that personal computers (PCs) and A/V equipment do not continue to evolve and develop on separate paths. This new collaboration will deliver a range of new and exciting in-home applications:

- Providing access to high-speed, Internet-based services for all members of a household
- Controlling A/V devices (such as VCRs, camcorders, and sound systems) from a single location in the house
- Home monitoring and security, which allows a monitor to show visitors approaching the house, babies sleeping in their cribs, and kids playing in the yard; as well as 24-hour perimeter surveillance
- Uniting all types of devices, including desktop PCs, DVD players, digital set-top boxes, stereos, VCRs, and high-speed Internet connections

HTI TOOLS AND EQUIPMENT

The tools and equipment used by the HTI range from general-purpose tools and equipment to the tools designed for specific tasks.

General-Purpose Tools and Equipment

Screwdrivers
Tape measure
Sturdy non-metallic ladder
Combination folding ladder
Tool belt
Safety goggles or glasses
Dust masks
Ear plugs
First-aid kit

Rough-in Tools

Measuring wheel
Cable caster
Fish tape
Hole saw
Long drill bits
Cable Joe
Stud sensor
Wire sensor
Drywall saw

Trim-out Tools

Stripping tools
Cutting tools
Termination tools
Crimping tools

Finish Phase Tools

Labeling machine
Voltage sensor
Multimeter
Metal sensor
Cable tester
Telephone test set
Cable certification meter

When selecting tools, select the highest quality tool that you can afford. Inexpensive tools break easily, and those with cutting edges become dull quickly. You will want your tools to last for many years. Good quality tools will be a good investment.

General-Purpose Tools and Equipment

There are general-purpose tools and equipment that a HTI finds necessary or useful for every project:

- **Sturdy non-metallic ladder:** It is important that you have a non-metallic ladder when working around power lines, so that if you accidentally come into contact with live power lines, the chance of being electrocuted is reduced. Purchase a ladder that is sturdy and is designed for industrial use rather than for light use.
- **Combination folding ladder:** A combination folding ladder can be used as a stepladder when folded and unfolded for use as a straight ladder.
- **Tool belt:** A tool belt, as shown in Figure 1–11, is an effective way of keeping tools and supplies handy. There are specialized tool belts for telecommunication technicians. Tool belts are generally leather or nylon. Choose one that has sufficient durability and capacity to carry all of the tools needed, but is light enough so that it is comfortable.

Figure 1–11
Typical HTI Tool Belt

- **Safety goggles or glasses:** Safety goggles, as shown in Figure 1–12, protect your eyes when working with power equipment.

Figure 1–12
Safety Goggles

- **Dust masks:** Dust masks protect your lungs from dust and debris.
- **Ear plugs:** Ear plugs protect your hearing when you work with power equipment. Figure 1–13 shows ear plugs on a cord that hangs around your neck when they are not in use. This keeps them handy when working with loud power equipment.

Figure 1–13
Ear Plugs

- **First-aid kit:** Keeping a first-aid kit on hand is important for every project. In addition to the usual bandages, make sure there is an eyewash cup. Dust and debris from ceilings can fall into your eyes and injure them.

Rough-in Tools

The rough-in phase of the installation process involves pulling cable. Therefore, the HTI needs the following specially designed tools.

- **Measuring wheel:** A measuring wheel is used to estimate the length of a cable run. The HTI rolls the reel down the intended path of the cable. A counter indicates the distance.
- **Cable caster:** A cable caster resembles a small fishing reel. It shoots a plastic dart attached to a fishing line through a duct or open space. You then attach a pull string to the end of the fishing line and reel the pull string across the space. After the pull string is in place, cable can be attached and pulled to the destination. This tool is particularly useful in new constructions, where the dart can be shot across the ceiling of an entire room so that you do not have to climb a ladder every few feet to pull cable.
- **Fish tape:** Fish tape is used to pull wires behind existing walls. It is used in retrofit constructions where holes in drywall are cut and the fish tape is inserted to pull wires to or from the area above the ceiling.
- **Hole saw:** Hole saws are used to make holes through wooden framing for cables and conduit.
- **Long drill bits:** Purchase drill bits that are at least 16 inches long to drill through inaccessible areas. Some installers have drill bits with holes drilled in them so that a pull string can be attached directly to the drill bit after it has drilled through the wall or stud and before it is retracted. The pull string is then pulled through the opening and can be attached to the cable for pulling.
- **Cable Joe:** A Cable Joe is a frame with a series of rollers that make it easy to pull multiple cables in bundles. See Figure 1–14.

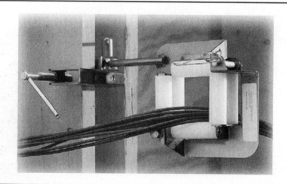

Figure 1–14
A Cable Joe Makes Pulling
Cable Easier

- **Stud sensor:** This sensor locates wooden studs and joists behind walls. This tool helps the installer make informed decisions about the best location to drill or saw when installing outlets or raceways.
- **Wire sensor:** A wire sensor finds wires hidden in the wall. This tool is designed for locating telephone lines. It should not be used on live electrical wires or on data circuits because it can damage the circuit.
- **Drywall saw:** Use a drywall saw to cut drywall for outlet boxes.

Trim-out Tools

The trim-out phase involves terminating and testing the cables. A variety of tools are needed for terminating the cables including stripping and cutting tools, punch-down or termination tools, and crimping tools.

Stripping Tools   Stripping tools are used to cut the cable jacket off the wires. Special stripping tools remove the outer jacket of twisted-pair cables such as CAT 5 (discussed in Chapters 3 and 8). These tools allow the HTI to ring the cable without damaging the insulation on the individual conductors. After the tool is used to ring the outer jacket, the jacket material can be pulled off by hand. See Figure 1–15.

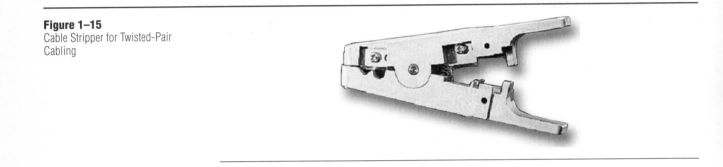

Figure 1–15
Cable Stripper for Twisted-Pair
Cabling

Stripping tools are also used to remove the jacket and dielectric of coaxial cables (another type of cable that is discussed in Chapters 3 and 8). The tools generally have two cutting blades. One blade slits the outer jacket of the cable while another blade cuts through the braid and dielectric. The blades are spaced properly to leave a finished end on the cable, ready for the installation of a connector. See Figure 1–16.

Figure 1–16
Cable Stripper for Coaxial Cable

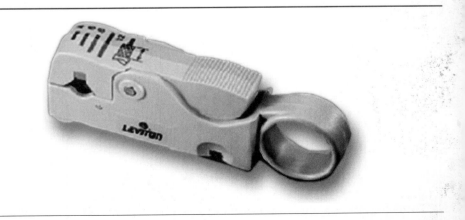

Cutting Tools   The most common cutting tools for cable installation are the six-inch diagonal cutting pliers, the electrician's scissors, and a cable cutter.

Diagonal cutting pliers, as shown in Figure 1–17, are used for cutting twisted-pair cable, coaxial cable, speaker cable, and alarm wires and are commonly found on a technician's tool belt. Avoid cutting anything harder than copper wire. Cutting steel conductors can create a notch in the cutting blades.

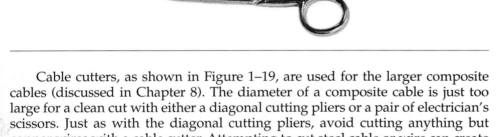

Figure 1–17
Cutting Pliers

The electrician's scissors are shown in Figure 1–18. They are one of the most versatile tools in a tool pouch. They are used for general cable cutting, cutting pull strings, and cutting pull ropes. They have notches on the back of the blade that can be used for stripping the insulation from 22-gauge and 24-gauge conductors.

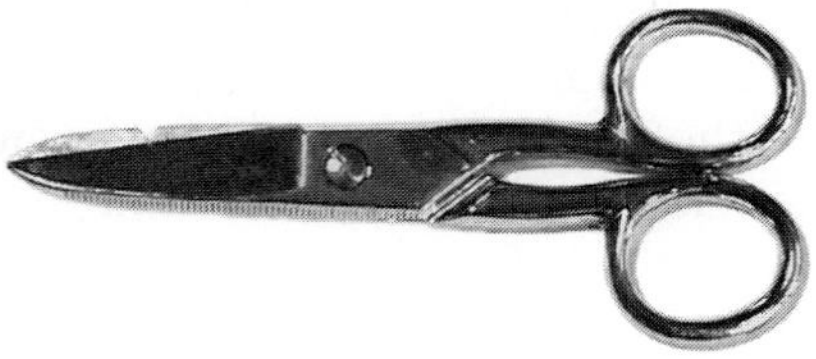

Figure 1–18
Electrician's Scissors

Cable cutters, as shown in Figure 1–19, are used for the larger composite cables (discussed in Chapter 8). The diameter of a composite cable is just too large for a clean cut with either a diagonal cutting pliers or a pair of electrician's scissors. Just as with the diagonal cutting pliers, avoid cutting anything but copper wires with a cable cutter. Attempting to cut steel cable or wire can create a notch in the cutting blades that will render the tool ineffective.

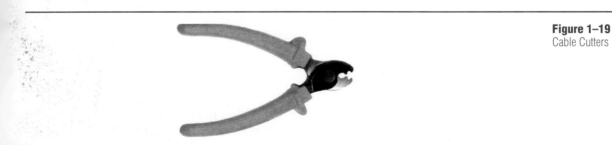

Figure 1–19
Cable Cutters

Termination Tools   A termination tool or punch-down tool is shown in Figure 1–20. It is designed to cut and terminate specific types of cable such as twisted-pair or coaxial cable. These tools are particularly useful for residential cabling infrastructure installations because the cut is clean every time. Additionally, they are safer to use because the cutting blade is recessed. This tool and other similar termination tools are designed to terminate and cut UTP cable and seat connecting blocks.

Figure 1–20
Termination (Punch-Down) Tool

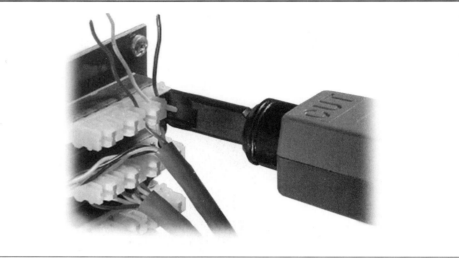

Some termination tools have interchangeable blades for terminating wires on 66 and 110 hardware. They are usually spring-loaded and adjustable, which is particularly helpful when working with wires of varying thickness.

Often overlooked, the torque wrench can be used for tightening F-type coaxial connectors. This is a small wrench with a preset torque setting at about 30 inch pounds. When the proper tightness is reached, the torque wrench emits an audible click, alerting the technician to stop tightening. Many cable television companies equip their installers with this tool. The torque wrench is used to achieve a uniform tightness to F-type connectors. Loose connectors can be a source of signal leakage. Overtightening can strip the threads.

Crimping Tools   A crimping tool is shown in Figure 1–21. It is used to attach a jack or plug at the end of a cable. Crimping tools come in different styles, depending on the manufacturer and the range of jobs they are intended to accomplish.

Figure 1–21
Crimping Tool

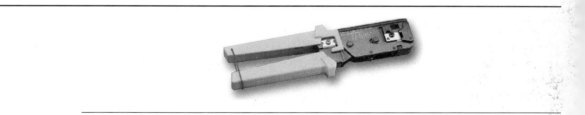

Crimping tools can be manual or battery-powered. Usually, manual crimping tools should be adequate for smaller installation jobs. For larger jobs, powered crimping tools can ease fatigue and increase productivity.

Crimping tools are generally designed for a precision crimp every time. Some of these tools have a replaceable die, which is the part of the tool that actually does the crimping. If the installer needs a different size connector crimped, the die can be swapped for the appropriate size. Some are generic, whereas others are designed to be used with the manufacturer's products.

When installing cable for video application, the snap-and-seal type of connector is recommended for coaxial terminations. After stripping, the actual connector is installed over the cable. The installation tool then forces the plastic sleeve up into the connector. Many CATV companies require snap-and-seal connectors for high-performance installations such as digital TV and cable modems. This type of connector creates a sealed hardware connection.

Finish Phase Tools   The finish phase involves labeling faceplates and installing them over outlets. To label faceplates, you should use a machine that makes professional-looking printed labels as shown in Figure 1–22. The installation of faceplates goes much faster and easier if you use an electric screwdriver.

Figure 1–22
Labeling Machine

Sensors and Testers

The remainder of the chapter focuses on the tools that test voltage or cable connections. There are several tools that you should have that will ensure your safety when working around high-voltage wiring (power lines) and other obstacles in the home. A voltage sensor detects electricity in circuits to avoid electrocution. A multimeter measures voltage. A metal sensor locates pipes behind walls so as not to drill into them. These tools can help avoid costly repairs on your part and possibly electrical shock.

In addition, you will need to test cable and certify it. You will need a cable tester, a telephone test set, and a cable certification meter to accomplish these tasks.

Voltage Sensors

When the home technology integrator installs systems in older homes, it is very important to be able to locate existing power circuits and distinguish power wires from low-voltage wiring. One of the easiest ways to identify live wires is to use a voltage sensor. A voltage sensor is shown in Figure 1–23. It provides a signal when voltage is present. One type of voltage sensor glows red or another color when placed near voltages between 24 and 90 volts. Use this tool for troubleshooting low-voltage circuits such as thermostats and lawn sprinklers. Another type of voltage sensor, a high-voltage AC tester that operates between 90 and 600 volts, is also available. This device can even be used to see if an extension cord is plugged in. The plastic-tipped tool does not have to come in direct contact with exposed metal wire or terminals to detect voltage.

Figure 1–23
Voltage Sensor

Multimeter

A multimeter, as shown in Figure 1–24, is a tool used for measuring voltage and other related parameters. This instrument can be used to ensure that there is no voltage on the telecommunications line before putting more sophisticated test equipment on the line. Also, a multimeter is a good tool to test power outlets during equipment installation. It is not unusual for some power outlets (especially in older homes) to have a voltage lower than the nominal 120 volts (U.S.) or no power. In such a case, plugging a device rated for the higher voltage can cause the device to malfunction or, in some cases, not to work at all.

Figure 1–24
Multimeter

A multimeter can measure AC/DC voltage, current, resistance, diode and transistor continuity, and can even test 9-volt or 1.5-volt batteries. Multimeters come with different feature sets, depending on the manufacturer and the range of functions or tests they are intended to perform.

Metal Sensor

It is important to determine whether there are metal pipes or rebar before starting installation because their discovery after starting can complicate a project. Before drilling for any cabling project, use a deep scanning metal sensor as shown in Figure 1–25. This tool can find metal studs, conduit,

Figure 1–25
Metal Sensor

copper piping, electrical lines, rebar, telephone lines, cable lines, nails, and other metal objects. The deep scanning metal sensor usually can scan through up to six inches of a nonmetallic surface such as concrete, stucco, wood, or vinyl siding. Using a metal sensor before starting any drilling or sawing greatly reduces the risk of broken drill bits, broken saw blades, or electrical shock.

Cable Tester

Another important category of tools is cable testers. Cable testers can determine existing and potential problems in a residential network cabling installation.

A cable tester is shown in Figure 1–26. This tool is used to test cables for opens, shorts, split pairs, and other wiring problems. After the cable installer has terminated a cable, the cable should be plugged into the cable tester to verify that the termination was done correctly. If a wire is accidentally mapped to the wrong pin, the cable tester indicates the wiring mistake. Similarly, it can test for problems with the cable such as shorts or opens. A cable tester should be a part of every residential cabling installer's toolbox.

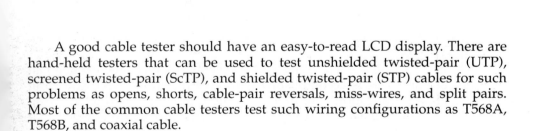

Figure 1–26
Cable Tester

A good cable tester should have an easy-to-read LCD display. There are hand-held testers that can be used to test unshielded twisted-pair (UTP), screened twisted-pair (ScTP), and shielded twisted-pair (STP) cables for such problems as opens, shorts, cable-pair reversals, miss-wires, and split pairs. Most of the common cable testers test such wiring configurations as T568A, T568B, and coaxial cable.

Telephone Test Set

The telephone test set is shown in Figure 1–27. It is another diagnostic tool that the home technology integrator may find useful. Also called a telephone buttset, this tool is used to listen for noise on a telephone circuit. It can also be used to detect voltage on the line. Most modern test sets include an LED that denotes line polarity. Technicians use the telephone test set for a variety of tests such as the capability to "break dial tone" when dialing out. They may also use the test set to call the central office for additional information or more tests. When connected to working telephone lines, the test set works like an ordinary telephone. When connected to nonworking lines, two technicians can connect a battery to the line and talk to one another over a pair of test sets.

Figure 1–27
Telephone Test Set or Buttset

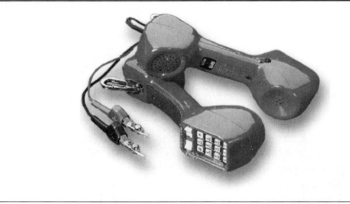

Caution should be exercised when using a test set on telephone lines. Ongoing conversations should never be interrupted. The test set has a monitor position that allows the technician to listen for conversation on the line before attempting to use the line to place a call. Common test sets work on only ordinary analog telephone lines. Care should be taken not to connect them to digital or data lines.

Sometimes, the need arises to access individual wires inside a telecommunications outlet or jack. The jack adapter is used to provide access to these wires. A common line cord is plugged into the adapters and then into the jack. A technician can then clip onto the individual wires or pairs of wires with a telephone test set. The technician can also use an ohmmeter or other test devices without having to disassemble the jack. Jack adapters, also known as Banjos, come in three- and four-pair configurations.

Cable Certification Meters

After all the faceplates have been attached and labeled, the HTI may perform certification tests on the cables. Certification tests are done after everything is in place in order to ensure that the cable or the connectors were not damaged in the finish phase.

After the cable has been tested and problems resolved, the cabling can then be certified. This is accomplished with a certification meter, shown in Figure 1–28. A certification test, unlike a simple test for a wiring segment, actually tests the whole circuit from the outlet to the distribution device; that is, it includes all the intermediate connectors and plugs. This is known as a channel test. The channel test verifies that end-to-end connectivity exists, that both ends of a cable are terminated in the proper pin configuration, and that no opens or shorts are present on the cable.

In addition to the channel test, you can perform a TIA certification test. Cable that is TIA-certified has met specific standards. The TIA certification test measures eight items including the following:

- **Length:** That the link length does not exceed 275 feet (90 meters)
- **Attenuation:** How much signal degradation occurs from one end of the link to the other
- **Near-end crosstalk (NEXT):** How much signal from one pair of wires is picked up by all other pairs

If the cable passes the TIA certification test, you can document that the cable is TIA-certified.

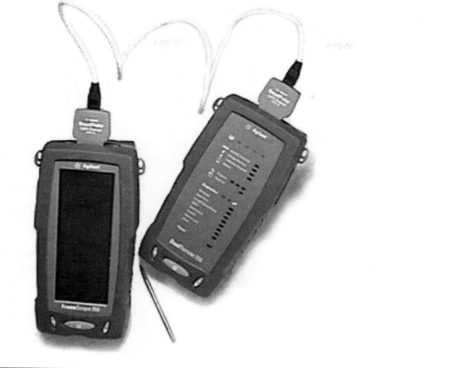

Figure 1–28
Certification Meter

SUMMARY

In this chapter, you learned the following:

- An integrated home network connects subsystems to each other and the Internet. The HTI is trained to design and install an integrated home network.
- Subsystems discussed in this course are computer networks, telecommunications, audio/video, security and surveillance, lighting, HVAC, water, access, and miscellaneous systems.
- There are three phases to the home network integration process. The design and engineering phase gathers requirements and designs the network. The installation phase consists of three steps: rough-in, trim, and finish. The customer support phase provides ongoing support and presents an opportunity for additional sales.
- The HTI and the homeowners are the key players in every project. Service providers, homebuilders, architects, manufacturers, distributors, retailers, and the media all play a role in the home network industry.
- Four components are needed for a home network: high-speed Internet access, residential gateways, home network technologies, and intelligent appliances.
- The ability to share Internet access, files, printers, games, and digital music within a home computer network is an important factor for homeowners.
- Home networks should be easy to use, reliable, scalable, compliant with industry standards, secure, private, and not require extensive new wiring.
- Tools are an investment for the HTI. High-quality tools produce professional results. Tools are classified as general-purpose tools, finish tools, and sensors and testers.

GLOSSARY

High-speed Internet access The capability to connect at speeds of 256 K, 512 K, 1 MB, or higher to the Internet via DSL or cable.

Home technology integrator An HTI is the person who designs, installs, and maintains an integrated home network.

Integrated home network The system that connects residential subsystems to each other and to the Internet.

Intelligent appliances Devices that have the capability to communicate and interoperate with other devices within the home network.

"No new wiring" technologies Technologies that do not require retrofitting a home with new wiring such as wireless communications, phone lines, and power lines.

Residential gateway The interface device that serves as the convergence point that bridges the Internet and computer network technologies in the home.

Service providers Provide services such as Internet, gas, electricity, and telephone for the home.

CHECK YOUR UNDERSTANDING

1. According to a study conducted by In-Stat/MDR, what is the primary reason for owning a home networking system?
   a. Audio/video control
   b. High-speed Internet access
   c. Home automation control
   d. Security

2. What is the second phase in the home network integration process?
   a. Customer support
   b. Design and engineering
   c. Installation
   d. None of the above

3. Which phase of the home network integration process documents how the network connects devices?
   a. Design and engineering
   b. Installation
   c. Customer support
   d. All of the above

4. One of the most important considerations when designing an integrated home network is to employ user-friendly controls and devices.
   a. True
   b. False

5. Which type of technology would be best for a retrofit home?
   a. Wireless
   b. Phone line
   c. Power line
   d. All of the above

6. Which of the following is the interface device that serves as the entry/exit point between the Internet and the computer network?
   a. Residential gateway
   b. xDSL
   c. Access
   d. Cable modem

7. A security system uses which type(s) of processing system(s)?
   a. Logical only
   b. Physical only
   c. Logical and physical
   d. Neither logical nor physical

8. Testing the network is done during which phase of the home integration process?
   a. Trim
   b. Finish
   c. Rough-in
   d. Design

9. A home network that has the ability to expand to accommodate future technologies and devices is:
   a. reliable.
   b. scalable.
   c. standards compliant.
   d. easy to use.

10. Interference from other home networking devices as well as common household appliances such as microwave ovens and cordless phones is NOT a common problem.
    a. True
    b. False

11. A lighting system uses which type(s) of processing system(s)?
    a. Logical only
    b. Physical only
    c. Logical and physical
    d. Neither logical nor physical

12. A computer uses which type(s) of processing system(s)?
    a. Logical only
    b. Physical only
    c. Logical and physical
    d. Neither logical nor physical

13. From 1999 to 2001, the home network business has:
    a. doubled.
    b. tripled.
    c. quadrupled.
    d. None of the above

14. The *"Archi-Tech Magazine"* magazine predicts that by 2005, the number of U.S. networked homes will reach:
    a. 1.7 million.
    b. 17 million.
    c. 170 million.
    d. 1.7 billion.

15. The professional who provides the customer with the right solution using the right technology is called a:
    a. Home Network Technician.
    b. Home Automation Specialist.
    c. Home Technology Integrator.
    d. None of the above

Standards, Codes, and Regulations

 **OBJECTIVES**

Upon completion of this chapter, the HTI will be able to complete the following tasks:

- Understand standards that provide specifications and guidelines to ensure proper installations and systems performance

- Understand codes that address the minimum safety requirements that ensure safety during the installations, use and disposal of materials, systems/components, fixtures, and other related subjects

- Understand regulations that are considered mandatory rules and are usually issued by government agencies

- Define new terminology related to the ANSI/TIA/EIA-570-A, Residential Telecommunications Cabling Standards

- Identify the organizations and associations that establish the standards, codes, and regulations that influence the HTI

- Determine the licensing requirements for the HTI

- Provide cabling and other materials that are certified by UL

INTRODUCTION

In this chapter, you will learn about standards, codes, and regulations that apply to integrated home networks. Many of the standards are written by industry organizations with the purpose of standardizing products, technologies, and techniques in the industry. Codes are primarily written to protect consumers and provide uniformity. Codes and proper code compliance protect home occupants from injury and death as a result of improper installation. Regulations that are pertinent for the HTI protect workers from being injured and provide the guidelines for the way work should be performed.

The various independent organizations, groups, and associations that establish and maintain standards, codes, and regulations affecting different aspects of home integration will be discussed in this chapter.

 **OVERVIEW**

Standards, codes, and regulations govern the requirements for residential telecommunications. They cover the installation, handling practices, and the type of materials used in the design and implementation of a home network. However, not even the best codes can ensure that systems are completely protected from intrusion, induced noise, or any other events that disrupt the flow of information.

Products are tested, rated, and certified before being sold. The organizations that do such work set or follow established standards and codes. Underwriters Laboratories (UL) is one such organization that ensures that products sold with its mark are safe for consumers.

Standards

The job of integrating residential systems components in a home network includes residential systems design, infrastructure wiring, and components installations. *Standards* provide specifications and guidelines for proper installations in order to allow a variety of devices from different manufacturers to perform together in an integrated installation. More specifically, standards are a collection of the requirements that embrace the properties of residential systems and components, establishing an accepted degree of functionality, interoperability, and longevity.

Because the HTI must have the knowledge and understanding of residential systems and their components to successfully complete a job, knowledge of both industry and important manufacturer product standards is crucial in giving the homeowner the proper recommendations. Standards are not mandatory rules. They are generally established as a basis to compare, judge, or measure the important aspects of a residential system, such as the following:

- Quantity and capacity
- Quality and value
- Performance and limitations
- Interoperability

Throughout this course, specific standards affecting the residential networking environment will be explored and explained for specific situations.

Codes

Although standards provide recommendations and guidelines for proper systems design and installation, codes address the minimum safety requirements that must be adhered to. More specifically, codes are intended to ensure safety during installations, use, and disposal of materials, systems/components, fixtures, and other related subjects. Codes are rules—they are usually invoked and enforced through national or local government regulations. Failure to comply with these codes can have dramatic results, including shutting down the job.

The HTI must be well informed about national or local government codes that can influence the decision-making process involving residential networking and systems integration.

There are two important terms contained in codes and standards that can have a major impact on how certain tasks are accomplished. These terms are defined as follows:

- **Shall:** A mandatory requirement
- **Should:** An advisory recommendation

Standards

The HTI or subcontractors must properly interpret this terminology in reading codes and standards.

Regulations

Regulations are considered mandatory rules issued by an official agency or organization. In the United States, an example of a government agency that issues regulations affecting the residential networking industry is the Federal Communications Commission (FCC). Since the elimination of most of the competitive regulations involving the telecommunications industry by the Telecommunications Act of 1996, regulations by government agencies are becoming less numerous in the United States. However, there are still some FCC and a significant number of state rules that apply to Local Exchange Carriers (LECs), or Access Providers (APs) and Competitive Local Exchange Carriers (CLECs), and Interexchange Carriers (IXCs).

In general, construction in just about every country is regulated by building codes, standards, and regulations. The enforcement of codes and laws in North America is normally assured by an Authority Having Jurisdiction (AHJ). Typically, installation methods, materials, electrical products, and many other specified products must conform to the requirements of the AHJ. The regulations are issued by a government agency, most often at the state or federal government levels.

An understanding of regulations helps an HTI to be better equipped in the design and implementation of a residential network and systems integration.

Ratings

In the United States, three categories are used to rate a product:

- **Listed:** When a product is rated as Listed, it has successfully completed a series of predetermined mechanical, electrical, and thermal characteristic tests, which simulate all reasonable and foreseeable hazards. However, this classification is exclusive to the product's particular application for which it was tested. It is therefore not valid for other applications.
- **Classified:** A product is Classified after evaluation and passing of tests for one or more of the following parameters:
  Its performance under specific conditions
  Specific hazards only
  Regulatory codes
  Other defined standards, including international standards

  The Classified rating is reserved for industrial or commercial products only.
- **Recognized:** A product is rated as Recognized after it has tested successfully for use as a component in a Listed package. However, the Recognized rating is more of a general-purpose approval than Listed is. The Recognized rating allows a product to be certified for a category of equipment uses. A common example is insulated wire, which is recognized as appliance wiring material whose category of uses includes the following:
  Telecommunications
  Instrumentation
  Data communication

An understanding of these rating categories greatly enhances the ability of the HTI to select the right products for home networking and system components integration projects.

Classification

Another rating that the HTI may encounter is the classification of cabling that has been established by two international organizations: the International Organization for Standardization (ISO) and the International Electrotechnical

Commission (IEC). These two organizations partnered together to create the ISO/IEC 11801 standard:

- **Class A:** Cabling is characterized up to 100 kHz
- **Class B:** Cabling is characterized up to 1 MHz
- **Class C:** Cabling is characterized up to 16 MHz
- **Class D:** Cabling is characterized up to 100 MHz
- **Class E:** Cabling is characterized up to 250 MHz
- **Class F:** Cabling is characterized up to 600 MHz
- **Optical Class:** Optical fiber links are characterized up to 10 MHz and above

 **STANDARDS**

ANSI/TIA/EIA-570-A

The primary standard that affects residential systems cabling infrastructure is *ANSI/TIA/EIA-570-A*, Residential Telecommunications Cabling Standard. This section will discuss this standard, as well as other relevant standards that the HTI needs to know in order to install cabling.

ANSI/TIA/EIA Standards

TIA

EIA

The American National Standards Institute/Telecommunications Industry Association/Electronic Industries Alliance (ANSI/TIA/EIA) is an important industry grouping that publishes a number of standards affecting residential, commercial, and industrial cabling systems. The *TIA* and *EIA* develop the standards and submit them to ANSI for approval. The standards are then published and made available to the industry. The significance of the ANSI/TIA/EIA is that it helps to standardize components and systems of multiple manufacturers that are installed in buildings.

Figure 2–1
ANSI, TIA, and EIA Logos

Of the many published standards, the ANSI/TIA/EIA-570-A, Residential Telecommunications Cabling Standard, focuses on residential cabling infrastructure, and it is the focus in this course. Other ANSI/TIA/EIA standards that can be a useful reference for the HTI and other systems installers in buildings are summarized as follows:

ANSI/TIA/EIA-568-A

- **ANSI/TIA/EIA-568-B:** Commercial Building Telecommunication Standard, 2001
- *ANSI/TIA/EIA-568-A:* Commercial Building Telecommunications Pathways and Spaces, 1998
- **ANSI/TIA/EIA-606:** Administration Standard for the Telecommunications Infrastructure of Commercial Buildings, 1993
- **ANSI/TIA/EIA-607:** Commercial Building Grounding and Bonding Requirements for Telecommunications, 1994

Only the ANSI/TIA/EIA-570-A has the residential environment as its primary focus. The following section details the ANSI/TIA/EIA-570-A standard.

ANSI/TIA/EIA-570-A

The purpose of 570-A is to standardize the requirements for residential telecommunications cabling. Within this standard, services are correlated to grades of cabling for residential units. The cabling infrastructure specifications within EIA/TIA-570-A are intended to include support for voice, data, video, multimedia, home automation systems, environmental control, security, audio, television, sensors, alarms, and intercom. This standard is intended to be implemented for new construction, additions, and remodeled single-family residential buildings. The practices detailed in EIA/TIA-570-A are also to be implemented for cabling systems designed for each individual unit of a multi-dwelling unit (MDU). The individual units of an MDU (typically townhouses, condominiums, apartment complexes, or assisted living facilities) have cabling needs similar to a single-family residence (read more about MDUs in Chapter 15).

Relevant specifics of this standard are cited throughout this course as the various residential systems are explored. It is important for the HTI and other residential systems installers to understand how and where to find information regarding the EIA/TIA-570-A standard.

The design requirements are the following:

- Use star topology for cabling configuration.
- Devices such as intercom, security system keypads, sensors, and smoke detectors may be wired in a star, loop, or daisy chain.
- A minimum of one outlet to each kitchen, bedroom, family/great room, den/study.
- Additional outlets should be provided on walls exceeding 3.7 meters (12 feet).
- Use 6- or 8-position modular outlets and connectors that comply with ANSI/TIA/EIA-568-A.
- Telecommunications outlets and connectors should be compatible with media.
- Application-specific electrical components such as splitters, amplifiers, and impedance matching devices are to be placed external to the telecommunications outlet or connector.
- 570-A recognizes the following cables:
  100-ohm 4-pair UTP
  Series 6 coaxial
  50/125 m multimode fiber
  62.5/125 m multimode fiber
  Single-mode fiber for special applications only

Section 4—ANSI/TIA/EIA-570-A   This section of the standard establishes a cable grading system based on the type of service to be provided. It defines Grade 1, which meets the minimum requirements for telecommunications services, and Grade 2, which is for more advanced cabling services. Figure 2–2 compares Grade 1 and Grade 2 standards.

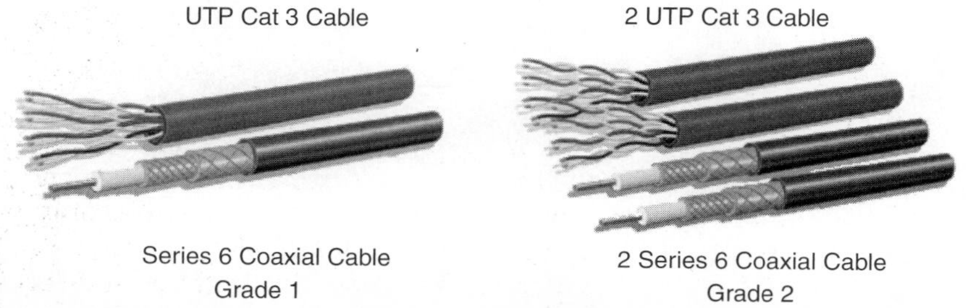

UTP Cat 3 Cable

2 UTP Cat 3 Cable

Series 6 Coaxial Cable
Grade 1

2 Series 6 Coaxial Cable
Grade 2

Figure 2–2
Comparison of Grade 1 and Grade 2 Cabling Systems

The grades are differentiated in the following manner:

- **Grade 1:** Provides a generic cabling system that meets the minimum requirements for telecommunications services. Typical applications consist of telephone, fax, and modem and CATV services—including cable modem and xDSL services.
- **Grade 2:** Provides a generic cabling system that meets the requirements for basic, advanced, and multimedia telecommunications services. Typical applications consist of telephone, fax, and modem and CATV services including cable modem and xDSL services.

Table 2–1 displays the requirements for Grade 1 and Grade 2 for telephone, television, data, and multimedia.

Table 2–1
Grade 1 and Grade 2 Requirements

Service	Grade 1 Requirements	Grade 2 Requirements
Telephone	One (1) 4-pair 100-ohm Category 3 (optional: One (1) 4-pair 100-ohm Category 5)	One (1) 4-pair 100-ohm Category 5 (optional: Two (2) 4-pair 100-ohm Category 5)
Television	One (1) 75-ohm Coax	One (1) 75-ohm Coax (optional: Two (2) 75-ohm Coax)
Data	One (1) 4-pair 100-ohm Category 3 (optional: One (1) 4-pair Category 5)	One (1) 4-pair 100-ohm Category 5 (optional: Two optical fiber)
Multimedia	Not accommodated	One (1) 4-pair 100-ohm Category 5 (optional: Two optical fiber)

Section 5—ANSI/TIA/EIA-570-A   In this section, you can find information on the description of single residential unit-cabling systems. You can also find information relating to the installation of the following:

- Patch cords (see Figure 2–3), equipment cords, and jumper cables
- Residential telecommunications outlet and connectors
- Auxiliary Disconnect Outlet (ADO) (Figure 2–4)
- Structured media center (SMC) or distribution device (DD): installation of input/output connections (see Figure 2–5)
- Outlet cabling, the recognized cables used, and maximum length requirements

Figure 2–3
Patch Cable

Figure 2–4
ADO

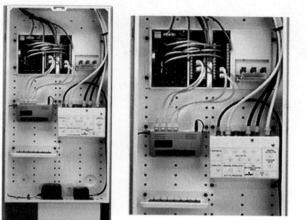

Figure 2–5
SMC or DD

These items are defined later in this chapter, in the section called "ANSI/TIA/EIA-570-A: New Terminology."

Section 6—ANSI/TIA/EIA-570-A   In this section of the standards document, you can find information about multi-dwelling and campus pathways, and spaces from the demarcation point to the MDU telecommunications room.

Section 7—ANSI/TIA/EIA-570-A   Here, you find information about the specifications for residential premises cabling components. In this section, the standard includes requirements for the following:

- Telecommunication outlet/connector terminations
- Connecting the hardware
- Equipment and patch cords
- Unshielded twisted-pair (UTP), coaxial, and fiber optic cables (Figure 2–6)

These cable types are discussed in detail in a later chapter.

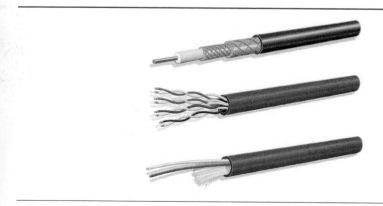

Figure 2–6
UTP, Coax, Fiber

Section 8—ANSI/TIA/EIA-570-A   This section covers general installation requirements. Specifically, it provides warnings (best practices) related to cabling operations. Included are the specifications for minimum bend radius, the maximum allowable pulling tension for different cable types, and so on. Table 2–2 lists the minimum bend radius and the maximum pulling tension for different cable types.

Table 2–2
Minimum Bend Radius and Maximum
Pulling Tension of Cables

Cable Type	Minimum Bend Radius	Maximum Pulling Tension (Measured in Pounds of Force [lbf] and Newtons [N])
Twisted-pair (100-ohm UTP)	Not less than 4 times the cable diameter	25 lbf (110 N)
Coaxial (75-ohm)	Not less than 20 times the diameter of the cable when pulling. Not less than 10 times the diameter of the cable when placing or dressing the cable.	35 lbf (150 N) for Series 6 cable; 90 lbf (400 N) for Series 11 cable
Optical fiber (2- and 4-fiber)	Not less than 10 times the diameter when there is no tension. Not less than 15 times the diameter when cable is subjected to tension.	50 lbf (222 N)

Annexes—ANSI/TIA/EIA-570-A   The ANSI/TIA/EIA-570-A also contains four important annexes or important additional material that provides specific and general information. These annexes are summarized in Table 2–3.

Table 2–3
ANSI/TIA/EIA-570-A Annexes

Annex	Description	Status
A	Optical Fiber Performance Specifications	Normative** and is part of ANSI/TIA/EIA-570-A
B	Field Test Requirements	Normative and is part of the standard
C	Cabling Residential Buildings	Informative, not considered part of the standard
D	Bibliography and References	Informative, not considered part of the standard

*Normative = Means a given provision of a standard is considered part of the standard as opposed to being an appendix.

Addenda—ANSI/TIA/EIA-570-A   Since ANSI/TIA/EIA-570-A was published in 1999, several addenda have been published to address issues such as security and audio cabling:

- **Addendum 1:** Security Cabling for Residences (ANSI/TIA/EIA-570-A-1-2002)
- **Addendum 3:** Whole-Home Audio Cabling for Residences (ANSI/TIA/EIA-570-A-3-2002)

New Terminology   Under the ANSI/TIA/EIA-570-A, some important new terminology is defined. These terms are considered to be specialized words for the industry and their mastery is necessary for the understanding of later chapters in this course.

- **Access provider (AP) or local exchange carrier (LEC):** This is the operator of any facility that is used to convey telecommunications signals to and from a customer's premise
- **Network interface device (NID):** The NID defines the demarcation or ending point from the AP or your local telephone company. It is usually a non-metallic box that is found on the side of a single-family residence.
- **Auxiliary disconnect outlet (ADO):** The ADO provides a means for the customer (homeowner or tenant) to disconnect from the service provider. The ADO is preferably located in the structured media center.

- **Structured media center (SMC):** A structured media center (also referred to as a distribution device or DD) is a convergence point in which outlet cables from various wiring points throughout the house are terminated. It is a cross-connect facility located within each home or tenant space. The SMC in effect is the center of the residential cabling network, similar to the telecommunications room (TR) or equipment room (ER) found in a large commercial or enterprise network.
- **Outlet cable:** An outlet cable provides the path of signal transmission between the structured media center and the outlet/connector or faceplate. When the cross-connect jumpers and patch cords are taken into account, the maximum length of such a cable is 100 m (328 ft).
- **Prewiring:** This represents the wiring installed before the walls are enclosed or finished. It also refers to any wiring that is put in place in anticipation of future uses or needs.
- **Crosstalk:** Unwanted coupling from signals transmitted on one circuit interfering with the signals on another circuit is referred to as crosstalk.
- **Twisted pairs:** These are pairs of copper conductors that are twisted together to become electrically balanced. The twisting of wire pairs maximizes signal strength and minimizes the effects of crosstalk.
- **Insulation displacing connection (IDC):** IDC is a connection method in which wires are forced between double metal contacts by means of a special punch tool to remove insulation and make contact. You will learn about some specific examples of IDCs in this course.

Underwriters Laboratories Inc. (UL)

Underwriters Laboratories Inc. (UL) is an independent, non-profit, product safety testing and certification organization. It has tested products for public safety for more than a century. Worldwide each year, more than 17 billion UL marks are applied to products worldwide.

Underwriters
Laboratories Inc.
(UL)

Figure 2–7
UL Logo

UL was founded in 1894 and has the reputation as the leader in product safety and certification. UL is an independent, not-for-profit product safety testing and certification organization. It is one of the most recognized, reputable conformity-assessment providers in the world. Today, UL's services extend to helping companies achieve global acceptance, whether it is an electrical device, a programmable system, or a company's quality process.

UL evaluates wire and cable products under more than 70 different product categories, utilizing more than 30 standards for safety. Although the UL focuses on safety standards, it has expanded its certification program to evaluate twisted-pair LAN cables for performance according to IBM and TIA/EIA performance specifications, as well as National Electrical Code

(NEC) safety specifications. The UL also established a program to mark shielded and unshielded twisted-pair LAN cables, which should simplify the complex task of making sure that the materials used in the installation are up to specification. Listing by UL denotes initial testing and periodic retesting to assure continuing conformance to standards.

Local fire-code and building-code regulators try to use standards like those of the NEC, but insurance groups and other regulators often specify the standards of the Underwriters Laboratories. UL has safety standards for cables that are similar to those of the NEC. UL 444 is the Standard for Safety for Communications Cable. UL 13 is the Standard for Safety for Power-Limited Circuit Cable. Network cable might fall into either category. UL tests and evaluates samples of cable and then, after granting a UL listing, the organization conducts follow-up tests and inspections. This independent testing and follow-through make the UL markings valuable symbols to buyers.

The UL Local Area Network (LAN) Certification Program addresses not only safety, but also performance. IBM authorized UL to verify 150-ohm shielded twisted-pair (STP) cable to IBM performance specifications, and UL established a data-transmission, performance-level marking program that covers 100-ohm twisted-pair cable. UL adopted the TIA/EIA-568-A performance standard, and through that, some aspects of the Anixter cable performance model. There is a small inconsistency, however. The UL program deals with both STP and unshielded twisted-pair (UTP) wire, whereas the TIA/EIA-568-A standard focuses on unshielded wire. The UL markings range from Level I through Level V. The TIA/EIA has Category 1 through Category 5e, but TIA/EIA has started to refer to levels, too. It is easy to become confused by the similarly numbered levels and types. The UL level markings deal with performance and safety, so the products that merit UL level markings also meet the appropriate NEC, MP, CM, or CL specifications, as well as the TIA/EIA standard for a specific category.

Companies whose cables earn these UL markings display them on the outer jacket as, for example, Level I, LVL I, or LEV I.

Some other organizations publish standards related to cabling for residential systems installation, which may impact installation work.

Web link: http://www.ul.com

Institute of Electrical and Electronic Engineers (IEEE)

Institute of Electrical and Electronic Engineers (IEEE)

The *Institute of Electrical and Electronic Engineers* (*IEEE*) is the world's largest professional engineering society. It publishes the following standards:

- Standards that rate the performance of various electrical and electronic equipment and materials
- Standards that govern the installation and maintenance of such equipment
- Training courses that allow engineers to keep abreast of new developments in the electrical and electronic engineering fields

IEEE 802.3

The HTI and other residential systems cabling infrastructure installers should be familiar with the *IEEE 802.3*, the IEEE equivalent for Ethernet.

Another IEEE standard is for a high-performance serial bus. Known as FireWire or IEEE 1394, this standard supports the high bandwidth requirements of devices such as digital video equipment and high-capacity storage.

Web links:
http://www.ieee.org
http://www.standards.ieee.org

Figure 2–8
IEEE Logo

International Organization for Standardization (ISO)

The International Organization for Standardization (ISO) is an international organization that sets standards for many industries. ISO and its partner—the International Electrotechnical Commission (IEC), an international organization responsible for developing global standards in the electrotechnical field—have defined cabling infrastructure standards that allow the interoperability of electronic networks and equipment.

International Organization for Standardization

Figure 2–9
ISO Logo

An example of such an ISO/IEC standard is the ISO/IEC Standard 11801, Information Technology—Generic Cabling for Customer Premises. This standard accepts both 100-ohm and 120-ohm twisted-pair wiring, and allows for the shielding of both of these cabling types as an option. The HTI should know this standard because many installations will use twisted-pair wiring.

Web link: http://www.iso.ch/iso/en/ISOOnline.frontpage

CODES

The cabling that is installed in the home is subject to rules and regulations governed by the National Fire Protection Association (NFPA) in the United States and the Canadian Standards Association (CSA) in Canada; as well as by local, county, and state building codes. This section discusses the NFPA and the CSA, and their respective electrical codes. Information on how to find out which local, county, state, or province building or fire codes apply to a particular project is also provided.

National Fire Protection Association (NFPA)

The *National Fire Protection Association* (*NFPA*) is an international organization that was established 100 years ago and has members in 70 countries. The NFPA develops, publishes, and distributes fire safety codes and standards intended to minimize the possibility and effects of fire and other risks. Many of these codes and standards are assembled in the National Electrical Code (NEC). Virtually every building, process, service, design, and installation is affected by NFPA documents. More than 300 NFPA codes and standards are used around the world. Figure 2–10 shows the NFPA logo. The

National Fire Protection Association (NFPA)

major fire codes relating to telecommunications installations are summarized as follows:

- **ANSI/NFPA-70:** National Electrical Code (NEC)
- **ANSI/NFPA-71:** Installation, Maintenance, and use of Signaling Systems for the Central Station
- **ANSI/NFPA-72:** National Fire Alarm Code
- **ANSI/NFPA-75:** Protection of Electronic Computer/Data Processing Equipment

More than 225 NFPA Technical Committees, each of which represents a balance of affected interests, develop NFPA documents.

Web link: http://www.nfpa.org

Figure 2–10
NFPA Logo

National Electrical Code (NEC), 2002 Edition

NEC

The *NEC* is revised every three years, and it is currently in its 2002 edition. The code is also known by its official alphanumeric designation of ANSI/NFPA-70. This edition of the NEC is arranged, for easy reference, by chapter, article, and section. You may find, for example, a citation that reads as "Chapter 8, Article 800, Section 800.49." Because of the significance of the safety of people and functionality of the integrated residential systems, it is useful for the HTI to verify essential information from the NEC. Table 2–4 summarizes some areas of interest of the NEC for quick reference.

The National Electrical Code of the United States (NEC) is intended to protect people and property from electrical hazards. It establishes the minimum requirements to safeguard people and property from electrical hazards. The National Fire Protection Agency (NFPA) sponsors, controls, and publishes the NEC in the United States jurisdictional area. Among other things, the NEC describes various types of cables and the materials used in them. The language of the code is designed so that it can be adopted as law through legislative procedure. The HTI should be prepared to see the term *NEC* used widely in cable catalogs, and should not confuse it with specifications from a major international equipment manufacturer with the same initials.

In general terms, the NEC describes how a cable burns. During a building fire, for example, a cable going between walls, up an elevator shaft, or through an air-handling plenum could become a torch that carries the flame from one floor or one part of the building to another. Because the outer coverings of cables and wires are typically plastic, they can create noxious smoke when they burn. Several organizations, including the Underwriters Laboratories (UL), have established standards for flame and smoke that apply to local area network (computer network) cables. However, the NEC contains the standards most widely supported by local licensing and inspection officials.

Table 2–4
NEC Reference Guide

NEC Domain	Title	Information Provided
Section 90.2	Scope	Information on what is found in the NEC.
Section 90.3	Code Arrangement	Notes that Chapter 8 covers communication systems and is independent of other chapters, except where they are specifically referenced therein.
Article 100–Part 1	Definitions	Terms that are not commonly defined in English dictionaries. Some examples include bonding, accessible, premises wiring, signaling circuit, and so on.
Section 110.26	Spaces About Electrical Equipment (600 Volts, Nominal, or Less)	Explains the spaces for working clearances around electrical equipment. Refer to this information when placing electronic devices and components on a communication rack or a terminal in an electrical closet.
Article 250	Grounding	Referenced from Article 800, provides information on communications grounding and bonding network requirements.
Chapter 3	Wiring Methods and Materials	Wiring and cabling installation in raceways. Referenced from Section 800.48.
Article 640	Audio Signal Processing Amplification and Reproduction Equipment	Covers the equipment and wiring for audio signal generation, recording, processing, amplification, and reproduction; sound distribution, electronic musical instruments, speech input systems, and so on.
Article 645	Information Technology Equipment	Covers equipment and power-supply wiring, equipment interconnection wiring, and grounding of information technology equipment and systems.
Article 725	Class 1, Class 2, and Class 3 Remote-Control, Signaling, and Power-Limited Circuits	Covers information on the requirements for security, control, and audio systems.
Article 760	Fire Alarm Systems	Covers the requirements for wiring and equipment of fire alarm systems.
Article 770	Optical Fiber Cables and Raceways	Requirements for listing of fiber optic cable, marking, and installation.
Article 800	Communication Circuits	Requirements for listing of copper communications cable, marking, and installation.
Article 810	Radio and Television Equipment	Covers the requirements for radio and TV receiving equipment.
Article 820	Community Antenna Television and Radio Distribution Systems	The requirements for community antenna TV, satellite, and radio distribution systems.
Article 830	Network-Powered Broadband Communication Systems	Contains information on network-powered broadband communications systems that provide combinations of voice, audio, data, and interactive services through a network interface unit (NIU).

Although the industry recognizes and generally conforms to the NEC standards, every individual municipality can decide whether to adopt the latest version of the NEC (currently the 2002 Edition) for local use. In other words, the NEC standards might or might not be a part of your local fire or building codes. The Occupational Health and Safety Administration (OSHA), a U.S. government agency that protects workers, has adopted the NEC as the code that businesses must comply with to protect workers inside buildings. When in doubt, it is prudent for the HTI or residential cabling subcontractor to select cable and materials that meet the NEC specifications for the given application.

The NEC is used by the following:

- Telecommunications distributions designers, including residential cabling installers and the HTI: Used to ensure a compliant installation
- Electrical inspectors and fire marshals: Used in loss prevention and safety enforcement
- Lawyers and insurance companies: Used to determine liability
- Residential systems manufacturers and suppliers: Used in warranty issues

Canadian Standards Association (CSA)

The Canadian Standards Association (CSA), chartered in 1919, was the first organization in Canada formed to develop industrial standards exclusively. The CSA is an independent, non-profit, standards-writing, certification, testing, and inspection organization. The association provides an open forum for the public, government, and businesses to voluntarily reach agreement through the consensus process on the criteria that best meet the community interest for materials, products, structures, and services in a wide variety of fields. The CSA has published more than 1,500 standards in eight major fields.

CSA standards cover many aspects, including materials, testing procedures, and construction. On behalf of the Standards Council of Canada (SCC), CSA represents Canada on various International Organization for Standardization (ISO) committees. CSA also works closely with the International Electrotechnical Commission (IEC) to develop standards.

Because CSA is not part of any government, it has no legislative powers. It is up to federal, provincial, and municipal governments to reference CSA standards in legislation. Many local governments have done so. The CSA Canadian Electrical Standard is generally adopted to become the Canadian Electrical Code, which is enforceable. The CSA logo is shown in Figure 2–11.

Web link: http://www.csa.ca

Figure 2–11
CSA Logo

Canadian Electrical Code (CEC), Part 1

Similar to the NEC, the Canadian Electrical Code (CEC), Part 1, is the standard for electrical wiring in Canada upon which all provincial codes are based. The CEC is derived from the Canadian Electrical Standards document created by the Canadian Standards Association (CSA). More specifically, the CSA sponsors, controls, and publishes the CEC, Part 1. It is a voluntary code that may be adopted by regulatory authorities of the various regions of Canada and enforced as the law.

CEC, Part 1 is the equivalent of the NEC in the United States. Table 2–5 displays the CEC, Part 1 Reference Guide, and summarizes those sections of the CEC, Part 1 that are of primary interest to the HTI.

CEC Reference	Title	Information Provided
2	General Rules	Marking of cables, permits, and flame spread requirements for electrical wiring/cables
10	Grounding and Bonding	Detailed grounding and bonding information Requirements for using and identifying grounding and bonding conductors
12	Wiring Methods	Requirements for installing: Raceway systems Boxes and other system elements
54	Community Antenna Distribution, Radio and Television Installation	Circuits that are used to distribute video and other information frequency signals
56	Optical Fiber Cables	Requirements for fiber optic cable installation
60	Electrical Communication Systems	Requirements for copper communication circuits installation
82	Closed-Loop Power Distribution	Requirements for controlling the signal between energy controlling equipment and utilization equipment

Table 2–5
CEC, Part 1 Reference Guide

Local, County, and State or Province Codes

The implementation of codes, such as building and fire codes, is often within the domain of city and/or county governments. This means that projects within the city are handled by the appropriate city agencies, whereas those outside the city are covered by county or state agencies. In some states, the implementation of codes varies from township to township. The HTI must be familiar with all of the different code-enforcement polices.

It is common for codes that require local inspection and enforcement to be incorporated into state or provincial governments and then possibly down to city and county enforcement units. Building codes, fire codes, and electrical codes are examples. Like occupational safety, these were originally local issues, but disparity of standards and lack of enforcement have led to national standards in many cases. When adopted by state or local authorities and enforced to appropriate levels, these standards are then turned over to the lower-level authorities for implementation.

However, many dynamics are applied to get local adoption and implementation of many codes. Insurance underwriters often set ratings for coverage based on implementation of various codes. Many states limit employer liability and employees' access to the courts as part of the enticement to businesses to not fight the implementation process.

The result is often an inconsistent application of the codes and requirements. It is therefore essential for the HTI to check the requirements for a particular job site. Some codes may be enforced by city, county, or state agencies. Fire codes are enforced by either the building permit department or the fire department. The specifics of how this is done in every single geographic region are beyond the scope of this course. The best way to identify the codes that apply to a specific project or area is to consult the local authorities, who can usually be found in city halls or municipal buildings. Alternatively, you can check the local governmental listings in the local telephone book under titles such as the following:

- Building inspector
- Local fire marshal or inspector
- Electrical inspector

When the HTI is subcontracting on a new construction project, the construction supervisor will be the main contact to verify all aspects of the job. When the project is a remodel for an existing home, the HTI will need to ensure that all of the necessary permits for the job have been obtained.

> **NOTE**   Understand the hierarchy of codes. Local codes always take precedence over state codes, which take precedence over national codes.

Building codes are based on accepted safety and fire-protection principles. In the United States, three private organizations have developed "model" codes:

- **The International Conference of Building Officials (ICBO):** Publishes the *Uniform Building Code*
- **The Building Officials and Code Administrators International, Inc. (BOCA):** Publishes the *National Building Code*
- **The Southern Building Code Congress International, Inc. (SBCCI):** Publishes the *Standard Building Code*

Typically, most government agencies adopt part or all of these model codes, making them mandatory and enforceable.

 REGULATIONS

The Federal Communications Commission (FCC) and the Occupational Safety and Health Administration (OSHA) regulate telecommunications wiring and workplace safety in the United States. An HTI who violates the regulations of either agency may be subject to criminal or civil action. In addition, some states or provinces regulate those who install cable by requiring licenses. Check with your state or province to determine whether you need to secure a license before your first project.

Federal Communications Commission (FCC)

Federal
Communications
Commission

The *Federal Communications Commission* is a Washington, D.C. based agency of the federal government that is in charge of monitoring and regulating telecommunications in the United States (Figure 2–12). It is a fairly complex administrative and regulatory body that oversees the activities of such companies in the United States as Sprint, AT&T, MCI, and so on. However, the FCC is also the regulatory body that oversees the manufacture, performance, installation, and operation of telecommunications equipment—including telephone headsets, modems, wireless devices, and so on—that is installed in individual residences. This equipment is referred to as "Customer Premises Equipment."

The FCC establishes policies and regulations that govern interstate and international communications by television, radio, wire, satellite, and cable. The FCC processes applications for licenses and other filings, analyzes complaints, conducts investigations, and develops and implements regulatory programs. Among other issues that the FCC deals with are wireless frequency auctioning, bandwidth, digital television, and home satellite dishes.

In February 1996, the Telecommunications Act of 1996 was signed into law, representing the first major overhaul of telecommunications policies in more than 60 years in the United States. Major FCC initiatives that influence the residential networking market are the following:

- Ensuring that all electronics equipment comply with FCC regulations
- Monitoring the rollout of broadband access and encouraging competition in the market
- Ensuring that third-generation wireless systems provide access, by means of one or more radio links, to a wide range of telecommunication services supported by the fixed telecommunication networks and to other services that are specific to mobile users

Figure 2–12
FCC Logo

FCC Part 68   *FCC Regulations Part 68* was developed to assure consumers, manufacturers, and carriers that terminal equipment and wiring could be connected without degrading the network. Part 68 describes the minimum requirements for the mechanical or physical properties of wiring devices, such as the amount of gold on jack contacts, electrical performance, dimensional integrity, material properties, and spring characteristics of contacts. Only equipment that meets FCC Part 68 standards may be manufactured and connected to the network. FCC Part 68 is of major importance and should be taken into account by the residential systems HTI when choosing or recommending products for homeowners.

FCC Regulations Part 68

Wiring Docket 88-57   Part 68 has been revised in light of the vast changes that are occurring in the networking industry in general and in the residential networking in particular. In early 2000, the FCC adopted an order (CC Docket No. 88-57) that set a new minimum quality standard for telephone inside wiring using Category 3 cable. The purpose of this was to eliminate the inferior cabling that was being installed, which caused crosstalk between telephone lines and negatively affected the service provider.

Occupational Safety and Health Administration (OSHA)

Most state and local government bodies recommend the establishment, implementation, and maintenance of safety programs for both on- and off-worksite locations. You must check with the regulations to ensure that you are following the required safety practices and documentation standards. Most nations have rules designed to protect workers against hazardous conditions. In the United States, for example, Congress in 1970 passed the Occupational Safety and Health Act in an attempt to ensure a healthful work environment for every working person in the country. Following this statute, the *Occupational Safety and Health Administration (OSHA)* was created within the U.S. Department of Labor. OSHA is the organization charged with worker safety and health in the United States.

Occupational Safety and Health Administration (OSHA)

Figure 2–13
OSHA Logo

OSHA is a federal and state program that sets guidelines and practices for safety. In the United States, if your state has an OSHA program, you are required to follow its guidelines; if it does not, however, you are required to follow the federal guidelines. Since the creation of this agency, workplace fatalities have been cut in half, and occupational injury and illness rates have declined 40 percent. At the same time, United States employment has nearly doubled: from 56 million workers at 3.5 million worksites to 105 million workers at nearly 6.9 million sites.

OSHA is responsible for protecting workers by enforcing United States labor laws. Technically speaking, OSHA is not a building code or building permit-related agency. However, OSHA inspectors have the power to impose heavy fines and/or shut down a job site if they find serious safety violations. Anyone who works on or is responsible for a construction site or business facility needs to be familiar with OSHA regulations; so does the HTI. The organization offers safety information, statistics, and publications on its web site.

Web links:
http://www.osha.gov
http://www.allaboutosha.com

Cable Installers' Licenses

The type of cabling infrastructure required for home technology integration is low-voltage. Currently, only a handful of states require licensing for residential installations involving low-voltage wiring, but the trend in the United States, Canada, and other nations is toward this type of licensing.

The HTI should verify the requirements for licensing in their area of operation. To determine whether licensing is a requirement in a specific state or region, contact the appropriate state or provincial government agency. This information is usually available in the local telephone book. The board of electrical inspectors should be able to provide the information or direct you to the appropriate agency.

SUMMARY

In this chapter, you learned about organizations and their standards, codes, and regulations that pertain to home integrated networks:

- Standards provide specifications and guidelines for proper installations in order to ensure systems performance. They are not mandatory rules; but they establish a basis to compare, judge, or measure the important aspects of a residential system.
- Codes address the minimum safety requirements to which products and installation procedures must adhere. The two important words in codes are "shall," which indicates a mandatory requirement; and "should," which is an advisory recommendation.
- Regulations are mandatory rules usually issued by government agencies. Installation methods, materials, electrical, and many other specified products must conform to the requirements of the Authority Having Jurisdiction (AHJ).

- Organizations, groups, and associations publish standards, codes, and regulations that specify the methods of installation of a residential environment. This information can be found on the Internet.
- ANSI/TIA/EIA-570-A, Residential Telecommunications Cabling Standard, is the standard that the HTI must follow when installing cabling infrastructure. It is implemented for new construction, additions to existing structures, and each individual unit of a multi-dwelling unit (condominiums, townhouses, and apartments).
- The National Fire Protection Association (NFPA) develops, publishes, and distributes fire safety codes and standards intended to minimize the possibility and effects of fire and other risks. Virtually every building, process, service, design, and installation is affected by NFPA documents. More than 300 NFPA codes and standards are used around the world.

- The NEC, sponsored by the NFPA, establishes minimum requirements to safeguard people and property from electrical hazards. The NEC contains the standards most widely supported by local licensing and inspection officials. Its equivalent in Canada is the CEC.
- The Occupational Health and Safety Administration (OSHA) uses the NEC as the code that businesses must comply with to protect workers inside buildings. The HTI should ensure that all cable and materials meet NEC specifications.
- Underwriters Laboratories Inc. (UL) evaluates wire and cable products for safety and performance. Companies whose cables earn these UL markings display them on the outer jacket, for example, Level I, LVL I, or LEV I.

GLOSSARY

ANSI/TIA/EIA-568-A A wiring standard for voice and data communications. Voice and data cable should be terminated using the 568-A standard.

ANSI/TIA/EIA-570-A A wiring standard for residential telecommunications cabling.

EIA (Electronic Industries Alliance) An organization that sets standards for interfaces to ensure compatibility between data communications equipment and data terminal equipment.

FCC (Federal Communications Commission) Federal government agency charged with regulating the communication industry in the United States.

FCC Regulation Part 68 A regulation that describes minimum requirements for the mechanical or physical properties of wiring devices.

IEEE (Institute of Electrical and Electronic Engineers) A professional organization whose activities include the development of communications and network standards. IEEE LAN standards are the predominant LAN standards in use today.

IEEE 802.3 The standard for Ethernet networks.

NEC (National Electrical Code) A set of rules and regulations plus recommended electrical practices that are published by the National Fire Protection Association and generally accepted as the building wiring standard in the U.S.

NFPA (National Fire Protection Association) A nonprofit organization that publishes the NEC.

OSHA (Occupational Safety and Health Administration) A federal and state program that sets guidelines and practices for safety.

Standards Agreed-on protocol principles that are set by committees working under various trade and international organizations.

TIA (Telecommunications Industries Association) An organization that develops standards relating to telecommunications technologies. Together, TIA and EIA have formalized standards, such as EIA/TIA-232, for the electrical characteristics of data transmission.

UL (Underwriters Laboratories Inc.) A private company that specializes in the testing of new products to make sure they meet safety standards. UL is also involved in the certification program for category-rated cable.

CHECK YOUR UNDERSTANDING

1. Which agency or organization administers OSHA?
   a. United States Department of Labor
   b. ANSI/TIA/EIA
   c. NFPA
   d. UL

2. The NEC governs the use of electrical wire installed in which of the following?
   a. Commercial buildings
   b. Businesses
   c. Homes
   d. All of the above

3. What is ANSI/TIA/EIA-570-A an example of?
   a. Regulation
   b. Code
   c. Standard
   d. Rating

4. Which government agency is charged with regulating the communications industry in the United States?
   a. NEC
   b. FCC
   c. ANSI
   d. IEEE

5. What is the National Electrical Code also known as?
   a. ANSI/NFPA-1
   b. ANSI/NFPA-70
   c. ANSI/NFPA-71
   d. ANSI/NFPA-72

6. Which standard or code should be followed when implementing cabling systems for multi-dwelling units?
   a. ANSI/EIA/TIA-570-A only
   b. IEEE 1394 only
   c. NEC only
   d. Both ANSI/EIA/TIA-570-A and NEC

7. Grade 1 cabling meets the minimum requirements for telecommunications services.
   a. True
   b. False

8. Which of the following provides information on communications grounding and bonding network requirements?
   a. NEC Article 250
   b. ANSI/TIA/EIA-570-A Annex A
   c. XCA
   d. ANSI/TIA/EIA-568-A Section 4

9. What do the UL level markings deal with?
   a. Performance only
   b. Safety only
   c. Both performance and safety
   d. None of the above

10. Two UTP Category 5 and two Series 6 coaxial cables are the minimum requirements for which of the following?
   a. Level 1 cabling systems
   b. Level 5 cabling systems
   c. Grade 1 cabling systems
   d. Grade 2 cabling systems

11. According to the EIA/TIA-570-A standard, devices such as intercoms, security system keypads, sensors, and smoke detectors may be wired in a:
   a. star topology.
   b. loop.
   c. daisy chain.
   d. All of the above

3

The Computer Network

OBJECTIVES

Upon completion of this chapter, the HTI will be able to complete the following tasks:

- Understand basic networking concepts

- Identify the layers of the OSI model

- Define the steps of the home network design process

- Know the advantages of both wired and wireless networks

- Understand the differences between dial-up, DSL, and cable modem connections to the Internet

- Develop connectivity documentation

INTRODUCTION

The focus of this chapter is the computer network as part of the integrated home network. The computer network is perhaps the most critical component in an integrated home network. The computer network is a data communication system that allows a number of independent devices to communicate directly with each other. The computer network connects devices such as PCs, laptops, printers, servers, and scanners. These devices are commonly referred to as *hosts*. Home technology integrators use cabling, such as CAT 5e, or wireless technology to connect hosts to the computer network. Other network devices include repeaters, bridges, switches, gateways, and routers. These devices help direct data flow from one device to another on the computer network.

The computer network provides the digital data access that enables other home subsystems to be linked and controlled throughout the integrated home network. Therefore, it is critical to carefully take into account some general considerations when designing and installing computer networks. First among these considerations are the homeowner's current and future needs. As an integrator, you need to be knowledgeable about industry trends, consider the growth in the home-owner's family size, and take into account the technology needs and level of technical experience possessed by the end-users. Knowledge of these key factors can greatly influence the extent to which potential future needs can be accommodated in the initial design and engineering of the computer network.

BASIC NETWORKING CONCEPTS

The purpose of a data network is to help an organization increase productivity by linking all the computers and computer networks so that people have access to the information, regardless of differences in time, location, or type of computer equipment.

In this section, you will learn about general data networking concepts. The knowledge you gain here will arm you with the basic tools needed to appreciate the significance of data networks in home network integration. You will learn about the OSI model and its central role in giving structure and organization to the communication process between computers within the same network or over an external network such as the Internet, for example.

Local-Area Network (LAN)

LAN

A local-area network, or *LAN*, is a physical interconnection of devices in a single building or group of buildings that communicate with each other using a specific protocol. The most common application of a LAN is the home computer network. However, home network integration connects other subsystem devices to the LAN to expand their capabilities and allow them to communicate with each other. Figure 3–1 illustrates a LAN.

Figure 3–1
Local-Area Network (LAN)

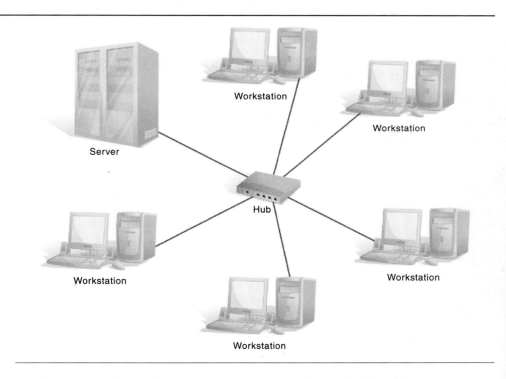

The LAN consists of the various devices (such as computers and printers), the method of connection among the devices (hardwired or wireless), and the protocol used to send and receive data along the connection to the devices.

Wide-Area Network (WAN)

A wide-area network (*WAN*), as illustrated in Figure 3–2, is a network that spans a larger geographic area than a LAN. It typically comprises two or more LANs.

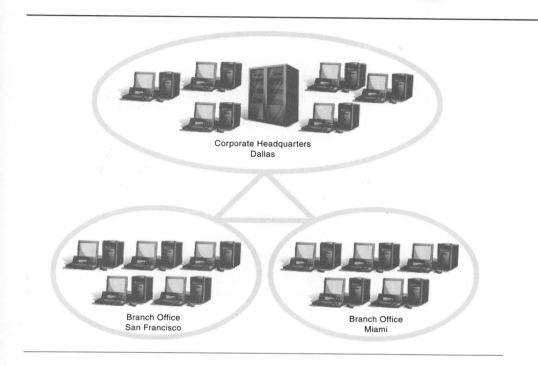

Figure 3–2
Wide-Area Network (WAN)

The Internet is the largest WAN in the world. With a high-speed Internet connection, data, voice, and audio/video can be quickly transmitted to and from the home network. Subsystems in the home that are integrated into the home network can then be remotely accessed, monitored, and controlled from any location in the world.

The connection of the LAN to the WAN via high-speed Internet access is a key element in home network integration. There are several high-speed Internet access technologies that will connect the WAN to the LAN. (We will explore these technologies in the following sections.) Each technology supports varying data-transmission speeds; however, a key characteristic is that they all allow the LAN to be always connected to the Internet. The customer does not have to dial up to the Internet and disrupt the traditional telephone service.

Open Systems Interconnection (OSI) Model

The International Organization for Standardization (ISO) developed the *OSI model* in the 1980s. The Open Systems Interconnection (OSI) reference model is an industry standard framework that is used to divide the functions of networking into seven distinct layers. Today, it is one of the most commonly used teaching and reference tools in networking. The OSI reference model provides a way to understand how a network operates; and it serves as a guideline for creating and implementing network standards, devices, and internetworking schemes.

There are several advantages of using the OSI model:

- It breaks down the complex operation of networking into simple elements.
- It enables engineers to specialize design and development efforts on modular functions.
- It provides the capability to define standard interfaces for "plug-and-play" compatibility and multi-vendor integration.

The OSI reference model has seven layers: physical, data link, network, transport, session, presentation, and application. The layers of the OSI model are shown in Figure 3–3.

Figure 3–3
OSI Model

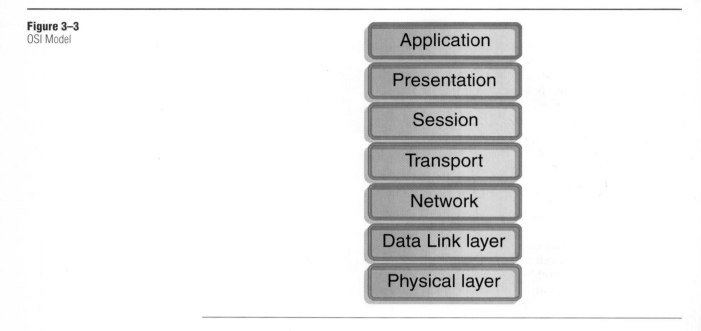

The four lower layers of the OSI model deal with data transport. The physical, data link, network, and transport layers define ways for end stations to establish connections to each other in order to exchange data. The three upper layers deal with applications. The session, presentation, and application layers define how the applications within the end stations will communicate with each other and with users. Figure 3–4 shows the functions of the layers of the OSI model.

Figure 3–4
Functions of the OSI Model

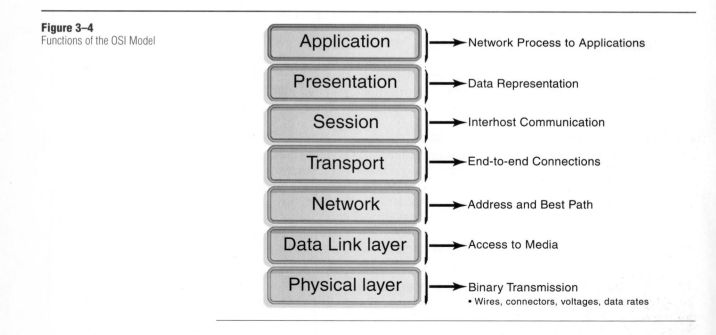

- **Application layer (Layer 7):** This is the highest layer of the model, providing services to application processes outside the model. It is the point where the user or application interfaces with the protocols to gain access to the network. For example, a word processor is serviced by file transfer services at this layer.
- **Presentation layer (Layer 6):** The presentation layer provides a variety of coding and conversion functions that are applied to application layer data. These functions ensure that data sent from the application layer of one system can be read by the application layer of another system. An example of coding functions is the encryption of data after it leaves an application. Another example is the JPEG and GIF formats of images displayed on web pages. This formatting ensures that all web browsers, regardless of operating system, can display the images.
- **Session layer (Layer 5):** The session layer is responsible for establishing, managing, and terminating communications sessions between presentation layer entities. Communication at this layer consists of service requests and responses that occur between applications located in different devices. An example of this type of coordination is that between a database server and a database client.
- **Transport layer (Layer 4):** The transport layer is responsible for delivery or network communication between end nodes. It uses protocols such as TCP or UDP to monitor whether a packet has been delivered; and if not, whether it should be sent again. TCP and UDP are discussed in detail later in this chapter.
- **Network layer (Layer 3):** This layer provides connectivity and path selection between two end systems. Routers and gateways are network devices that function at this layer. They take logical addresses (also known as Internet Protocol [IP] addresses) that are contained in data packets to determine the route the packet will take to reach its destination.
- **Data Link layer (Layer 2):** Provides the reliable transfer of data across media using physical addressing and the network topology. Bridges and switches are network devices that function at the data link layer. They use data link layer information, namely the Media Access Control (MAC) addresses of network interface cards (NICs) that are contained in data packets, to direct the data to the correct destination. The MAC address is also known as a physical address.
- **Physical layer (Layer 1):** The physical layer is made up of the physical media, or the cable and wire, which transmit data from a source computer to a destination computer. Hubs and repeaters are networking devices that also belong to this layer.

Encapsulation

Encapsulation is a function of the OSI model that wraps data with control information and unwraps data for display at the destination. Figure 3–5 illustrates the process of encapsulation. Each layer of the OSI model allows encapsulated data to pass across the network. These layers exchange information to provide communications between the network devices. To exchange information, the layers use protocol data units (PDUs). A *PDU* is the technical name of a frame of data transmitted over the data link layer (Layer 2) in a communications network. PDUs are more commonly referred to as *packets*, and they include control information and user data needed to get the packet to its destination. Control information resides in fields called headers and trailers.

PDU

Figure 3–5
Encapsulation

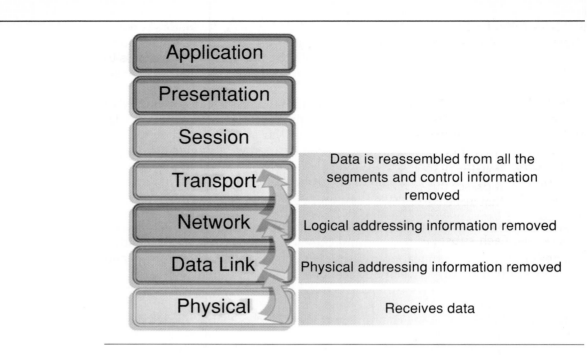

Application

Presentation

Session

Transport — Data is reassembled from all the segments and control information removed

Network — Logical addressing information removed

Data Link — Physical addressing information removed

Physical — Receives data

Encapsulation

To attach control information to a packet, the layers use a process called *encapsulation*. When a layer receives a packet, it encapsulates it with a header and trailer, and it then passes the packet down to the next layer. The control information that is added to the packet is read by the peer layer on the remote device. For example, the transport layer receives a packet from the upper layers. It adds control information such as the application from which the packet was generated. It then passes the packet to the network layer. The network layer encapsulates the packet with its own header information. The packet is then passed to the data link layer. The data link layer encapsulates the network-layer information in a packet called a frame. The frame header contains information required to complete the data link functions. When the physical layer receives the frame, it encodes the frame into a pattern of ones and zeros for transmission, usually on a wire.

Although encapsulation seems like an abstract concept, it is actually quite simple. Imagine that you want to send a book to a friend in another city. How will the book get there? Basically, it will be transported on the road or through the air. You can't go outside and set the book on the road or throw it up in the air and expect it to get there. You need a service to pick it up and deliver it. So, you call your favorite parcel carrier and give them the book. But that's not all. You need to provide the parcel carrier with an address and then send the book on its way. But first, the book needs to be packaged. Here's the complete process:

Step 1   Pack the book.
Step 2   Place an address label on the box.
Step 3   Give the box to a parcel carrier.
Step 4   The carrier drives down the road.

This process is similar to the encapsulation method that protocol stacks use to send data across networks. After the package arrives, your friend has to reverse the process. He takes the package from the carrier, reads the label to see who it's from, and finally opens the box and removes the book. The reverse of the encapsulation process is known as *de-encapsulation.* De-encapsulation is further explained in the next section.

De-Encapsulation

When a remote device receives a sequence of bits, it passes the sequence to the data link layer for frame manipulation. When the data link layer receives the frame, it does the following:

Step 1   It reads the control information provided by the peer source device.
Step 2   It strips the control information from the frame.
Step 3   It passes the frame up to the next layer, following the instructions that appeared in the control portion of the frame.

This process is referred to as de-encapsulation and is illustrated in Figure 3–6. Each subsequent layer will perform this same de-encapsulation process.

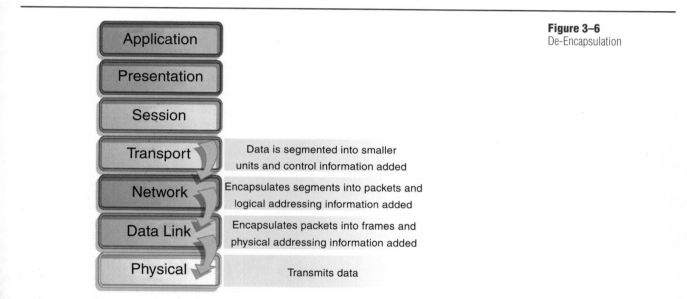

Figure 3–6
De-Encapsulation

TCP/IP Protocol Stack

Transmission Control Protocol/Internet Protocol (*TCP/IP*) is a set of protocols used to communicate across interconnected networks. A *protocol* is a controlled sequence of messages that are exchanged between two or more systems to accomplish a given task. Protocol specifications define this sequence together with the format or layout of the messages that are exchanged. In coordinating the work between systems, protocols use control structures in each system that operate like a set of interlocking gears. This occurs so that computers can precisely track the protocols' points as they move through the sequence of exchanges.

The TCP/IP protocol stack is an industry standard. It is the protocol of choice for the Internet. TCP/IP is used for such common applications as e-mail, remote login, terminal emulation, and file transfer. Figure 3–7 shows how TCP/IP relates to the OSI model.

Figure 3–7
The OSI Model Compared to TCP/IP

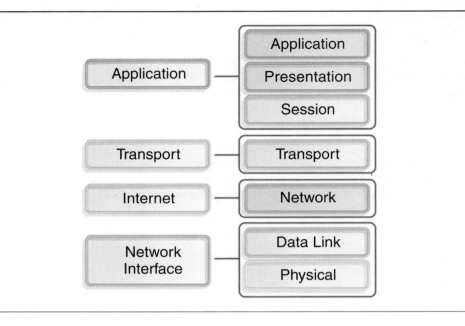

The TCP/IP protocol suite is composed of four layers. The lowest layer, the network interface layer, corresponds to the physical and data link layers of the OSI model. The second layer, the Internet layer, corresponds to the network layer of the OSI model. The third layer, the transport layer, corresponds to the transport layer of the OSI model. Finally, the fourth layer, the application layer, corresponds to the three upper layers of the OSI model: session, presentation, and application.

The Internet layer of the TCP/IP protocol suite is the part of the protocol that provides addressing and path selection. It is the layer at which routers operate in order to identify paths in the network.

The transport layer performs two functions: flow control provided by sliding windows, and reliability provided by sequence numbers and acknowledgments. Flow control is a mechanism that allows the communicating hosts to negotiate how much data is transmitted each time. Reliability provides a mechanism for guaranteeing the delivery of each packet.

The application layer of the TCP/IP protocol suite defines the application protocols for the following operations:

- **File transfer:** Trivial File Transfer Protocol (TFTP), File Transfer Protocol (FTP), and Network File System (NFS)
- **E-mail:** Simple Mail Transfer Protocol (SMTP)
- **Remote login:** Telnet and rlogin
- **Network management:** Simple Network Management Protocol (SNMP)
- **Name management:** Domain Name Service (DNS)

TCP versus UDP

Two protocols are provided at the transport layer: TCP and UDP.

Transmission Control Protocol (TCP) is a connection-oriented, reliable protocol. In a connection-oriented environment, a connection is established between both ends before transfer of information can begin. TCP is responsible for breaking messages into segments, reassembling them at the destination station, resending anything that is not received, and reassembling messages from the segments. TCP supplies a virtual circuit between end-user applications.

User Datagram Protocol (*UDP*) is a connectionless and unacknowledged protocol. Although UDP is responsible for transmitting messages, no checking for segment delivery is provided at this layer. UDP depends on upper-layer protocols for reliability. The differences between TCP and UDP are shown in Figure 3–8.

Figure 3–8
TCP and UDP

TCP

Using TCP is similar to using FedEx for sending a letter. FedEx tracks the letter from when it is picked up to when it is delivered.

UDP

Using UDP is similar to sending a letter by regular post. The letter is not tracked and delivery is not guaranteed.

TCP and UDP use protocol port numbers to distinguish multiple applications (described in the sections that follow) running on a single device from each other. The port number is part of the TCP or UDP segment, and is used to identify the application to which the data in the segment belongs. There are well-known, or standardized, port numbers assigned to applications, so that different implementations of the TCP/IP protocol suite can interoperate. Examples of these well-known port numbers include the following:

- File Transfer Protocol (FTP): TCP port 20 (data) and port 21 (control)
- Telnet: TCP port 23
- Trivial File Transfer Protocol (TFTP): UDP port 69

THE REASONS TO NETWORK

The following sections describe the most popular reasons for choosing to have a home network installed. These reasons will vary with each homeowner, but the benefits are obvious. The capability to share files, a printer, and media components saves time and money in the home, just as it does in the office. Video surveillance is an emerging technology that allows the homeowner to monitor their property. These topics are discussed in the next few sections.

File and Print Sharing

For the homeowner, there will always be a need to share resources like printers and files on the home network. These services can be made easier by the presence of some sort of centralized resource location, such as a home server. Luckily, the current residential gateway market trend is the inclusion of a built-in home server. This is typical of most multi-service residential gateways. If the home server cannot support print sharing, another option is to consider including printers that use the Data Link Control (DLC) protocol and can be connected directly to the network. Residential gateways are explained later in the chapter. Figure 3–9 shows how using one printer can save costs.

Figure 3–9
Shared Network Printer

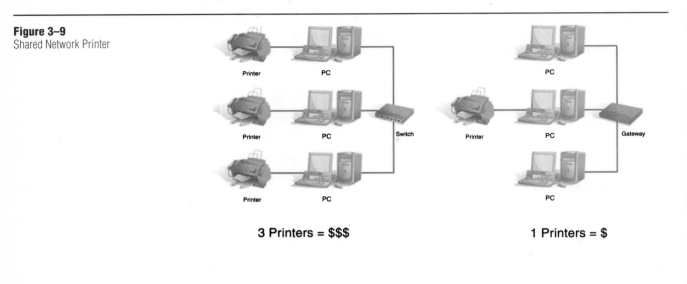

3 Printers = $$$ 1 Printers = $

One printer can serve the entire household. Sharing printers saves homeowners money.

Media Sharing

Data storage is a common practice to any network user. For a home network, this may be the most important feature in the entertainment and audio/video subsystem, especially if large files have to be downloaded from the Internet to be shared by family members. For this reason, a residential gateway with a built-in server having a large hard drive would be preferable. Figure 3–10 shows a server, a gateway, and a PVR.

Figure 3–10
Server, Gateway, and PVR

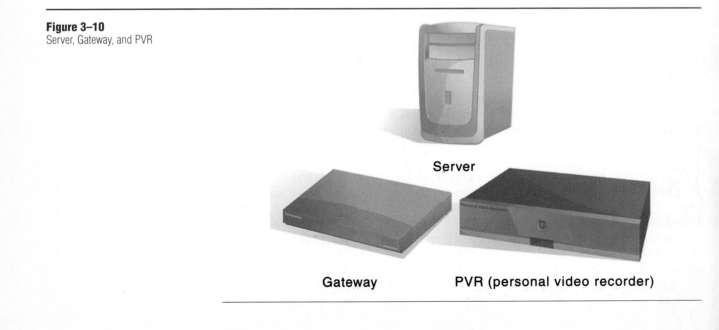

Server

Gateway   PVR (personal video recorder)

In addition, if the homeowner's budget permits, the HTI can deploy additional components such as personal video recorders (for example, Replay TV and TiVO). These devices usually have large hard drives that can hold hours of programming.

E-Mail and Data Storage (ISP)

E-mail and data storage are important services for home network users. Many people with high-speed Internet service to their home today tend to want to host their own "@home" e-mail service. In making a decision to go this route, the user has to be well-advised about the need for a home server and also the increased security demands that will go with such a service. As mentioned earlier, many homeowners do not necessarily possess the network management skills that will guarantee the security of such a system.

Currently, many home users still outsource their @home e-mail service for hosting by their Internet service provider (ISP). Although many keep a certain capacity for data storage at home, users also prefer to rent data storage space from their ISP for a small monthly fee. The ISP has MIS personnel whose job it is to manage both e-mail and data storage for its clients, home network users. This tends to be a win-win situation for both the home network user (e-mail and data security) and the ISP (source of revenue). Figure 3–11 shows examples of different ISPs.

Figure 3–11
Internet Service Providers—ISPs

Home servers can be used as dedicated tools to perform more specific tasks. For instance, music servers can be used to download and store music from the Internet and then stream the music to multiple speakers throughout the home.

Video Surveillance

Video surveillance is a common practical use of the residential network. Many mini digital cameras are available today that can be connected to the home network and used for real-time monitoring of all or parts of the residence. Figure 3–12 shows a video camera that is part of a home video surveillance installation.

Figure 3-12
Video Camera for Home Surveillance

The images taken can be stored on a computer and accessed any time. The one big advantage of video surveillance is that the camera can be accessed remotely via the Internet by the homeowner from his or her place of work. One typical application of such an in-house service is the home portal, which allows the private virtual viewing of the home's interior via the Internet.

Define the Scope of the Network

The first step in the network design process is to gather the homeowner's technical requirements so that you can determine what functionality the homeowner needs. Therefore, before designing a computer network, it is important to understand what the homeowner wants to accomplish. In addition, it is important to document where equipment is to be located throughout the house, as well as the budget constraints of the project. Time constraints are also a concern. Develop a timeline of work to be completed in order to stay on track. The HTI needs to meet with the homeowner to gather all this information, and must continually update the homeowner on the progress of the network design and installation. Doing so ensures that the requirements are being met and enables you to take into account any changes that need to be made.

Determine Network Services

Some homeowners will be very specific about their requirements. Others will not understand what home networks can offer them. For the latter, be prepared to give a small presentation of the possibilities that a home network can provide. The presentation should be tailored to the type of project (retrofit versus new construction). A customer who is having a new home built will be able to incorporate and install the computer network infrastructure during the electrical phase of construction. As an HTI, you will be invaluable in making sure that all rooms are set up with the appropriate wall jacks as well as a central location for the router or switching equipment. A homeowner with an existing structure needs to retrofit the infrastructure for a computer network. This process includes running cable through the attic and down the walls. Although this may cause some damage, it is easily repaired and is preferred to running cable along baseboards. A retrofit can also be accomplished with wireless technology. (Wireless technology is discussed later in this chapter.) As the HTI, it is your job to make sure that the homeowner is aware of all the options that are available for installing a computer network. Use the following list to determine the homeowner's network requirements:

- Services such as video, e-mail, Internet, and virtual private network (VPN) (an illustration of a VPN is shown in Figure 3–13)
- Home office/telecommuter
- Lifestyle: business use, casual use, games, music, surfing the Internet
- Single person, couple, family
- Security, child protection/control (sensitivity to security issues)
- Technical competency of homeowner: A first-time or technologically shy user requires a very user-friendly graphical user interface (GUI) and may need customer support material.
- Professional needs of parents and educational needs of the children; e-mail type and volume
- Need to share devices (printer, CD-ROM, files, digital cameras, and so on)
- Remote access needs: This includes the ability to connect to a home network via a computer or phone line.
- A management information systems (MIS) contact person for home-to-office connection

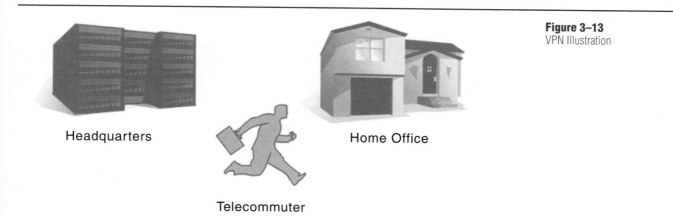

Figure 3–13
VPN Illustration

Headquarters

Telecommuter

Home Office

A VPN is a private network that is constructed within a public network infrastructure such as the Internet. A telecommuter can access the company's network via the Internet by using VPN. A VPN is a service that offers secure, reliable connectivity. A VPN is the most cost-effective method of establishing a point-to-point connection between remote users (telecommuters and home office users) and a company's LAN.

Starting at the first meeting with the homeowner, carefully documenting detailed requirements will help you design and install a network that is tailored to the homeowner's needs. Additionally, detailed notes of what the homeowner needs to accomplish and the physical layout of the home will help in the design phase. You will also need to inspect the home to document the physical layout of the building and the existing equipment. The size of the home will affect the cost of wiring as well as the placement of devices.

Document Customer's Equipment

As an HTI, you will be responsible for ensuring that existing equipment will function on the home network and that it is compatible with new equipment that the homeowner plans to acquire at a later date. Understanding these fundamentals will help to determine the location of the wall jacks for the cable connections.

There is a wide variety of computing and computer-networking products to choose from on the market today. The determination of which product to buy or recommend to the customer/homeowner should always be guided by the expressed and/or evaluated needs of the user, both for the present time and the future. Very often, the customer will say what they currently need, but it is up to the HTI to use this information—together with his or her knowledge of technological trends—to advise the customer on the best way to invest in equipment and products that ensure future compatibility. You should keep current with the latest technology in order to give advice about possibilities that may not have been considered by the homeowner. Some of the key pieces of information to take into consideration include the following:

- The number of computers and the peripherals that will be included on the network. The family's lifestyle and computing attitudes, such as their use of multimedia content, computer games, graphical applications, and so on can affect the type of equipment to buy. Computers with faster processors, larger memory, and more storage may be needed to accommodate the customers.
- Homeowners who telecommute or have a home office will need products that enhance availability and reliability of high-speed Internet access. They may require high-speed Internet access and may also need dial-up backup links for mission-critical applications.
- The prevalence of laptops may require that a hybrid wireless and wired system be used if portability of computers emerges as an expressed need by the customer. Wireless technologies are ideal for a home network that includes laptop computers in addition to desktop computers and other peripherals.

You should discuss the type of equipment that the homeowner will be adding or needing in the future in order to design a network for growth. The homeowner might want or need the following for the network: new computers, printers, scanners, and so on. Discuss the type of equipment that the homeowner is likely to purchase in the next few years. Also, you need to forecast how much additional equipment will be purchased in the next 5–10 years. Because ownership of personal computers and peripherals has skyrocketed in the past 10 years, this trend can be expected to continue. For instance, if a family now has a computer in the home office and one in the children's room, plan on the family wanting another computer in the den or kitchen, as well as one in the master bedroom. If you plan ahead, the homeowner will not need to upgrade the network during the foreseeable future.

Determine Location of Outlets

The location of existing and future equipment should be discussed with the homeowner. All rooms in a house can be connected in the integrated home. The possibilities are endless. As an HTI, it is up to you to determine the options available based on the location. Different types of outlets used in a home integration project are shown in Figure 3–14.

Figure 3–14
Examples of Outlets

For a retrofit computer network project, consider setting up a wireless computer network so that no new wires have to be run and the homeowner can move equipment or add equipment later. The construction of the home, however, is a factor. If the interior walls are made of solid masonry such as brick or concrete blocks, a wireless network may experience difficulties. A similar problem can occur with stucco houses made with wire mesh. Radio and television signals have difficulty going through wire mesh and, therefore, two devices can have trouble communicating. If this is the case, wiring will need to be installed.

In new construction, running network cables as the electrical wiring is being installed is the current standard. Most new homes will provide the outlets for computers and other devices. In established homes, in which wireless is not an option due to the construction restraints, the retrofit can be accomplished by running the cables in an attic, in a basement, or along baseboards.

WIRED SOLUTIONS

A major consideration of designing and installing the computer network has to do with the wiring infrastructure. The options for the homeowner are wire or non-wire (wireless) systems. The choice of one infrastructure technology over the other is usually dependent upon the construction type (new, remodel, retrofit). Before homeowners can select an infrastructure, they need to have a clear understanding of what is available.

Wired computer networks use either copper or fiber-optic cables to transmit data between devices. The way in which computers, printers, and other devices are connected is known as *topology*. Physical topology describes the layout of the wires and devices as well as the paths used by data transmissions. Logical topology is the path that signals travel from one point to another. Topologies include bus, ring, star, hierarchical, and mesh, as shown in Figure 3–15.

Figure 3–15
Topologies

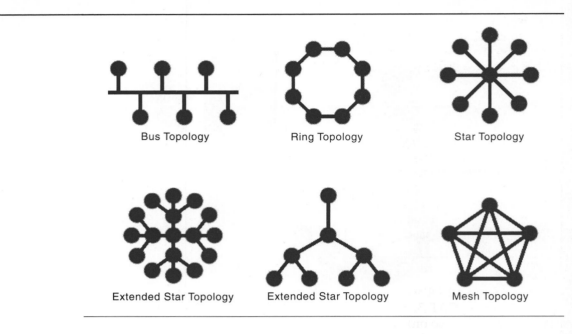

The most common type of wired network for businesses is Ethernet. There are also types of wired networks available that don't require running cables in the house. For example, the phone-line system uses existing telephone wiring. Power-line systems use existing power lines and outlets. These types of technologies are advantageous because no new wiring is needed throughout the home, but they do have limitations that will be discussed in a later section.

Ethernet

Ethernet is a type of LAN, and an Ethernet network is easy to set up in a home that is under construction. In an existing home, it requires running the cable through the attic, a basement, or along the baseboards. The data transmission rate is measured in megabits per second **(Mbps)**. Ethernet transmits data at a rate of 10 Mbps. Newer versions of Ethernet technology are Fast Ethernet, which supports data-transmission up to 100 Mbps, and Gigabit Ethernet, with up to 1,000 Mbps of data-transmission. Ethernet specifications are defined under IEEE 802.3 (the Institute of Electrical and Electronic Engineers standards that are specific to Ethernet).

Ethernet operates on a technology known as carrier sense multiple access collision detect (CSMA/CD). Each device on the wire listens for data signals that have been transmitted by other devices. If one device has data to send to another device and the wire is clear, it will transmit the data in a special format called an Ethernet frame. If the wire is busy, the device waits until it is clear before trying to transmit.

If two or more devices try to transmit at the same moment, the signals will collide. When the collision is detected, everything stops; and after a random waiting period, one of the devices will try to transmit again. If necessary, the process will repeat. Ethernet's efficiency, simplicity, and flexibility make it a popular choice for networks that do not have many users. This system works well for networks with low-to-moderate activity levels, such as a home network.

An Ethernet LAN can take several forms, depending on the type of cabling used. The two forms that use coaxial cable attach all devices along a single cable, a topology known as bus.

Mbps

Standard Ethernet, also known as thicknet or 10Base5, is an old type of wiring rarely used now. Originally, it was used to span long distances, up to 500 meters. Thinnet, or 10Base2, uses thinner cable that is about one-quarter inch in diameter, and can run up to 185 meters long.

> **NOTE** The designations 10Base5, 10Base2, 10BaseT, 100BaseTX, and 100BaseFL are used to indicate the maximum data transfer rate (10 or 100 million bits per second), the type of transmission (baseband uses the whole capacity of the channel for a single signal; broadband splits the channel to accommodate both transmitting and receiving signals), and then a number or letter combination to further distinguish them (the numbers 5 and 2 correspond to the maximum number of meters that the signal can travel for thicknet and thinnet cable; T and TX indicate twisted-pair cabling; and FL indicates fiber-optic cabling)

10BaseT and 100BaseTX use inexpensive unshielded twisted-pair (UTP) cabling, such as CAT 3, 5, or 5e, in a star topology. Each device (computer, printer, server, and so on) has its own cable (up to 100 meters long) that is connected to a central device that is either a hub or a switch.

According to TIA/EIA 570-A, the minimum acceptable twisted-pair cabling for Grade 1 residential networks is CAT 3. (See Chapter 2 for Grade 1 and 2 specifications.) Although CAT 3 is acceptable, it is by no means desirable. CAT 3 has a maximum carrier frequency or signaling rate of only 16 megahertz (MHz) compared to 100 MHz for CAT 5 and 5e. Therefore, CAT 5 is recommended for Grade 1 installations. For Grade 2 installations, CAT 5 is the minimum, whereas CAT 5e is recommended.

Finally, 100BaseFL is the designation for fiber-optic cables that have a range of up to 2 kilometers. Installing fiber-optic cabling has many advantages (discussed in Chapter 8), but it is considerably more expensive to install.

Phone-Line Systems (HomePNA)

Existing telephone wiring is an excellent medium for networking home PCs without adding new wires. The Home Phone Network Alliance (*HomePNA*) system uses the existing phone line. HomePNA technology uses a method called frequency-division multiplexing (FDM) to transmit voice and data signals at the same time. FDM works by allocating channels of specified bandwidth and frequency to each signal. Implementing HomePNA is fairly straightforward. The HomePNA hardware is installed on a computer or some other network-enabled device. A patch cord is then used to connect the device to the phone line via a telephone outlet.

HomePNA

Using telephone wiring reduces the complexity and inconvenience of rewiring a home. The average multiple-PC household in the United States has four to five telephone jacks, located in places where many people set up their computers: a home office, bedroom, family room, or kitchen. Other advantages include the following:

- Easy to install
- Inexpensive
- Reliable
- Equipment and protocol are standardized
- Operates at a constant 10 Mbps, even when the phone is in use
- Requires no additional networking equipment (for example, hubs or routers)
- Supports up to 25 devices

- Fast enough for bandwidth-intensive applications such as video
- Compatible with other networking technologies
- Works on Windows, Macintosh, and older PCs
- Provides a moderately secure environment for data transmission

HomePNA does have some drawbacks. A phone jack must be close to each computer; otherwise, phone extension cords or new wiring need to be run. Even though the latest version, HomePNA 2.0, operates at a very reasonable speed of 10 Mbps (using the Ethernet CSMA/CD framing and transmission protocol), this is still 10 times slower than fast Ethernet (100 Mbps). If the homeowner will be sending huge amounts of data between computers, this solution may not be adequate.

The main problem encountered with HomePNA systems relates to noise being generated on the cabling and interfering with the data signal. Some sources of noise include analog communication devices (telephones and fax machines using the same wire) or electrical power sources. Adding low-pass filters to the system can mitigate this problem. Such filters transmit the lower frequencies used by telecommunications systems while blocking other interfering frequency ranges.

HomePNA technology-based computer subsystems do not require traditional Ethernet hubs for signal distribution. Instead, each PC attached to the phone line network is connected via a phone network adapter (PNA). These phone line LAN adapters come in external parallel versions or internal Industry Standard Architecture (ISA) or Peripheral Component Interconnect (PCI) bus cards. Leading manufacturers of HomePNA equipment include 3Com, D-Link, Diamond, Intel, and Linksys.

Important considerations for a HomePNA system include the following:

- A maximum of only 25 networking devices can be on the system.
- The cable length between HomePNA devices cannot exceed 500 feet (150 meters).
- The overall area of coverage should not exceed 10,000 square feet (1,000 square meters).
- HomePNA technology may not function properly in residences that use a home private branch-based telephone system (home PBX).

HomePNA works with the majority of existing wiring in homes. In the United States, fewer than one percent of the homes cannot use the existing phone line.

Power-Line Carrier Systems (PLC)

Power-line is a LAN distribution method that uses the existing power-line cabling in the home to send signals to devices. Because most devices in the home, including lights and appliances, draw power from the electrical system, it is possible to communicate with and control these devices through the power-line system. In most homes, power receptacles are already positioned in multiple locations within each room. As with phone lines, power lines are laid in a bus topology. Typical applications for PLC systems are computer-networking distribution, audio/video distribution, telephone distribution, appliance control, security, lighting, HVAC, and energy-management control.

The PLC solution works well for many telecommunication requirements. Integrated home-control systems are usually bursty telecommunications systems. (*Bursty* refers to data that is transferred in uneven, short intervals.) Voice, data, and video are mostly continuous and transmit larger files.

Burst mode bandwidth and reliability requirements are typically low because missed instructions can simply be retransmitted by the control mechanism. However, power-line is the lowest-performing LAN because of the high tendency for interference between devices that are obtaining AC voltage from the power line and devices that are transmitting data along the same line. Before installing PLC, consider these disadvantages and how they relate to your project:

- In larger homes using multiple electrical panels, a typical two-phase electrical system can block telecommunications between devices on different phases of the same power grid.
- Surge protectors are particularly problematic for these systems, but may be resolved by the application of a whole-home surge-suppression system.
- PLC has limited bandwidth, and there are limits to the distance communicating devices can be from each other.
- Devices used by neighbors can interfere with signals.
- Electrical disturbances such as power surges, lightning strikes, and so on can lead to reliability issues.
- Power usage in the home can also negatively impact the performance of the system.

Monitoring and control are addressed by power-line protocols such as X-10, CEBus, HomePlug, or LonWorks. X-10 and CEBus are considered the industry standards, and are the most commonly used PLC technologies in the U.S. residential networking market.

X-10 *X-10* is a communications protocol that allows compatible home networking products to talk to each other via the existing 110-volt electrical wiring in the home. No costly rewiring is necessary. X-10 transmitter devices send a coded low-voltage signal that is superimposed over the 110V AC current. Any X-10 receiver device plugged into the household power supply will see this signal. By using X-10, it is possible to control lights and almost any other electrical device from anywhere in the house with no additional wiring. Virtually all the home automation tools have relied on this protocol. Typical commands include on, off, all on, all off, and dim.

X-10 has several limitations. It cannot handle more sophisticated monitoring and commands, and has a relatively slow transfer rate of information. It is unreliable due to noise in the circuitry and its one-way communication. If you send a "turn off" message to a lamp, there will be no return communication that says, "The light is off." Furthermore, one-way X-10 is not intelligent enough to detect whether a device is already on or off. So, if the lamp was already turned off and you sent the command "turn off," the device would then be turned on because the original state of the lamp is not detectable to X-10. Newer two-way X-10 technology offers the capability to know the current state of a device. It is able to detect when a lamp is off, and will turn it on.

Web link: http://www.x10.org

CEBus   In the late 1980s, the EIA (Electronics Industry Association) and the CEMA (Consumer Electronics Manufacturing Association) sponsored the development of a new standard, the Consumer Electronics Bus protocol (CEBus). The CEBus Standard was developed around Common Application Language (CAL), which is a universal communications language for home network products. CAL is designed to be understood by most home electronic products. The mission of the CEBus Industry Council (CIC) is to provide information to the design and development community about CEBus and CEBus Home Plug & Play. Home Plug & Play is an industry specification that describes the way consumer products will cooperate with other products. Many of these devices can be installed without the need for a central controller, which enables CEBus Home Plug & Play devices to be used for many simple automation problems. CEBus Home Plug & Play devices can also be networked with a central controller for more extensive automation projects.

X-10

CEBus products are an ideal solution for an existing home network that uses existing power lines because they do not require any additional wiring to be installed.

Although few products have been developed, the specifications are open-sourced or free for anyone to develop products. More information can be found at

Web link: http://www.cebus.org

HomePlug   The mission of the HomePlug Powerline Alliance is to "enable and promote rapid availability, adoption and implementation of cost-effective, interoperable and standards-based home power-line networks and products." It provides a forum for the creation of open specifications for high-speed, home power-line networking products and services. It also seeks to accelerate the demand for these products and services through the sponsorship of market and user education programs.

The vision is to deliver Internet and multimedia from every home power outlet. Member companies are building products—primarily Ethernet and Universal Serial Bus (USB) bridges as of the date of this writing—that seamlessly connect to each other through your customer's power outlets.

Web link: http://www.homeplug.org

LonWorks   Echelon has developed its own protocol, called LonTalk, to automate controls in commercial and residential buildings. Used only in some European residential markets and in some large U.S. homes, its proprietary nature has made some in industry reluctant to embrace it.

Web link: http://www.echelon.com

Broadband Coaxial Systems (Series 6)

Like power lines and phone lines, coaxial cabling is already found in many homes, particularly those built within the last 20 years. Coaxial cabling is used to provide cable TV service (known as community antenna television or CATV in the industry). Cable TV providers now offer video programming, telephone service, and high-speed broadband Internet services (requiring a cable modem) to residential customers using the existing coaxial cabling within their homes.

Because each telecommunication service transmits and receives signals at a different frequency, the same coaxial cabling can be used to both distribute television programming and transmit high-speed broadband services. Using this same principle, manufacturers are working on products that can be used to distribute voice and data (that is, computer) information over the same cabling infrastructure by using a different frequency range. With multiple signal capacity, data transmission can be broken into channels and allocated for simultaneously sending and receiving data.

The type of data transmission that can simultaneously carry multiple signals is called broadband. It uses frequency multiplexing to allow the combination of the various signals for transmission and then signal recovery at the receiving end. Baseband, on the other hand, is a type of digital data transmission in which each wire or other medium carries only one signal or channel at a time. Data can be sent or received, but not sent and received at the same time.

The Society of Cable Telecommunications Engineers (SCTE) is currently involved in a collaborative initiative with the Cable Television Laboratories, Inc. (CableLabs) to develop an open cable standard providing interoperability for advanced terminal devices, including set-top CATV boxes. CableLabs is a nonprofit cable CATV industry organization. The ongoing initiative is slated to deliver a blueprint of advanced telecommunications services to consumers and define a family of digital cable services.

High-Performance Serial Bus—IEEE 1394

Also known as FireWire, the IEEE 1394 is a high-performance serial bus standard designed to support high-bandwidth requirements of devices such as digital video equipment and high-capacity mass storage. Supported by the Institute of Electrical and Electronic Engineers (IEEE), the IEEE 1394 standard was developed to meet the need for high data-transfer rates between computer components, and between the computer components and connected devices such as video cameras. Currently, data-transmission rates of up to 400 Mbps can be achieved, with even higher data rates in development. Figure 3–16 shows an example of FireWire.

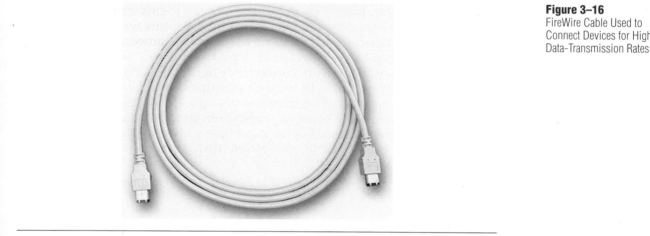

Figure 3–16
FireWire Cable Used to Connect Devices for High Data-Transmission Rates

If you include IEEE 1394 as part of the infrastructure technology in a home network, the following design criteria need to be considered:

- To achieve the associated high speeds, the cable used to connect two devices cannot be more than 4.5 meters (14.8 feet) in length.
- A maximum of 17 devices can be daisy-chained to each other over 16 links, for a maximum end-to-end distance of 72 meters (236 feet).
- Following a tree topology (a combination of a bus and star topology), a maximum of 63 devices can be interconnected using the IEEE 1394 technology.

In addition to its high-speed characteristic, FireWire is ideal for applications viewed in real time because it offers a guaranteed data delivery rate, called *isochronous* (pronounced "eye-sock-ra-nuss"). An example of isochronous is a video watched on the Internet. It must be delivered quickly in order for the video to play, and the audio and video must match.

WIRELESS SOLUTIONS

Wireless is a practical alternative in hard-to-cable areas of an existing home, or for such applications where the portability of computing devices is crucial. The advantages and disadvantages of wireless are shown in Table 3–1.

Wireless Advantages	Wireless Disadvantages
Ease of installation	Limited signal distance
No ugly wiring throughout the home	Signal interference is a possibility
Computers are portable	Signals are not always secure

Most wireless LANs use the 2.4 GHz (gigahertz) frequency band, which is the only portion of the RF spectrum reserved around the world for unlicensed devices. Others use the 5 GHz band, and still others use infrared light to transmit information.

Several types of wireless technologies are on the market today, but the three most widely accepted are IEEE 802.11, Bluetooth, and HomeRF.

Infrared (IR)

Infrared

Most of us are familiar with everyday devices that use *infrared* technology, such as remote controls for TVs, VCRs, and CD players. IR transmission is categorized as a line-of-sight wireless technology. This means that the workstations and digital devices must be in direct line to the transmitter in order to operate. An infrared-based network suits environments in which all the digital devices that require network connectivity are in one room.

There are, however, new IR technologies that can work out of the line of sight. IR home networks can be implemented quickly. A disadvantage is that people walking across the room or moisture in the air can weaken data signals. IR in-home technology is promoted by an international association of companies called Infrared Data Association (IRDA).

Web link: http://www.irda.org

IEEE 802.11

Wireless LANs conforming to the Institute of Electrical and Electronics Engineers (IEEE) 802.11 specification can be integrated into wired Ethernet networks using an access point. The devices on the wireless network (PCs, laptops, printers, and so on) can then communicate with devices on a wired network. Installing an 802.11 wireless network alongside an Ethernet network may be the best solution for homes in which the homeowners have both desktop and laptop computers and want to use the laptops anywhere in the house.

The 802.11 standard allows for transmission over different media, including infrared light and two types of radio transmission within the unlicensed 2.4 GHz frequency band, frequency hopping spread spectrum (FHSS) and direct sequence spread spectrum (DSSS). Spread spectrum is a modulation technique that spreads a transmission signal over a broad band of radio frequencies. This technique is ideal for data communications because it is less susceptible to radio noise and creates little interference. FHSS is limited to a 2 Mbps data transfer rate, and is recommended for only very specific applications. For all other wireless LAN applications, DSSS is the better choice.

The IEEE 802.11 standard is continually evolving and adapting to meet the needs of the industry as new technology is developed. The current wireless LAN standard is the 802.11b, and it offers speeds of up to 11 Mbps. However, there is a wide range of newer versions in the making, all competing to be the successor to IEEE 802.11b. Each attempts to address the issues of speed and/or security to a different extent. IEEE 802.11a significantly increases the speed up to 54 Mbps while using Orthogonal Frequency Division Multiplexing (OFDM). IEEE 802.11g doubles the speed of 802.11b up to 22 Mbps, and is backward-compatible with it.

IEEE 802.11a

Instead of using the 2.4 GHz frequency band that the other 802.11 standards use, the 802.11a standard uses the unlicensed 5 GHz frequency band. The need for a different frequency band is due to the use of the 2.4 GHz band by wireless phones and other wireless devices around the home. The devices on the wireless network have to compete with other devices in the home and perhaps even with wireless devices used by neighbors. This congestion can hamper the network's speed. In addition, the radio signals from these other wireless devices can interfere with the network. Because the 5 GHz frequency band is unique in the house, only the transmissions from devices on the wireless network will be on that band. Thus, the different band allows speed up to 54 Mbps, nearly five times that of 802.11b. 802.11a is ideal for wireless networks that require support for transmitting video, voice, and other large files.

The 802.11a network is not always the solution. Increased speeds are accomplished only if the networking devices are closely grouped. According to testing done by Atheros Communications data-transmission rates of 802.11a decreased from the optimal 54 Mbps when devices were more than 20 feet apart. Table 3–2 shows the maximum data-transfer rate between wireless components.

Web link: http://www.atheros.com

Distance between Wireless PCs, Laptops, and Printers (Feet)	Maximum Data-Transfer Rate (Mbps)
0–20	54
20–39	48
40–74	36
75–84	24
85–134	18
135–174	12
175–224	6

Source: J. Chen, *Measured Performance of 5-GHz 802.11a Wireless LAN Systems*, Atheros Communications, Inc. (http://epsfiles.intermec.com/eps_files/eps_wp/AtherosRangeCapacityPaper.pdf), 2001.

Table 3–2
Maximum Data-Transfer Rate of 802.11a Wireless LANs Based on the Distance between Wireless Devices.

IEEE 802.11b

The IEEE 802.11b standard is the most commonly used wireless technology. 802.11b transmits at 2.4 GHz and sends data up to 11 Mbps using direct sequence spread spectrum modulation (DSSS). Fewer access points are needed to maintain the maximum transfer rate. Whereas with 802.11a, the maximum transfer rate begins to decline at 20 feet, the maximum transfer rate of 11 Mbps is retained at 100 feet (Source: J. Chen, *Measured Performance of 5-GHz 802.11a Wireless LAN Systems*, Atheros Communications, Inc., [http://epsfiles.intermec.com/eps_files/eps_wp/AtherosRangeCapacityPaper.pdf], 2001).

Although the 802.11a solution is faster, the higher operating frequency equates to relatively shorter range. 802.11b, as shown in Figure 3–17, is best for large homes or multi-dwelling units in which cost is a factor because 802.11b needs far fewer access points than 802.11a. 802.11b may also be the best choice if the homeowner already has wireless cards in PCs, laptops, and printers because most are 802.11b-compliant.

Figure 3–17
802.11b

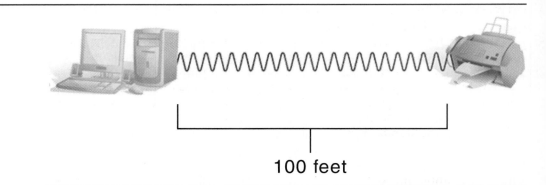

100 feet

802.11a and 802.11b are not interoperable because they have different radio frequency and modulation types. A device with an 802.11a network interface card (NIC) cannot connect with an 802.11b access point. Therefore, all equipment must be purchased as one standard.

Equipment that is compliant with the 802.11b specification receives a seal of approval, indicated by the word "Wi-Fi," by the Wireless Ethernet Compatibility Alliance (WECA). Wi-Fi stands for "wireless fidelity," similar to Hi-Fi for "high fidelity."

HomeRF

The HomeRF Working Group was formed in 1997. The key goal of the group is to enable interoperable wireless voice and data networking within the home at a reasonable price.

A HomeRF is a radio frequency (RF) networking technology that uses a protocol known as Shared Wireless Access Protocol (SWAP). It includes six duplex voice channels in addition to the IEEE 802.11 Ethernet specification for data transmission. Using the frequency range 2.4 GHz, HomeRF 1.x has a maximum transfer rate of 1.6 Mbps, whereas the newer version 2.0 has a maximum transfer rate of 10 Mbps.

Up to 127 devices can be installed on a HomeRF network, but the typical data-transmission distances are limited to 75–125 feet (23–37 meters) and can be extended up to 150 feet (45 meters) as illustrated in Figure 3–18.

Figure 3–18
Home RF

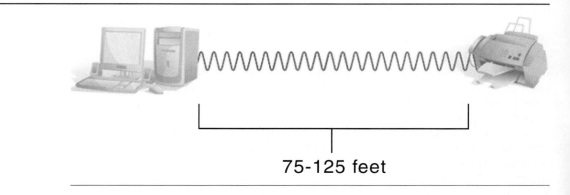

75-125 feet

When designing this technology, the HomeRF Technical Committee chose to reuse proven RF networking technology for data and voice communications, and added simplifications where appropriate for home usage. With this approach, SWAP inherited native support for Internet access via TCP/IP networking, and for voice telephony via the Public Switched Telephone Network (PSTN) and Voice over Internet protocol (VoIP).

HomeRF has struggled in its acceptance. The earlier versions (1.0–1.2) were relatively slow compared with 802.11b, and the standard was not interoperable with 802.11b, the most common wireless standard that is commonly used for laptop computers. Today, leading IT manufacturers such as Intel, Motorola, and Compaq offer HomeRF-compatible equipment to network PCs, laptops, and Internet access. In addition, HomeRF Group is petitioning the FCC to permit the standard to use the 5 GHz channel that 802.11a uses.

Bluetooth

Bluetooth is another industry group developing a specification for low-cost, short-range radio links between mobile computers, cameras, and other portable in-home devices. It enables home networking users to connect a wide range of computing and telecommunications devices easily and simply, without the need to buy, carry, or connect cables.

Bluetooth is a short-distance RF communications technology operating for distances of up to 10 meters (33 feet) as illustrated in Figure 3–19. This is a major drawback if devices are not placed near each other.

Bluetooth

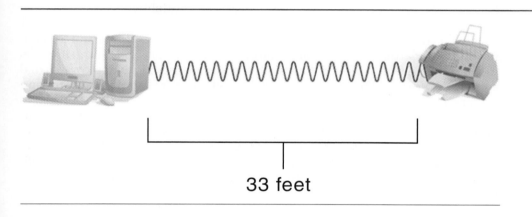

33 feet

Figure 3–19
Bluetooth Short-Distance RF Communication

Bluetooth operates on a frequency of 2.45 GHz. Because Bluetooth uses the same frequency as 802.11b and HomeRF, interference could occur if other wireless networks or wireless devices (cordless phones, garage door openers, and so on) are being used at the same time.

This technology uses spread-spectrum frequency hopping (SSFH), and transmitters change frequencies 1600 times per second. Bluetooth can support both voice and data. Voice can be transmitted at 64 Kbps. The maximum data-transfer rate is 721 Kbps for one direction and 57.6 Kbps for the other direction. In other words, the homeowner can transmit files at 721 Kbps while downloading a web page at 57.6 Kbps.

Bluetooth is intended to be an RF solution in home networking to replace the infrared remote controls used by many devices today—such as televisions, video cassette recorders, and so on. The technology is being promoted and has been adopted by a group of companies called the Bluetooth Special Interest Group (SIG). The group includes 3Com, Ericsson, IBM, Intel, Lucent, Microsoft, Motorola, Nokia, and Toshiba.

Selecting a Wireless Technology

Selecting a wireless technology depends on the range between devices, the speed desired, and cost. Table 3–3 summarizes and compares the four most common wireless technologies.

Table 3–3
Comparison of Wireless Technologies

Technology	Frequency	Technique Used	Data-Transfer Rate	Range
802.11a	5 GHz	Orthogonal Frequency Division Multiplexing (OFDM)	54 Mbps	Short-range (maximum transfer rate operable only up to 20 feet)
802.11b	2.4 GHz	Direct Sequence Spread Spectrum Modulation (DSSSM)	11 Mbps	Medium-range (maximum transfer rate operable up to 100 feet)
HomeRF	2.45 GHz	Shared Wireless Access Protocol (SWAP)	10 Mbps	150 feet
Bluetooth	2.45 GHz	Spread-Spectrum Frequency Hopping (SSFH)	721 Kbps	30 feet

Securing the Wireless Network

Wireless networks need security. Neighbors with wireless LANs or computer hackers who are located within range can access your network. Hackers have cracked the standard encryption keys for 802.11 networks, although the 802.11 Group has tried to stay one step ahead. Because these encryption and authentication standards are vulnerable, stronger encryption and authentication methods should be deployed. One such security measure is to deploy a virtual private network (VPN), particularly if the homeowner is a telecommuter and is working on sensitive information from home.

A VPN uses the Internet to connect an individual, such as a telecommuter, to a LAN located elsewhere, such as at his or her company. A VPN can employ strong authentication and encryption mechanisms between the access points (the telecommuter) and the network. Despite these vulnerabilities, encryption and authentication remain essential elements of wireless LAN security.

INTERNET ACCESS

Internet access is the entry and exit point linking a home network to the World Wide Web. Several factors go into determining the type of Internet service that would be brought to the home. Some of these include the following:

- Availability of the desired service in your area
- Cost of service and equipment
- Location of the home—presence of mature trees, around a hillside, etc. with respect to the service provider central office (CO)
- Comfort level of user with the type of technology

The usual alternatives from which to choose may include digital subscriber line (DSL), cable broadband services, integrated services digital network (ISDN), and two-way satellite. Each of these services requires the use of some sort of device, typically a modem to interface with the home network.

Regardless of the type of Internet connection, when setting up e-mail services, you will need the Simple Mail Transfer Protocol (SMTP) and Post Office Protocol (POP3). SMTP is used for the transmission of e-mail over computer networks. POP3 is the standard protocol for receiving e-mail. You will also have to make sure that they are configured so that ports 25 and 110, respectively, are open on the firewall. Firewalls will be discussed later in this chapter.

Dial-up Modem

Dial-up modem access is not high-speed, but it is the most common access method used, and every HTI should understand how it works. A modem is a device that allows data transmission between computers over the Internet. The modem is a device that allows a digital computer to communicate through a telephone line using an analog signal. It converts the digital signal used by the computer to an analog signal that can be used by the telephone line. The word modem comes from its function, **mo**dulator and **dem**odulator.

There are different types of modems in use today. A 56K dial-up modem is the one most commonly used by homeowners. The 56K speed is dependent on the type of service that is being used and the traffic on the lines when dialing up. In reality, it is much slower. Once connected to the Internet, speed can be slowed by graphics, audio, or video, as well as by the amount of traffic that the telephone company is dealing with at that time. In homes in which an Internet connection is critical 24/7, a 56K dial-up modem may be used to back up other connections such as DSL or cable. Figure 3–20 shows a network layout with a dial-up modem connection to the Internet.

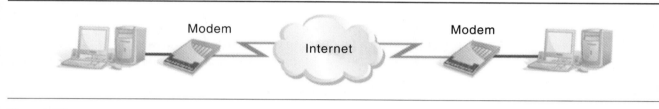

Figure 3–20
Network Layout Using a Dial-up Modem

Although a modem can use a variety of different protocols to translate signals, protocols are specified at particular rates of data transmission. In addition to the protocol designation, the following characteristics distinguish modem capabilities:

- **Bits per second (bps):** The speed at which the modem can transmit and receive data
- **Voice and fax capabilities** (in addition to data transmission): Used for expanded Home Network options
- **Data compression:** Enables faster data-transfer rates

NOTE   The benefits of the modem speed, or bps, are dependent upon the speed at which the signal is received. The signal must be received from the wide-area network (WAN) at a speed equal to or greater than the modem speed in order to transmit at the maximum speed of the modem.

The sections that follow explore each of these technologies further, explaining their suitability as a gateway of the home network to the outside world (the Internet).

Digital Subscriber Line (DSL)

DSL takes advantage of the frequencies that go unused on the telephone wire and moves significant quantities of data without affecting the voice portion of the line. A telephone uses frequencies on the phone line below about 20 kHz. DSL uses the frequencies starting at 25 kHz to just above 1 MHz. This allows the phone and the DSL to share the line between the home and the nearest central telephone office.

DSL modems are very suitable gateway devices for boosting the Internet bandwidth available to appliances running on an in-home network. DSL modems support data transmission over standard telephone lines as much as 50 times faster than the analog modems—up to 1.5 Mbps.

When a homeowner orders DSL service, the ISP will either provide or sell a modem designed to work with DSL signals. DSL modems can be either external or internal. The telephone R-11 jack plugs into the modem, and there is a line out to accommodate a telephone on the same line.

A digital subscriber line (DSL) circuit consists of two DSL modems connected by a copper twisted-pair telephone line. To maintain backward compatibility with the standard telephone system and to avoid disruption of service due to equipment failure, the voice part of the frequency spectrum is separated from the digital modem circuitry by means of a passive filter called a plain old telephone system (POTS) splitter. This means that if the DSL modem fails, the POTS service is still available. Under this configuration, users are able to simultaneously make voice calls and transmit Internet data over the same broadband DSL pipe. When a DSL transmission is received at the central office, a more advanced POTS splitter is used to send the voice traffic to the public telephone network and data to the Internet. Figure 3–21 illustrates a network layout using a DSL modem to connect to the telephone company's local exchange.

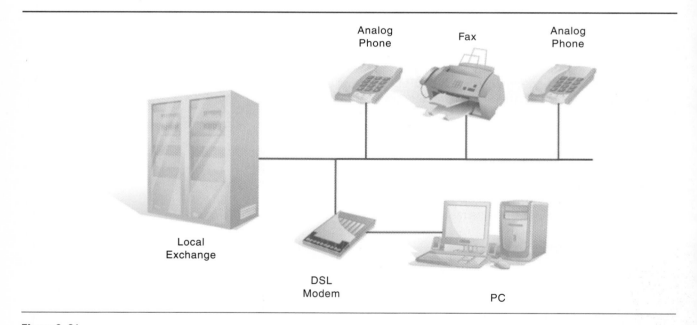

Figure 3–21
Network Layout Using a DSL Modem

DSL can transmit voice, data, and video traffic to and from the home network. Distance is a major drawback when considering DSL as a high-speed Internet solution. The DSL customer must be located fairly close to the telephone central office, known as the telco. This is because the DSL signal cannot travel as far down the lines, due in part to its higher frequency and also due to the presence of load coils and other line-conditioning equipment that keep the telephone network optimized for the voice frequencies it was designed to carry.

Other problems can be a hum on the phone line and lower voice quality. A low-pass filter on the phone line keeps the data signals from interfering with standard telephone calls. These low-pass filters should be attached to all phone jacks that are not using the DSL connection.

DSL advantages:

- The DSL signal is an always-on connection. After the connection has been established, the user is always connected to the Internet.
- DSL services provide high-speed access compared to that of a telephone 56K modem.
- The voice portion of the service remains unaffected when the computer is online.

DSL disadvantage:

- The DSL customer and the telco or service provider's central office must be in close proximity.

DSL is typically asymmetrical, meaning that it provides a higher data rate for the downstream path into the home than for the upstream path out of the home. DSL is available in many forms, including asymmetric DSL (ADSL), symmetric DSL (SDSL), rate-adaptive DSL (RADSL), high-bit-rate DSL (HDSL), very-high-bit-rate DSL (VHDSL), and ISDN DSL (ISDL).

- **ADSL:** Used by most homes, asymmetric DSL line has a download data transfer rate that is about 3–4 times faster than its upload rate, that is, 1.5 Mbps downstream and 640 Kbps upstream. Users therefore can download information fairly quickly from the Internet, but the upload time will be much slower.
- **SDSL:** Used mainly by small businesses, symmetric DSL doesn't allow you to use the phone at the same time, but the speed of receiving and sending data is the same.
- **RADSL:** Rate-adaptive DSL is a variation of ADSL, but the modem can adjust the speed of the connection depending on the length and quality of the line.
- **HDSL:** High-bit-rate DSL receives and sends data at the same speed, up to 1.5 Mbps, but it requires two lines in addition to the regular home phone line.
- **VHDSL:** Very-high-bit-rate DSL is the fastest connection with speeds of up to 52 Mbps, but it works only over a short distance.
- **ISDL:** ISDN DSL is for existing users of Integrated Services Digital Network (ISDN) who want to use their existing equipment. Another advantage is that it can work the farthest from the telephone company. The drawback is that ISDL is slower than most other forms of DSL, operating at a fixed rate of 144 Kbps for both download and upload data transfers.

Table 3–4 displays the different DSL technologies, compares the upload and download rates, and shows the maximum distance from a telephone company.

Table 3–4
DSL Technologies

DSL Technology	Maximum Download Data-Transfer Rate	Maximum Upload Data-Transfer Rate	Maximum Distance from Telephone Company (in Feet)
VHDSL	52 Mbps	16 Mbps	4,000
RADSL	7 Mbps	1 Mbps	18,000
SDSL	2.3 Mbps	2.3 Mbps	22,000
ADSL	1.5 Mbps	640 Kbps	18,000
HDSL	1.5 Mbps	1.5 Mbps	12,000
ISDL	144 Kbps	144 Kbps	35,000

DSL requires a DSL transceiver, often called a DSL modem. The service provider will most likely refer to the equipment as an ADSL Termination Unit–Remote (ATU-R). It connects to the computer's USB port or RJ-45 port, depending on the type of transceiver purchased. The cable must have the appropriate plugs for the ports.

Cable Modem

Cable modem high-speed Internet service is a broadband technology that is provided by cable television (CATV) service providers, and runs on coaxial cable. Cable modem data transmissions can operate between 320 Kbps and 10 Mbps, depending on different factors such as the modulation scheme of the signal and the type of coaxial cable. The actual speed of the Internet access service coming to the home depends on what the service provider offers. Typically, the bandwidth is shared between users in the neighborhood.

Cable modems operate over two-way hybrid fiber and coaxial lines. Some cable operators that have not upgraded their networks to provide full bidirectional services use their hybrid fiber coaxial (HFC) network for high-speed downstream data transfer and use the local telephone network for the return path or upstream data.

A standard cable modem, as shown in Figure 3–22, will have two connections; one port is connected to the TV outlet on the wall and the other to the subscriber's PC. The cable modem will then communicate over the cable network to a device called a Cable Modem Termination System (CMTS). The speed of the cable modem also depends on traffic levels and the overall network architecture. Cable modems are capable of receiving and processing multimedia content at 30 Mbps, which is literally thousands of times faster than a normal telephone connection to the Internet. In reality, however, subscribers can expect to download information at between 0.5 Mbps and 1.5 Mbps because the bandwidth is shared by a number of homes throughout the neighborhood.

Figure 3–22
Cable Modem

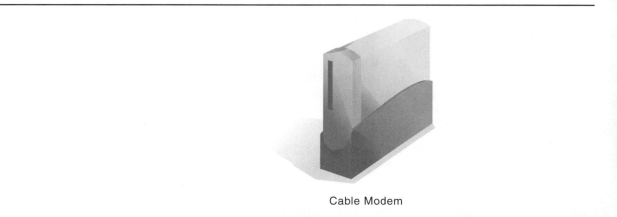

Cable Modem

The advantage of using cable modem high-speed Internet access is that coaxial cable wiring offers greater bandwidth for the home network. In addition, many cable lines are already in place, making cable modem Internet access more readily available than DSL.

If the homeowner purchases a cable modem from a retailer, they need to be certain that the modem is compatible with the provider's system. The cable industry has set standards for cable modems. In North America, most cable modems will be compliant with a standard called Data Over Cable Service Interface Specification (DOCSIS). DOCSIS specifies that the minimum data transmission rate is 500 Kbps and the maximum is 30 Mbps. In Europe, Asia, or Australia, the gateway device would be compliant with a specification called EuroModem.

Integrated Services Digital Network (ISDN)

Another type of high-speed Internet access over telephone lines is Integrated Services Digital Network (ISDN). ISDN provides a faster data-transfer rate for computers than can be obtained from modems by using a bearer channel (B channel) of 64 Kbps. It differs from other high-speed Internet services because it is a dial-up service rather than an always-on or dedicated connection.

ISDN in concept is the integration of analog or voice data together with digital data over the same network. Although the connection is digital, ISDN still supports analog telephones and fax machines. When a telephone call comes in while a computer connection is in progress, the computer automatically scales back to allow 64 Kbps until the phone call is terminated.

A network terminator (NT1) and ISDN terminal adapter (TA) are required to connect an ISDN to a PC. The NT1 plugs into the two-wire line from the telephone company with an RJ-11 connector and provides four-wire output to the TA. Within the U.S., the NT1 is typically built into the TA; but they are separate devices outside the United States.

An external TA is plugged into the serial port whereas an internal TA is plugged into an expansion slot. The TA can hook into a parallel port for higher speed. The TA also has the capability to automatically switch between analog and digital, depending on the type of call.

ISDN requires adapters at both ends of the transmission, so the access provider and the end-user both need an ISDN adapter. The ISDN adapter translates the digital signal received from the Internet into a digital signal that can be distributed on the home LAN, and vice versa. Figure 3–23 shows a network layout using an ISDN adapter to connect to the telephone company's local exchange.

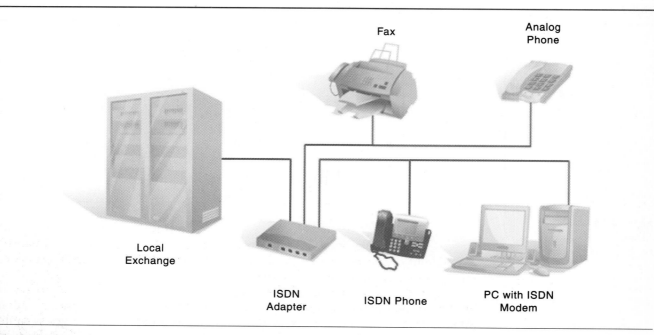

Figure 3–23
Network Layout Using an ISDN Adapter

There are two levels of service:

1. Basic Rate Interface (BRI) is intended for the home and small enterprise. The BRI consists of two 64 Kbps B-channels and one 16 Kbps D-channel. A Basic Rate user can have up to 128 Kbps service.
2. Primary Rate Interface (PRI) is for larger users. The Primary Rate consists of 23 B-channels and one 64 Kbps D-channel in the United States, or 30 B-channels and 1 D-channel in Europe.

Both rates include a number of channels called Bearer channels (B-channels) and Delta channels (D-channels). Each B-channel carries data, voice, and other services. Each D-channel carries control and signaling information.

ISDN is generally available from the phone company in most urban areas in the United States and Europe. ISDN is at a competitive disadvantage with the DSL and cable modem forms of high-speed Internet access because of the lower transmission rates and higher charges from local telephone carriers.

Satellite

Communication satellites orbit above the surface of the earth. They are used to link two or more earth-based microwave stations. Communication satellite providers typically lease some or all of a satellite's channels to large corporations, which use them for long-distance telephone traffic, private data networks, and distribution of television signals. Leasing these huge communications "pipes" can be very expensive; therefore, they are not suitable for the mass residential marketplace. Consequently, a new suite of services, called the direct broadcast satellite (DBS) system, has been developed to provide residential networking consumers with a range of high-speed Internet access services.

A DBS system consists of a mini-dish that connects in-home networks to satellites with the capability to deliver multimedia data to a home network at speeds in excess of 45 Mbps. However, this speed can be achieved only when downloading content. To upload or send information to the Internet, the system uses a slow telephone connection. Figure 3–24 shows a satellite receiving a signal from the Internet.

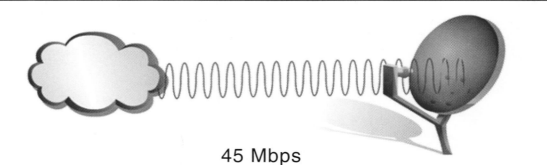

45 Mbps

Figure 3–24
Satellite Transmission

Satellite systems normally use the Quadrature Phase Shift Keying (QPSK) modulation format to transmit data from the dish in the sky to the mini-dish on the roof. QPSK refers to the four angles used that are typically out of phase by 90°. In assessing satellite as the high-speed Internet choice for a home-integration project, one important consideration is where to mount the satellite dish. Another consideration is the location of the home in terms of the line of sight with the earth-based relay stations or antennae.

The advantage of satellite as a high-speed Internet option to the home (cost not being prohibitive) is that satellite provides a wireless high-speed Internet

access method and has great availability worldwide. In the past, high-speed Internet access via satellite was at a disadvantage because of the lack of interactive data service. The data path used point-to-multipoint transmission, or one-way data transmission, from the service provider to the home. Therefore, an additional method of data transmission had to be used to send data from the LAN out to the WAN. This made satellite a complicated, expensive, and least-desirable choice for Internet access. Current expected home-transfer rates can be 1.5 Mbps down and 512K up. Cost will range from U.S. $50–150 per month. The latest satellite Internet services provide two-way data transmission, eliminating the need for a secondary Internet service.

Fiber-to-the-Home

Optical fiber cabling may prove to be the most popular medium for delivering information to the home of the future. Fiber-optic cable can transmit enormous bandwidth over very long distances. Fiber-optic cabling systems are also immune to electromagnetic interference (EMI). These systems are also highly reliable, which tends to lower long-term cost of maintenance.

Newly installed fiber-optic cabling from the service provider to home can provide the highest possible data rates and support such intensive applications as high-definition television (HDTV). These potential high data rates are the main benefit of fiber optic in the cabling system, and constitute the reason why fiber-to-the-home has the potential to be a great alternative to high-speed Internet broadband to the home.

COMPUTER NETWORK EQUIPMENT

The following sections describe the devices needed to properly design and build a home network. These devices include a network interface card, servers, switches, hubs, gateways, and routers. In addition, wireless access points that distribute signals are discussed.

Network Interface Card (NIC)

A *network interface card* (*NIC*) is a circuit card that plugs into one of the expansion slots in the back of a personal computer. The NIC is the most vital link between the computer processor and the network. It interprets information between the computer and the network, and feeds information in and out of the computer so that both the computer and the network can accept it. Figure 3–25 shows a typical NIC.

Network interface card (NIC)

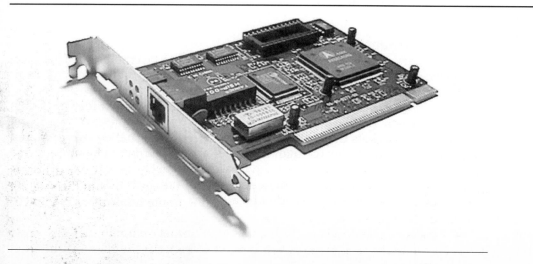

Figure 3–25
Network Interface Card (NIC)

MAC

A NIC comes with a unique identifier known as the Media Access Control (*MAC*) address that is built into the circuit board. A second address, known as the Internet protocol (IP) address, is assigned and binds to the NIC. The MAC address is known as the physical address, and the IP address is known as the logical address. The combination of these two addresses enables a computer fitted with a NIC to interface and communicate with other devices on the home network.

Servers

Server

A *server* is a computer normally designed to provide specific services to the network, such as storing files or running networked programs. Figure 3–26 shows a network server. It is set up with specialized software that allows it to serve as the repository for files and programs. All computers on the network can be logically connected to the server with a router, hub, or switch. A server has a keyboard and monitor that enable access to the system to perform maintenance.

Figure 3–26
Network Server

Server

A major concern in a home network is security on the Internet. When a server is used as a gateway, it provides one point of access that can be secured with hardware or specialized firewall software.

A common location for a server is the structured media center (SMC), where it stays close to other networking devices. In many cases, the server may not be a stand-alone device, but instead a built-in component of the residential gateway, discussed later in this section.

Switches and Hubs

The computers and peripherals on a network need to be connected. This is accomplished with the use of a switch or a hub to attach cables from all computers and peripherals.

Switch

Hub

A *switch* is a network management device that filters and forwards data packets between LAN devices. This allows only the recipient device to receive the data packets that are specifically addressed to that device. Switches are generally more expensive than hubs, but are similar in appearance. A *hub* differs from a switch in that all devices on the network can see all data packets at any given time. The data packets are not directed to only the intended recipient device, as provided by a switch. Figure 3–27 shows a typical network hub.

Figure 3–27
Network Hub

One advantage of using a switch instead of a hub is that devices connected to a LAN using a switch do not compete for bandwidth. Instead, devices have dedicated bandwidth that is used only at the time of signal transmission. In a busy home, the use of a hub involving shared bandwidth might not be enough. Including a switch or a residential gateway with switching capabilities in the network design improves the network responsiveness and is more convenient to the users.

Residential Gateways

A *gateway* is also connected to the switch or hub. It is used to connect to the Internet.

Gateway

The residential gateway (RG) is a preassembled and preconfigured package of a modem, switch, or hub, and router or bridge devices to make the home network installation easy and convenient. Physically, the residential gateway can consist of a full or partial set of these devices to handle signal connection, translation, and direction. They are wired networking distributor products (wireless distributor products are discussed later in this chapter). Figure 3–28 illustrates a residential gateway.

Figure 3–28
Residential Gateway

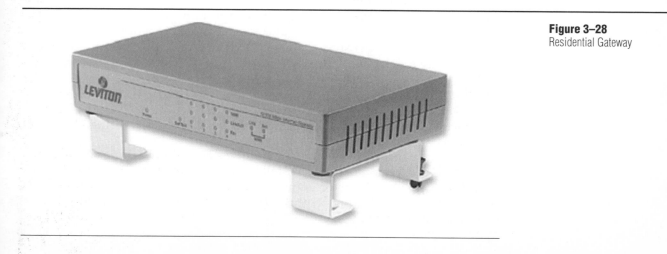

The selection and configuration of the devices in the residential gateway depend on the complexity of the network. Before selecting any particular residential gateway for a network-integration project, consider how the customer plans to use the network. For instance, if the customer has cable service and cable modem is the chosen high-speed Internet access method, a cable modem should be included in the RG. Review the specifications for the manufacturer's residential gateway to ensure that it supports the specified Internet service and is scalable for future attachments as the customer's home network expands and as technology evolves.

Routers

Routers

Routers are the most sophisticated internetworking devices discussed so far. They are slower than bridges and switches, but make "smart" decisions on how to route (or send) packets received on one port to a network on another port. Each port to which a network segment is attached is described as a router interface. Routers can be computers with special network software installed on them, or they can be other devices built by network equipment manufacturers. Routers contain tables of network addresses along with optimal destination routes to other networks. A router is shown in Figure 3–29.

Figure 3–29
A Router

In researching network-distribution products during the home network design process, it is important to understand that some residential gateways are simply called routers by their manufacturers.

Gateways, routers, switches, and hubs are usually installed on a distribution panel, which is located in the telecommunications room (see Chapter 10 for more on the distribution panel). It is important that these devices are not subjected to extreme heat or cold, and are not placed in a dusty environment. These devices are expensive, sophisticated pieces of computer equipment that can be easily damaged by the environment into which they are placed.

Wireless Access Points

Wireless networks consist of radio transceivers and special-purpose network interface cards (NICs) that distribute signals around the home. These are known as wireless access points or distributors.

Wireless computer subsystems within the home networking space require a wireless backbone hub to be installed on a conventional Ethernet LAN to distribute the signals to the computers equipped with wireless Ethernet cards. A wireless base station can be plugged into the broadband modem to serve as the signal distributor.

Wireless computer subsystem components are usually more expensive than their wired counterparts, and their data-transfer rate is usually slower. It is expected, however, that wireless LANs will become more reliable and affordable. Another wireless computer subsystem option is using HomeRF products. These products were developed especially for the residential market and are priced accordingly.

The location of wireless access points within the home network is of particular interest in the home network design because of the way wireless works. It was mentioned earlier in this chapter that wireless uses radio frequencies (RF) or infrared (IR) waves to transmit data between devices on the LAN. The type of wireless technology is therefore an important factor in the location of the wireless access point or distributor. For example, most of the existing infrared technology is described as line-of-sight, which means that the workstations and digital devices must be in direct line to the transmitter (access point) in order to operate.

Most wireless-based home networks also suffer from distance limitations. This means that the placement of an access point or wireless distributor must take into account the distance limitations imposed by the specific technology being used.

CONNECT THE NETWORK

Connecting the network devices properly is crucial for network performance and stability. Several factors come into play when it comes to network connectivity. There are different types of outlets, depending on the nature of the conductor. Common examples of wall outlets include the phone outlet, cable outlet, data outlet, power outlet, and the universal outlet type. After the equipment is installed, it must be tested to verify that it interfaces with existing equipment. Connectivity documentation is developed that provides valuable information regarding the installation.

Network Design and Termination Points

The network termination point is the weakest point of the cable. The operation of any wires installed as part of the home networking wiring depends on proper terminations. There are a wide variety of network termination points in any network installation. Terminations occur at the telephone and data modules of the structured media center (SMC) and at the wall outlets, as shown in Figure 3–30.

Figure 3–30
Telephone and Data Modules for the SMC

The following are important terms to understand:

- **Terminate:** To terminate a conductor means to connect the wire to something, typically a piece of equipment.
- **Terminal:** A terminal is a screw or quick-connect device where a conductor is meant to be connected.

At the structured media center (SMC), the common network termination points include punch-down blocks and patch panels. Four main types of blocks are common in telecommunications:

- 110-type termination block—IDC termination block used in both voice and data cabling applications, as shown in Figure 3–31.

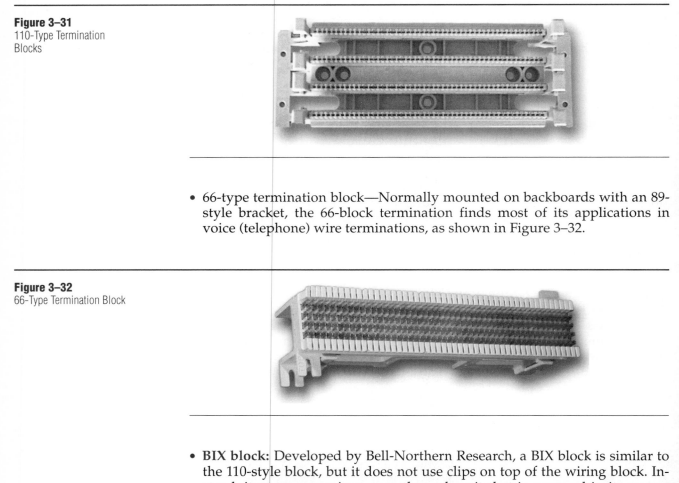

Figure 3–31
110-Type Termination
Blocks

- 66-type termination block—Normally mounted on backboards with an 89-style bracket, the 66-block termination finds most of its applications in voice (telephone) wire terminations, as shown in Figure 3–32.

Figure 3–32
66-Type Termination Block

- **BIX block:** Developed by Bell-Northern Research, a BIX block is similar to the 110-style block, but it does not use clips on top of the wiring block. Instead, it uses a one-piece, pass-through unit that is reversed in its mount after termination of the cable to expose the opposite side, which provides a cross-connect field.
- **Krone block:** Widely used in Europe and Australia, the Krone block provides silver-plated contacts at a 45-degree angle with the conductor being held in place by tension in the contacts. It is also available in patch panels.

IDC is the recommended method of copper wire termination recognized by the ANSI/TIA/EIA-568-B.1, *Commercial Building Telecommunication Cabling Standard, Part 1: General Requirements,* for twisted-pair cable terminations. This method removes the conductor's insulation as it is seated in the connection.

Jacks, Outlets, and Data Ports

Wiring in the home network is either directly connected to a component or it is connected to a jack at a wall outlet location. The location of jacks and data ports (outlets) is mainly determined by the homeowner's placement of computers and other network-enabled devices within the home. Some typical types of wall outlets include the following:

- **Phone outlet:** Connects a phone line using a Registered Jack (RJ)-style modular connector. Figure 3–33 shows a phone outlet.

Figure 3–33
Phone Outlet

- **Cable outlet:** Connects the coaxial cable using a coaxial series 6 (RG-6)-style connection. Figure 3–34 shows a cable outlet.

Figure 3–34
Cable Outlet

- **Data outlet:** Connects Category 5 wiring to Category 5 data jacks.
- **Power outlet:** Connects the power line to the power connector.
- **Universal outlet:** Customizable to the types of jacks required at the outlet. Coaxial, phone line, Category 5/Category 6, and power line can connect at a universal outlet.
- **Type I wall outlet:** Most common type of outlet. It connects one Category 5/Category 6 cable and one coaxial cable.

- **Type II wall outlet:** Outlet that connects two Category 5/Category 6 cables and two coaxial cables.

Where to locate computers is a big issue in many homes, especially those with children. Safe computing is an important aspect of residential networking. Parents want to make sure that their young ones aren't being exposed to dangerous content from the Internet when out of their sight. As a result, some parents don't want a computer hidden away in the child's bedroom; instead, they prefer public placement of the computers. This doesn't mean that older children should not have a computer in their room, however. The question therefore is "Where to have data outlets?" Generally, it is good practice to plan for jacks and ports/outlets in all the rooms, even those for the teens. However, depending on the choice of your customer, all data wiring for such locations in which placement of data networking devices is undesired for the time being can be blind terminated. *Blind termination* refers to an outlet that is installed but not seen. Blind terminations may become useful in future upgrades or when the occupants mature and can be trusted with a computer in the privacy of their bedroom.

Install Equipment and Test Hardware

The final phase of connectivity is equipment installation and hardware testing. After identifying the existing equipment, developing connectivity documentation, putting in place all the network termination points, and figuring out all the interface issues with legacy equipment (where applicable), proceed with the equipment installation. Most of the equipment that you order comes with comprehensive installation manuals. Where this is not the case, the manufacturer typically directs you to its website to find the specifications for installation. Sometimes, there are even installation demos on these sites. Drivers and all necessary software are also normally included in the shipment or are available at manufacturers' websites.

Follow the manufacturer's instructions to the letter because some products are certified for residential use only as per a specific method of installation and configuration. Therefore, even if you have installed similar products many times from other manufacturers, do not assume anything. Also, make sure to follow the instructions of the product manufacturer to test the product when you have finished with your installation work.

Do a test run at the end of the installation, and have your customer watch it. This can be anything from more comprehensive tests to a simple test such as sending a document to a newly installed printer for test printing.

Interface with Existing Equipment

In retrofits, it is almost certain that some legacy (existing) equipment and devices will be an integral part of a home network design and implementation. The immediate challenge in such a context is to understand how any existing equipment will interface with new equipment. Secondly, you must figure out either from existing manuals, manufacturer websites, or other resources any interoperability issues between legacy and new equipment.

There are several ways to approach this problem:

- Upgrade software such as computer operating systems, the basic input/output system (BIOS), and device drivers on legacy equipment to make them compatible with current equipment.
- You may need to run additional cabling to avoid high-traffic volumes or intermittent failures in signal transmissions.
- In some instances, legacy and new equipment may have incompatible device interfaces. A solution to this is to research any available adapters that the

manufacturers might have developed to take care of backward-compatibility issues.
- Finally, explore the possibility of using hybrid wired and wireless solutions to overcome interface issues with legacy equipment.

Develop Connectivity Documentation

The next important task in connecting network devices is to develop connectivity documentation. One of the most important documents is the wire detail chart. It shows the devices, the outlets, and the types of wiring used; as well as the location of the devices. It also gives an indication of how each device should be treated during the rough-in and trim phases of the wiring (rough-in and trim are discussed in a later chapter).

This chart also includes a legend that illustrates the wire types used in the signal flow diagrams, which demonstrates the importance of being consistent. You should use the same color, abbreviation, and so on for a device or cable throughout your documentation. It will be much easier for you to identify it if you are consistent.

As illustrated in the wiring chart, the computer and telephone subsystems consist of a device symbol, a device description, and outlet prefix. The outlet prefix is used to establish a convention for labeling cables. Location is where the device is placed and the destination is where it is connected to the distributor of the computer or telephone subsystem. This chart also indicates the type of wire needed to connect devices and the color of wire to make it easier to recognize it among other wires.

Several types of documents may be included as part of the connectivity documentation. They may include the following:

- **Signal flow diagram:** This diagram shows the path that signals move throughout the system; it illustrates internal and external devices and signal distribution to all components within the network.
- **Schematic block diagram/Equipment layout:** This diagram shows the theoretical connection of the gateway, hardware, and components.
- **Floor plan:** The floor plan incorporates the integrated network into the original floor plan design. This can be developed from hard copies or electronic CAD files of blueprints, or from sketches and measurements of the home layout created using a diagram program such as Visio.
- **Wire detail chart:** The wire detail specifies the media type and connections for each icon on the floor plan. It details the outlet type, location of the outlet, wiring destination, manufacturer of the wiring, wiring color code, and the connectors at the first two stages of installation (rough-in and trim).

CONFIGURE AND TEST THE NETWORK

To configure a network, you will need software. Just as PCs use computer operating systems, network servers use operating systems called network operating systems. The software version you acquire for the network's and the way it is installed and configured will completely determine the network functionality, responsiveness, and availability.

Procure and install only the latest versions of software, and pay attention to software that is upgradeable. It is also useful to look for software from well-established manufacturers in order to be sure that upgrades and patches will always be available and technical support for optimal configurations will mostly be assured as well. An open-source software such as LINUX is also a good choice because help with configuration issues can be found in many places on the Internet.

Also included with most home servers is a suite of software programs to support the following capabilities:

- IP address sharing, automatic IP configuration
- Web presence
- Home information database

The home-networking software suite normally includes the following components:

- **Network Address Translation (NAT):** NAT enables all devices on a home network to share a single IP address to access the global Internet. The NAT program makes all the necessary address translations. In addition, NAT provides inbound access security so that access to appliances and devices within a home can be controlled.
- **DHCP server:** Dynamic Host Configuration Protocol (DHCP) is a standard method for computers to automatically obtain configuration information from a server.
- **Micro-Web server:** This server program provides a point of presence on the World Wide Web for such things as remote management and administration.

Internet Protocol (IP) Addressing

IP address

An *IP address* is the means by which a device on a home network identifies and communicates with other devices within the network and with hosts on the Internet. IP addresses are crucial for the proper configuration of the residential network. Each device on the network must be configured with a unique IP address within the network. An IP address is described as public or private. The terms *public* and *private* are very important in home network IP addressing. A public IP address is located on the WAN (Internet) side of the residential gateway, whereas the private IP address is located on the internal (home) LAN side. Additionally, a static address is fixed as opposed to a dynamic address, which changes every time a user logs onto the network. Therefore, you also have public static and private static IP addressing.

There are a number of considerations when assigning IP addressing. If a device needs to be visible from the outside (i.e., remotely accessed), it must be assigned a static IP address.

Generally, IP addressing can be achieved in two ways:

- You can assign a public static IP address; this is a less secure option because your device will be visible to the outside world.
- You may instead move the device to the internal home LAN and then assign it a private static IP address. Next, designate a port on the RG to point to the assigned private IP address. Typically, ports 5634 and 5637 are used. This is a more secure option because your device is not visible to the outside world. It can only be reached through network address translation (NAT) by the residential gateway.

Network Address Translation (NAT)

NAT enables all devices on a home network to share a single IP address provided by an ISP to access the Internet. The NAT program makes all the necessary address translations as illustrated in Figure 3–35. In addition, NAT provides inbound access security so that access to network devices and appliances within the home can be controlled. Most residential gateways available in the residential networking products market today have a built-in NAT program.

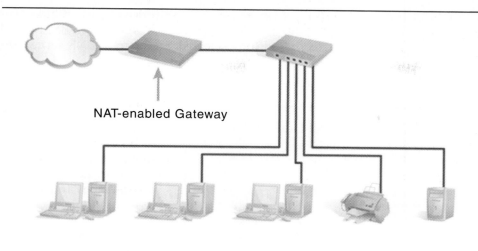

NAT-enabled Gateway

Figure 3–35
Network Address Translation
(NAT)

A home network only needs one public IP address if a NAT-enabled
gateway is used.

Public versus Private Address Configuration

There are some important implications in going with a private versus a public
addressing scheme. More specifically, the following issues need to be taken
into consideration:

- The ISP provides a dynamic or static IP address.
- The dynamic IP is cheaper in price to obtain, but has two major weak-
  nesses, including the following:
  Does not support remote access control (e.g., VPN, home automation
    control)
  Not a good candidate for upgrade (i.e., the capability to install and
    integrate additional subsystems to the computer network)
- A static IP is higher-priced, but has clear advantages over dynamic:
  Supports remote access control (e.g., VPN, home automation control)
  Good candidate for upgrade

Usually one static IP is sufficient if all the home network access goes
through the residential gateway. However, if the customer has a home control
super-system that requires remote control, multiple IPs are desirable. A home
super-system that needs to be remotely accessed via the Internet is assigned a
separate static IP outside the residential gateway.

Dynamic Host Configuration Protocol (DHCP)

Dynamic Host Configuration Protocol (DHCP) is a standard method for computers to automatically gain configuration information from a server.

DHCP is a software utility that runs on a computer and is designed to assign IP addresses to clients (PCs and other networked devices) logging onto a
TCP/IP network. This eliminates the need for manual IP address assignments. For the home network design, it is desirable to consider having this
service home-hosted. There are residential gateway products in the market
that have the DHCP service built in, which eliminates the need for a separate
DHCP server and also facilitates setup and management. It is important to
understand that DHCP is an internal service to the home LAN, and is very
useful for reducing the overall cost of ownership of a home network, espe-

cially in large homes. It also greatly reduces the burden of network management or service calls by the homeowner. Figure 3–36 shows the IP Protocol properties dialog box with the buttons selected to obtain a DNS server address automatically.

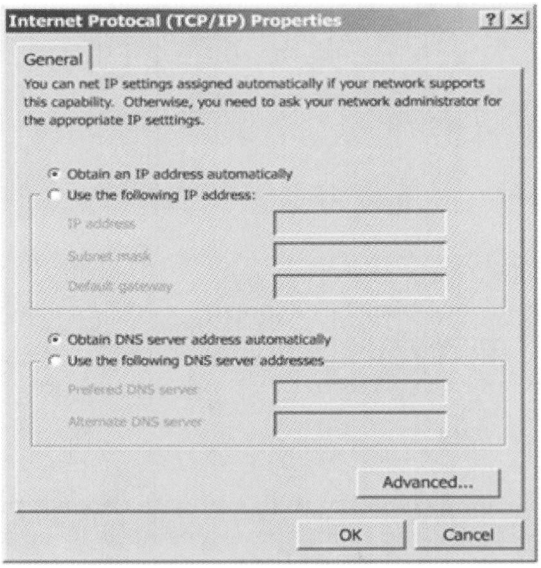

Figure 3–36
IP Protocol Properties Dialog Box

To obtain a DNS server address automatically, select the following two buttons in the Internet Protocol (TCP/IP) Properties window:

1. Obtain an IP address automatically

2. Obtain DNS server address automatically

Address Conservation

The use of address-conservation techniques in designing the IP addressing scheme for a home network is an important skill required for the home integrator. One such method is the use of network address translation. Because of the shrinking stock of public IP addresses, Internet service providers (ISPs) are usually reluctant to commit a large amount of address space to one homeowner. Therefore, one or two public addresses can be used with a private IP address via NAT translation to access the Internet. Private addresses are known as Request For Comments (RFC) 1918 addresses. The RFC internal private address ranges that anyone can use include the following:

- 10.0.0.0 to 10.255.255.255
- 172.16.0.0 to 172.31.255.255
- 192.168.0.0 to 192.168.255.255

Firewall

Today, more and more people are beginning to take advantage of the always-on feature of broadband technologies. This makes a lot of sense, but users need to be aware of the risks associated with connecting in-home network devices and appliances to the outside world. First and foremost, they are exposing their private network to a variety of security risks. Breaches on a home network can result in a hacker stealing your stored data or even taking over your network and using it to hack into other networks outside the reach of the home.

There are a number of steps that may be taken to minimize the security risk of connecting a private in-home network to the public Internet. The implementation of a home firewall is a key step in preventing hacker attacks on a home network. In the world of corporate IT, firewalls are the most popular method of protecting local-area networks from outside intruders. A *firewall* is a hardware or software system that is used to protect sensitive data from being accessed by unauthorized people. A firewall in the context of a home-networking environment is normally located at the interface between the local devices and the high-speed broadband connection. All data, video, and voice traffic between these two networks is examined by the in-home firewall. Other features of a typical home firewall system include the following:

Firewall

- Automatically closes your broadband connection on the detection of attempts to hack into one of your digital appliances
- Allows different members of your family to set their own levels of security
- Records all broadband Internet access events

An in-home firewall normally consists of a variety of hardware and software components. The exact combination of components chosen to build a firewall depends on the level of security required. A typical home firewall comprises a standard desktop PC running a personal firewall system, which examines all information coming in from outside the home network as illustrated in Figure 3–37. The home-networking security market has matured in the past few years, with software companies offering consumers a variety of sophisticated security products.

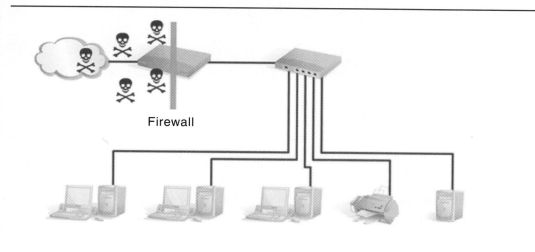

Figure 3–37
Illustration of How a Firewall Stops Intruders

Firewall

A hacker will try to gain entry into the home network. A good firewall will stop the hacker.

New firewall products continue to be developed and put on the market. It is therefore recommended that you consult magazines and other related industry publications for the most up-to-date information on the best product suitable for a given situation. Look for buying guides and "head-to-head" comparisons done by many industry magazines.

After installing a firewall, you will need to open certain ports on the firewall in order to use e-mail and the Internet. SMTP (explained in the DSL section) uses port 25 and POP3 uses port 110 to transmit and receive e-mail. In addition, you will need to open port 80 Hypertext Transfer Protocol (HTTP), the standard protocol that supports the exchange of information on the World Wide Web.

Test Connectivity

After installing the computer network, like any technical project, proceed with some basic tests to make sure the system is functional. Testing the system can be as simple as verifying that all computers on the network connect to the Internet. At each PC, click on the web browser icon. The start or home page should display. Each computer will need to be set up independently to enable file sharing and printing.

Ping

Devices on the network can be tested for connectivity by using *ping*. Ping is a simple but highly useful command-line utility that is included in most implementations of TCP/IP. Ping can be used with either the host name or the IP address to test IP connectivity. Ping works by sending an Internet Control Message Protocol (ICMP) echo request to the destination computer. The receiving computer then sends back an ICMP echo reply message. This message is shown in Figure 3–38. ICMP reports errors and provides other information relevant to IP packet processing. *Ping* is derived from *pack internet groper*.

Figure 3–38
Ping Response Example

```
c:\>ping cisco.com
Pinging cisco.com [17.254.3.183] with 32 bytes of data:
Reply from 17.254.3.183: bytes=32 time=430ms TTL=90
Reply from 17.254.3.183: bytes=32 time=371ms TTL=90
Reply from 17.254.3.183: bytes=32 time=370ms TTL=90
Reply from 17.254.3.183: bytes=32 time=371ms TTL=90

Ping statistics for 17.254.3.183:
    Packets: Sent = 4, Received = 4, Lost = 0 <0% loss.,
Approximate round trip times in milli-seconds:
    Minimum = 370ms, Maximum = 430ms, Average = 385ms
```

If you cannot ping another device, you have to troubleshoot the problem. First, ensure that the active equipment is working properly by checking on a known-good circuit and other stress points in the system. Use the appropriate test equipment for accuracy. You may proceed by replacing one item at a time, resetting after each element is replaced. Some examples of so-called stress points you need to test may include the following:

- Line cords
- Patch cords
- Wall plate components
- Wiring closet components
- Installed cable (use certified cable tester)

More advanced testing techniques are discussed in later chapters.

SUMMARY

Designing and installing a computer network requires the HTI to gather information from the customer, conduct a site survey, determine the best type of network technology to use for the project, design the network layout, install the media and equipment, configure the network equipment, test it, and teach the customer how to use the network.

This chapter focused on the following topics:

- A home network forms a LAN that can connect to the Internet, a WAN.

- The OSI reference model has seven layers: physical, data link, network, transport, session, presentation, and application.
- The first step in the design process is determining what your customer needs.
- In older homes, power-line, phone-line, or wireless networks are the best choice.
- CAT 5/5e is the standard cabling for home installations.
- Bluetooth uses radio waves to transmit information between network devices situated no more than 33 feet apart.

- X-10 uses existing power lines to control devices.
- A residential gateway is used to connect the network to the Internet.
- DHCP automatically configures all network settings, including obtaining an IP address.

- A firewall is a hardware or software system that is used to protect sensitive data from being accessed.
- NAT enables all devices on a home network to share a single IP address.

GLOSSARY

Bandwidth   The measurement of how fast data travels from source to destination. It is often measured as millions of bits per second (Mbps, where the M stands for *mega*) or thousands of bits per second (Kbps, where the K stands for *kilo*).

Bluetooth   A wireless technology that uses short-distance radio signals operating for distances of up to 10 meters (33 feet).

De-encapsulation   Stripping data packets of addressing information.

Encapsulation   Wrapping data packets with addressing information to ensure the best chances of delivery to their intended destination.

Ethernet   The most widely installed local-area network (LAN) technology. It typically uses copper wire, yet there are also wireless and fiber Ethernet networks.

Firewall   A hardware or software system that is used to protect sensitive data from being accessed by unauthorized people.

Gateway   A device that acts as an entrance to another network.

HomePNA   A technology that uses the existing phone line. The technology supports up to 25 devices and delivers a constant 10 Mbps, regardless of whether or not the phone is in use.

Hub   A networking device into which all devices are physically connected by a cable and share bandwidth.

Infrared   A variety of wireless control applications, including home entertainment remote controls, wireless mouse devices, cordless modems, motion detectors, and fire sensors.

IP address   The means by which a device on a home network identifies and communicates with other devices within the network and hosts on the global Internet.

LAN   Local-area network. A physical interconnection of devices in a single building or group of buildings that communicate with each other using a specific protocol.

MAC address   A unique identifier assigned to the NIC that enables communication between devices.

Mbps   Stands for *megabits per second,* and is a measure of the total information flow over a given time, called bandwidth, on a medium such as copper or fiber-optic cable.

Network Interface Card (NIC)   A computer circuit board that provides an interface to connect to a network.

OSI model   Open System Interconnection model. Used to divide the functions of networking into seven distinct layers.

Ping   A simple but highly useful command-line utility that is included in most implementations of TCP/IP.

Protocol   A controlled sequence of messages that are exchanged between two or more systems to accomplish a given task.

Protocol data unit (PDU)   Controls information and user data contained in data packets used by OSI layer hardware and software to encapsulate packets.

Router   A device that determines the best route that data should take to reach its destination. A router is also used in the home network as a gateway for Internet access and may provide network security features.

Server   A computer or computer program that provides services to other computers on the network such as storing files and managing printing services.

Switch   A device on a computer network to which all devices are connected either directly or via another device. A switch selects a path for the data to travel, segments the traffic, and makes intelligent decisions.

TCP/IP   Transmission Control Protocol/Internet Protocol. A set of protocols used to communicate across interconnected networks.

UDP   User Datagram Protocol. A connectionless and unacknowledged protocol.

WAN   Wide-area network. A network that interconnects LANs that span a larger geographical area.

X-10   A communications protocol that allows compatible home networking products to talk to each other via the existing 110-volt electrical wiring in the home.

CHECK YOUR UNDERSTANDING

1. Which of the following is not an advantage of the OSI reference model?
   a. Creates faster transfer rates
   b. Breaks down the complex operations into simple elements
   c. Allows for specialized design and development of modular functions
   d. Provides the capability to define standard interfaces for PnP and multi-vendor integration

2. What does encapsulate refer to?
   a. Final delivery of a packet
   b. Determining correct address information
   c. Adding header and trailer information to a packet
   d. Establishing communication between networks

3. A controlled sequence of messages that are exchanged between two or more systems to accomplish a given task is known as a:
   a. control application.
   b. protocol.
   c. data link.
   d. PDU.

4. What is the first step in the design process?
  a. Take inventory of existing equipment
  b. Determine what your customer needs
  c. Determine location of equipment
  d. Determine type of network technology to use

5. Which type of equipment is used to support file or printer sharing?
  a. Switch
  b. Router
  c. Server
  d. Hub

6. CSMA/CD stands for:
  a. computer sense multiple access with collision detect.
  b. carrier sense multiple access with collision detect.
  c. common sense multiple access with collision detect.
  d. control sense multiple access with collision detect.

7. To connect to a network, every computer device needs which of the following built into it?
  a. NIC
  b. Switch
  c. Hub
  d. Gateway

8. The type of NIC purchased depends on the type of media used.
  a. True
  b. False

9. IEEE 802.11, Bluetooth, and HomeRF are widely accepted forms of:
  a. modems.
  b. ISPs.
  c. wired technologies.
  d. wireless technologies.

10. Wireless computer networks can support data transmission speeds from 1 to:
  a. 12 Mbps.
  b. 24 Mbps.
  c. 56 Mbps.
  d. 1.544 Mbps.

11. What is an acceptable Ethernet cable for wiring a home computer network?
  a. Thicknet
  b. Thinnet
  c. CAT 5
  d. Homeplug

12. HomePNA, even when the phone is in use, operates at a constant:
  a. 1.5 Mbps.
  b. 10 Mbps.
  c. 100 Mbps.
  d. 1000 Mbps.

13. What is the maximum operating distance of Bluetooth?
  a. 10 feet
  b. 33 feet
  c. 100 feet
  d. 512 feet

14. What is the major limitation of using DSL?
  a. Slow data transmission rate
  b. Expensive equipment
  c. User must be near telco
  d. Equipment must be within line of sight of each other

15. A cable modem communicates over the cable network to a device called:
  a. Cable Modem Terminal Service.
  b. Cable Modem Termination System.
  c. Cable Modem Termination Service.
  d. Computer Modem Termination System.

16. A wire detail chart provides important information for an HTI. Explain why.

17. Explain the importance of a firewall on a home network.

18. What are the benefits of a server on a home network?

19. List the determining factors involved in recommending the type of Internet connection for a home network.

20. What is "ping" used for?

Telecommunications

OBJECTIVES

Upon completion of this chapter, the HTI will be able to complete the following tasks:

- Identify the many different components in a home telephone system

- Understand PBX and how it works

- Design and configure a home telephone system

- Determine proper low-voltage wiring for telecommunications

- Document the network and troubleshoot the system

INTRODUCTION

Telecommunications in the home began when the first telephones to residences were installed near the dawn of the previous century. The advent of multiple lines allowed business numbers to be installed as well as lines for private use, and this may have constituted the beginning of the home office and telecommuter movement that is growing towards maturity today.

Residential telecommunications and connectivity is at the center of the emerging Small Office/Home Office market (SOHO). The HTI installer must be well-grounded in the theory, use, installation, and troubleshooting of this important technology.

TRADITIONAL AND IP TELEPHONY

This chapter begins with a discussion of both traditional telephone systems and the latest voice delivery technology. Traditional telephony systems rely on the *Public Switched Telephone Network* (PSTN), also known as *Plain Old Telephone Service* (POTS). PSTN is one huge network composed of all the telephone companies and their lines in the world. Local telephone companies are part of the PSTN so that anyone can place a call from Los Angeles to San Diego or Paris or Singapore. The Internet also uses the PSTN to send and receive data across a city or the world. In addition, a new technology called Simultaneous Voice and

Public Switched Telephone Network

Plain Old Telephone Service

Data Technology (SVD) is being employed to carry both voice and data over the PSTN, where high-speed Internet access is not readily available or is cost-prohibitive.

Although the PSTN has been around since the early 1900s, it is increasingly coming under pressure from the Internet and data technologies since they are better equipped to carry multimedia communications. Internet Protocol (IP) Telephony, otherwise known in the industry as Voice over Internet Protocol, or VoIP, is the latest standard for delivering voice over data networks. How VoIP works will also be discussed later in this section.

Public Switched Telephone Network (PSTN)

The telephone company is responsible for deploying, supporting, and maintaining the PSTN infrastructure, which has evolved into a very robust infrastructure over time. This infrastructure is very reliable, highly available, redundant, fault-tolerant, and has set a high standard to meet in terms of replicating this level of service.

For more than 100 years, people have depended on the telephone to communicate. In the early days of the telephone, people communicated by first contacting the operator, who would then connect the caller to the appropriate receiver. This was accomplished with the use of a switchboard. Each caller was identified on the switchboard by wires that terminated at a plug. The operator inserted these plugs into the patch panel to make the connection.

In recent years, the telephone has become even more important as a tool used to fax documents and connect to the Internet. When using PSTN for an Internet connection, however, it is restricted to about 56 Kbps. The telephone works by sending analog signals through the telephone wires. It starts when a person speaks into the phone. The voice data is converted into an *analog* signal, and it is transmitted to the local telephone office. From there, it is strengthened and routed to a network of telephone offices toward its destination. At the other end, the signal causes the phone to ring, indicating that voice data is on the line. Ringing is the result of a 90-volt, AC wave at 20 Hz, which causes the sound or ring. Because the transmission is not robust, it is subject to interference that causes line noise. This noise is the main problem with analog signals.

Analog

Today, calls go to an electronic device in a junction box called a concentrator, located close to the home. The concentrator converts the dialing and voices into digital pulses, collects all the digital pulses from all the phones in your neighborhood, and sends these pulses over a common wire to the exchange. The exchange has many inputs from concentrators that are mixed together and put on a trunk that connects all the exchanges in a city.

NOTE *Circuit-switched* is a type of network in which a single connection is made between two devices for the duration of the connection. PSTN is a circuit-switched network. The telephone company reserves a specific physical path to the number you are calling for the duration of your call. During that time, no one else can use the physical lines involved. Circuit-switched is often contrasted with *packet-switched*. Packet-switched networks break communication down into packets that are then sent to their destination. This allows the same data path to be shared among many users in the network. Most traffic over the Internet and most LANs uses packet-switching.

Signaling is used to tell the telephone company the status of a phone line. There are two types of signaling: off-hook and on-hook. Off-hook is simply the indication that the handset has been lifted off the cradle. That means the user is getting ready to place a call or is trying to answer an incoming call. The on-hook condition is indicated when the handset is in the cradle. On-hook simply tells the telephone company that the phone is ready to receive a call or that a call has ended.

> **NOTE** Current phone systems use *tone dialing*, which sends a short burst of a tone with a different pitch for the different numbers dialed, rather than the series of disconnects for every digit, as in *rotary dialing*.

For outside lines, the telephone company checks the status of the destination telephone. If a destination phone is off-hook, the telephone company sends a busy signal to the caller. If the destination phone is on-hook, the telephone company connects the caller and the destination phone rings.

Simultaneous Voice and Data Technology (SVD)

One of the latest telecommunications technologies to emerge is Simultaneous Voice and Data Technology (SVD). SVD provides low-cost communication of high-quality voice, data, fax, and images simultaneously using PSTN. Data can be transmitted at 28 Kbps when no voice is present on the line, and 16.8 Kbps when voice is transmitted simultaneously. Although not widely used, SVD is gaining popularity where high-speed Internet access is unavailable.

Voice Over Internet Protocol (VoIP)

VoIP, also known as IP telephony, is a technique for sending real-time voice over data networks, such as the Internet, or an internal IP network using IP datagrams. A *datagram* is a unit of data, or packet, that is transmitted in a TCP/IP network. Each datagram contains the source and destination addresses and data. As discussed in Chapter 3, each device on the network has an IP address, which is attached to every packet for routing. VoIP packets are no different. To place a call from one device to another, users may enter the IP address of the called party using a voice-conferencing software package. In VoIP, the digital signal processor (*DSP*) segments the voice signal into frames that are then transmitted over data networks such as the Internet.

VoIP is typically used as an internal business communication tool or as a secondary communication tool in conjunction with PSTN line access. Additional analog copper-line connections are not needed for VoIP networks. IP telephony works on a computer network, and operates on pairs 2 and 3 of a CAT 5 cable. Standard analog phones operate on pair 1 of a CAT 3 cable.

VoIP can be used to communicate solely between computers or telephones or a combination:

- **PC to PC:** This is the ideal solution. Long-distance calls are free if using VoIP between two computers. It is also the easiest way. All the homeowner needs is VoIP software, a microphone, speakers, a sound card, and a high-speed Internet connection so that conversations can take place in real time.
- **PC to/from a phone:** The PC must have VoIP software installed in order to place or receive phone calls. If the call is initiated from a phone, the caller dials a VoIP service that will connect him or her to the computer. The caller is charged for this service.

VoIP

Datagram

DSP

• **Phone to phone:** The caller dials a VoIP service. The call is then conducted over a data network rather than the PSTN. The caller is charged for this service.

TELEPHONE SERVICES

An understanding of the telephone services that are available will help you design and install a telephone network that will best serve the customer. Telephone services provide features that many homeowners find beneficial. Call extensions to provide multiple lines, voice mail, and call conferencing allow the homeowner to take advantage of the services available.

Call Extension

A home can have multiple lines or extensions. This allows a user to call out from one phone while a fax is delivered in the home office and another user receives a phone call. Homeowners who have a home office or who telecommute may consider multiple lines a necessity. In addition, homeowners who wish to have one type of DSL service, HDSL (High DSL), will need two phone lines.

An example of a home with four incoming lines:

• Line 1 is the primary line
• Line 2 is for the business
• Line 3 is for the fax
• Line 4 is for the children

The layout can be similar to the following:

In the home office, the business line appears on position 1 of the jack and the primary line appears on position 2. The fax appears on position 1 of another jack within the room.

In the children's bedrooms, the child's line appears on position 1 of that jack while the primary line appears on position 2.

Finally, in the master bedroom, the primary line appears on position 1, the children's line appears on position 2, and the business line appears on position 3.

Voice Mail

Most telephone service providers offer voice-mail services. For multiple lines, there can be different voice-mail accounts. A stutter dial tone is a common signal used to notify the user that a message is waiting. Stutter dial tones and regular dial tones are similar, but the stutter dial tone does not have a continuous tone. Phones are available that will notify the user when a call is waiting by a flashing light. Without this feature, the user must pick up the phone and listen for the stutter tone. Retrieving voice mail requires that the user call a designated number and enter a password. Each account will be able to access voice mail with options to replay, save, or delete. Each account will also have access to administrative features that determine the number of rings before voice mail answers, specify how long saved messages are kept, and perform password maintenance.

Call Waiting

Another feature offered by a telephone provider is call waiting, which is used to notify the user of an incoming call while the phone is in use. The flash button, located on the telephone keypad, is pressed to open the incoming line and pressed again to return to the original call.

Call Conferencing

When a conference call involves a small number of participants, such as three or four callers, it can be accomplished with a normal telephone system. Conference calling is a service provided by the telephone company. A conference call is started by calling the first participant and pressing the conference button, then the second participant is called and the conference button is pressed, and so on. Call quality will be degraded when the number of participants increases. It is recommended that you use a conferencing service bridge provided by the local telephone company when there is a large number of participants.

PRIVATE BRANCH EXCHANGE (PBX)

A *PBX* creates a private telephone network in the home. Each phone in the home has a different extension. It's the equivalent of giving the homeowner multiple lines from the phone company for the cost of a single line. Calls made within the PBX do not use the phone company's lines and therefore do not incur any charges.

PBX

> **NOTE** A variation of the PBX is called *Centrex* (*Central Office Exchange Service*). All the communications equipment necessary to implement the PBX is owned and managed by the telephone company, who sells the various services to the homeowner.

Centrex (Central Office Exchange Service)

The Home PBX unit is a simplified version of the Office PBX System. The features of the Home PBX include the following:

- **Call transfer:** This allows for calls received on one line to be switched to another line, for example, in another room.
- **Call restriction:** This feature will block certain numbers from being connected if dialed.
- **Conference calling:** Allows multiple parties on different lines to join a conversation.
- **Music-on-hold:** This feature will play pre-recorded music on the line when the caller is put on hold.

How a PBX Works

A private branch exchange, or PBX, becomes the central switch that connects incoming calls to the appropriate extension or room. When calling outside, users connect to the PBX by dialing "9" and then the number. The PBX shares a certain number of outside lines to make telephone calls external to the PBX. A PBX can support either digital or analog signals.

A PBX system can be used for security. A phone at a gate or door will alert the homeowner to visitors. When a person comes to the door or gate, they pick up this phone. The PBX sends the call to all phones in the home. Different rings for the door, gate, another phone in the home, or an outside line can be programmed to alert the homeowner to the location of the call.

PBX Installation

The PBX is installed in the structured media center. The wires coming from the demarcation point of the home are fed into the PBX unit as shown in

Figure 4–1. Lines are then run to each individual room (using the star topology) that is designated for telephone or data use. A star topology for the phone wiring is required to install a Home PBX. The systems should be installed on a separate unswitched electrical circuit. When using an uninterruptible power supply (UPS), make sure there is enough space. There are different models available that are small or can be mounted on the wall.

Figure 4–1
PBX Box

On-Net and Off-Net Calling

Off-net calling is the term used to indicate when a call is placed from within the PBX to outside the PBX. The homeowner will dial a number, usually "9," to obtain an outside line first. Then the telephone number is entered.

The other type of call that a homeowner can make with a PBX is an on-net call. By dialing an extension, the homeowner can call any phone within the PBX network. The phone call never goes to an outside line. The advantage of on-net calling is that the homeowner does not pay the telephone company for these calls because the telephone company's lines are never accessed.

Call Restriction

There are several features that a PBX can offer in addition to those offered by the telephone company. One such feature is call restriction. Phones on the system can be restricted from making outbound long-distance and specific 1-900 number calls. They can also be set up to receive only inbound calls.

Intercom/Paging System

Intercom/paging systems can be incorporated into the home PBX. They can also be set up as a stand-alone system. The intercom goes to specific locations throughout the home. The paging system uses the system to broadcast throughout the home or specific zones. In addition to being incorporated into the intercom, the paging system can use speakers in the phone or home speakers.

 THE HOME TELEPHONE SUBSYSTEM

The telephone subsystem delivers telephone services throughout the home network. The telephone subsystem can act as a conduit for signal distribution of other subsystems, such as security and computers that require telephone lines, when it is integrated into the network. By using the structured wiring scheme in an integrated home, the features and functionality of the telephone subsystem can be augmented with the addition of a PBX unit.

Design considerations for telephone and data connections should be based on the local codes and standards for the area, site limitations, and the specifications of the homeowner.

Considerations include:

- Overall distance to determine the amount of wire or cable
- The services that will be provided
- Flexibility when requirements change
- Scalability in order to expand the system as needed

Demarcation Point (Demarc)

The point at which telephone service enters the home is known as the demarcation point, or *demarc*, as shown in Figure 4–2. From this point, the service is distributed throughout the home. The demarc is where the service provider's responsibility stops and the homeowner's responsibility starts in terms of any problems that might occur. The line from the service provider's facility to the demarc is the responsibility of the service provider.

Demarc

Figure 4–2
Telephone Demarc

Distribution Equipment

Distribution equipment is used to distribute telephone signals throughout the telephone and computer subsystems in the home. Up to four telephone lines can be connected to a standard telephone hub device. More than one hub can be used when additional phone lines are coming into the home. In these situations, a PBX should be considered to meet the needs of the homeowner.

The distribution equipment is housed in a structured media center (SMC) or distribution panel, as shown in Figure 4–3. The figure shows the telephone line distribution modules circled in the upper left corner. The current ANSI/TIA/EIA-570-A standard defines 24 inches as the minimum height of a structured media center (SMC). Although a 14-inch panel does not meet the current standards, it is expected to be part of the next standard, ANSI/TIA/EIA-570-B, which is expected to be ratified in 2003. The smaller panel is used in smaller, simpler installations.

Figure 4–3
SMC with Telephone Line
Distribution Modules

The decision regarding where to put certain equipment throughout the home is based on different location factors. The SMC should be located in a clean dry area in which it will be easily accessible. If a closet does not provide enough room, consider a location in the basement, or some other area where there is adequate power, air conditioning, and ventilation. A laundry room is usually not a good location because of the added humidity and contaminants. Additionally, a garage does not typically maintain a consistent temperature and can be dusty. Building a small enclosure to maintain the equipment is one option if it is warranted.

The benefits of designating a central distribution device in the home include organization, the ability to hide the hardware and wiring, and easy access. There is more flexibility when the installation is for a new home. Central distribution in a retrofit will depend on the availability of designated space.

Termination Blocks

Termination blocks are used to terminate telephone wires on the home network. They incorporate 110-type connections to provide connectivity between circuits. In commercial applications, 110-type blocks come in two different configurations: 100-pair and 300-pair. The 100-pair configuration is the most common, with four rows of 25-pair terminations. Figure 4–4 shows the 110-type block, which includes wire management troughs that act as spacers between the blocks. The 110-type connection provides a high-density, high-performance insulation connection. As this insulation is displaced, it forms around the conductor to prevent corrosion.

Figure 4–4
110-Type Block

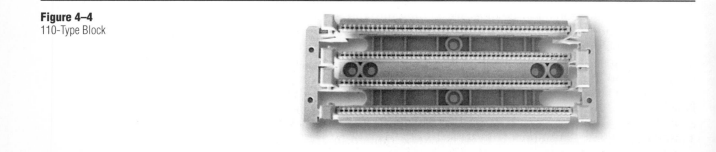

A variation of these blocks can be used in residential applications as shown in Figure 4–5.

Figure 4–5
Voice and Data Module

110-type blocks use a special punch-down tool, as shown in Figure 4–6. It is able to punch down up to five pairs of wire at a time. Punch-down refers to the insertion of wires into connectors such as the blocks, clips, and jacks. Using the correct punch-down tool is recommended. A screwdriver can damage the contacts, resulting in an inferior connection.

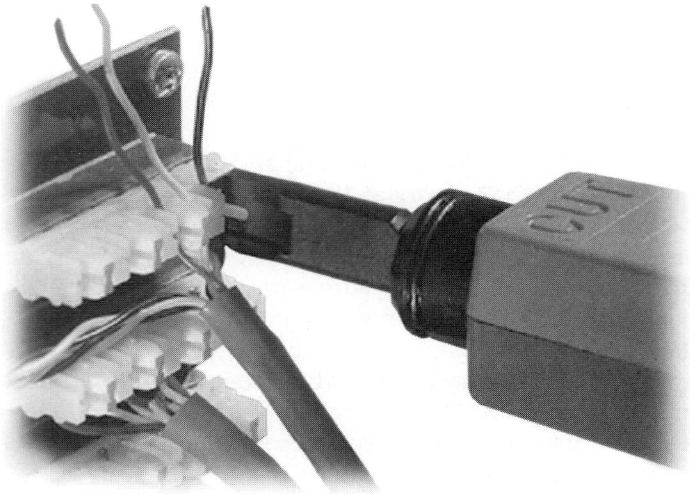

Figure 4–6
Wire Punch-down and
Termination Tool

The connectors in the home are terminated as follows:

- The first connector is terminated to the demarcation point from the service provider.
- The next eight connectors are terminated at the outlet locations within the home using twisted-pair wiring.
- The last connector can be used for terminating to another bridging module that can distribute more telephone lines.

Key System Units

A key system unit (KSU) or Key Telephone System (KTS) is used when a homeowner wants additional features for the telephone system. KSUs do not use switches, unlike a PBX. A single-line phone behind a key system can gain access to only one outside line. The key telephone sets do not terminate in an intelligent switching device, so it does not have the ability to make decisions about the use of pooled, or shared, circuits to connect to the Public Switched Telephone Network (PSTN).

A KSU can provide flexibility when the home network has multiple phone lines. It is able to connect different incoming phone lines to the various room extensions in the home, put conference calls together, play music to incoming calls put on hold, or act as an intercom. Key systems are less expensive than PBX system and many do not require professional installation. A KTS requires special phone equipment, but its features are valuable to a homeowner who has a high volume of calls and requires the security it offers.

Phone Jacks

Telephone jacks should be installed in every room that allows for Internet connection. Additional jacks should be installed if additional telephone lines will be used or the homeowner wants more than one telephone or fax in the room. The communications outlet should be located at least 16 inches away from an AC outlet to prevent signal interference.

Handicapped users might require special access for telephone, fax, and caller line ID equipment. TDD (Telecommunication Device for the Deaf), a text device that allows hearing-impaired people to communicate over telephone lines, is easier to connect when the phone jacks are above the counter instead of below. In every instance, adjust the height of the phone jack to best accommodate the user.

 **LOW-VOLTAGE MEDIA**

CAT 3

CAT 5

The subsystem wiring starts at the telephone and uses a phone wire (*CAT 3*) or *CAT 5* wire. CAT 3 is suited to work at 10 MB, was designed to work with Ethernet at 10 MB, and is the FCC-required minimum wire rating for phone lines. An older home will typically have CAT 3 already installed whereas newer constructions typically use CAT 5. CAT 5 wire is recommended by ANSI/TIA/EIA 570-A because it is scalable, which means that it can be expanded and can be used for VoIP.

The wiring is laid for every room that will use a phone connection and an Internet connection. In the past, a daisy chain wiring topology was used. New installations should use a star topology where all the wires are run from a central point, typically the distribution panel.

Tip and Ring

In the early days of the telephone, switchboard plugs were inserted into the proper receptacle or jack in order to complete a call. The tip of the plug has one conductor attached, whereas the other conductor is fastened to a ring around the plug. This is referred to as "tip and ring." The colors of the ring are blue, orange, green, brown, and slate. The colors for the tip are white, red, black, yellow, and violet. Figure 4–7 illustrates an old telephone switchboard plug.

Figure 4–7
Old Telephone Switchboard Plug

Color Codes for Twisted-Pair

In a four-pair cable, such as CAT 5, the primary color is the tip, which is usually white with a tracer, or stripe, that is the same color as the pair's solid color wire, which is the ring. The white wire is designated as the tip because this wire is normally the first wire of each pair to be punched down to a punch block, particularly when working with larger pair-count cables. Most twisted-pair cables have the color-coding scheme shown in Figure 4–8.

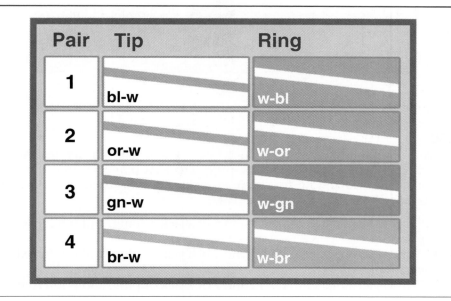

Figure 4–8
Color-Code Scheme for a
Four-Pair Cable

Telephones using only two pairs of wire are less common. The first pair has one green wire ("tip") and one red wire ("ring"). The second pair has one black wire ("tip") and one yellow wire ("ring"). For a single phone line, only the green and red pair is normally used. The black and yellow pair is normally a spare and available to install a second phone line.

> **NOTE** Category 3 cables are laid out as follows: pair 1 is white-blue, pair 2 is white-orange, pair 3 is white-green, and pair 4 is white-brown. White is always the tip side of the line, and the color is always the ring side of the line.

Although the four-pair cable is the most common, it is not the only possible configuration. The same general color-code scheme is used in multiple-pair cables. The four-pair wiring scheme is a subset of this larger color-coding system. Most four-pair wiring follows this standard. Some manufacturers do not put a colored stripe on the tip wire. Others do not use a solid color, but rather a translucent shade that tints the color of the wire that shows through it. Finally, some manufacturers distinguish wires by using periodic splotches of the companion color. Figure 4–9 shows a color-code scheme for 25-pair cable.

USOC

Figure 4–9
Color-Code Scheme for
25-Pair Cable

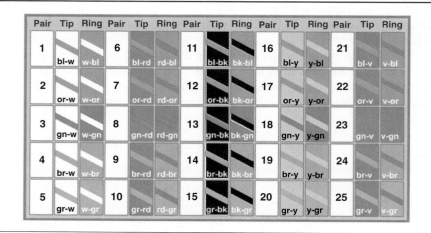

Pair	Tip	Ring	Pair	Tip	Ring	Pair	Tip	Ring	Pair	Tip	Ring	Pair	Tip	Ring
1	bl-w	w-bl	6	bl-rd	rd-bl	11	bl-bk	bk-bl	16	bl-y	y-bl	21	bl-v	v-bl
2	or-w	w-or	7	or-rd	rd-or	12	or-bk	bk-or	17	or-y	y-or	22	or-v	v-or
3	gn-w	w-gn	8	gn-rd	rd-gn	13	gn-bk	bk-gn	18	gn-y	y-gn	23	gn-v	v-gn
4	br-w	w-br	9	br-rd	rd-br	14	br-bk	bk-br	19	br-y	y-br	24	br-v	v-br
5	gr-w	w-gr	10	gr-rd	rd-gr	15	gr-bk	bk-gr	20	gr-y	y-gr	25	gr-v	v-gr

USOC

The Bell Telephone Universal Service Order Code (*USOC*) defines the way wires are organized into a modular plug. The first pair goes into the center two pins. The rest follow from left to right, splitting each pair down the middle that separates the data wire pairs:

- Pair 1 is terminated in pins 4 and 5 of a jack or plug, and is referred to as position one.
- Pair 2 is terminated in pins 3 and 6 of a jack or plug, and is referred to as position two.
- Pair 3 is terminated in pins 2 and 7 of a jack or plug, and is referred to as position three.
- Pair 4 is terminated in pins 1 and 8 of a jack or plug, and is referred to as position four.

Figure 4–10 shows the patterns.

Figure 4–10
USOC Wiring Patterns

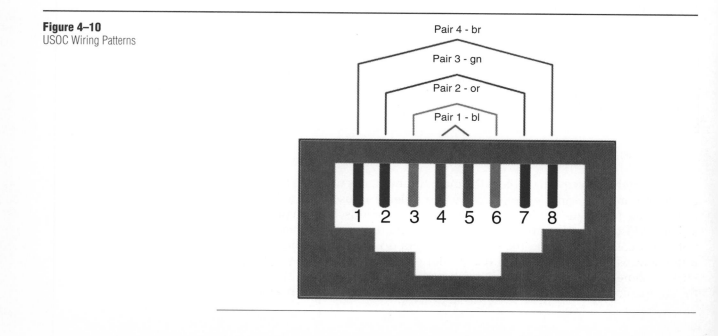

For homes with a single line, pins 4 and 5 are used to send and receive voice transmissions. These pins are the tip and ring located in the center of the plug. For two-line phones, the first line is in the center on pins 4 and 5 of the jack, and the second line is on pair 2 (pins 3 and 6). Because pins 4 and 5 are used for voice transmission, Internet data is sent on pins 1 and 2 and received on pins 3 and 6.

This wiring scheme is used for T1 circuits. The wiring scheme has been modified by ANSI/TIA/EIA to keep pairs apart to avoid crosstalk or the unwanted coupling of signal from one or more circuits to other circuits.

ANSI/TIA/EIA T568A and T568B

ANSI/TIA/EIA developed two wiring schemes: *T568A* and *T568B*. These two wiring schemes indicate the order that pairs should be mounted in modular plugs and jacks, as shown in Figure 4–11. Do not confuse these wiring schemes (T568A and T568B) with the TIA/EIA standards that specify them (TIA/EIA-568-B).

T568A

T568B

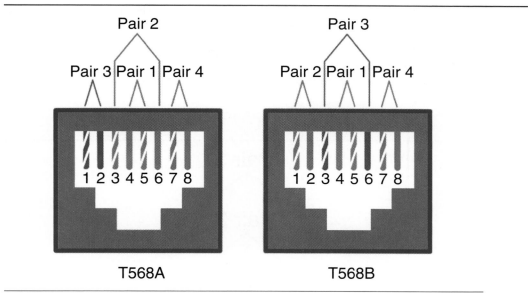

Figure 4–11
T568A and T568B Wiring Patterns

For the T568A standard:

- Pair 1 is terminated on pins 4 and 5 of a jack or plug, and is referred to as position one.
- Pair 2 is terminated on pins 3 and 6 of a jack or plug, and is referred to as position two.
- Pair 3 is terminated on pins 1 and 2 of a jack or plug, and is referred to as position three.
- Pair 4 is terminated on pins 7 and 8 of a jack or plug, and is referred to as position four.

For the T568B standard:

- Pair 1 is terminated on pins 4 and 5 of a jack or plug, and is referred to as position one.
- Pair 2 is terminated on pins 1 and 2 of a jack or plug, and is referred to as position three.

- Pair 3 is terminated on pins 3 and 6 of a jack or plug, and is referred to as position two.
- Pair 4 is terminated on pins 7 and 8 of a jack or plug, and is referred to as position four.

If you are working on an existing network, use the wiring scheme already employed. If it is a new project, use T568A. This is the recommendation set forth by the ANSI/TIA/EIA 570-A, the standard for residential networking.

It is useful to note that with USOC, T568A, and T568B, pair 1 is always the blue pair, pair 2 is always the orange pair, pair 3 is always the green pair, and pair 4 is always the brown pair. One way to remember these colors is to go from high to low: Blue sky, orange sun, green grass, brown dirt.

6P6C (RJ-11) Jacks and Plugs

As shown in Figure 4–12, Category 3 cable uses 6P6C (six-position six-contact) connectors for terminating. The number of positions (6P) refers to the connector's width. The number of contacts (6C) refers to the maximum number of conductors the connector can terminate. This common connector has six pins, as follows:

- Pair 1 (white/blue) is terminated on pins 3 and 4.
- Pair 2 (white/orange) is terminated on pins 2 and 5.
- Pair 3 (white/green) is terminated on pins 1 and 6.

Figure 4–12
6P6C Connectors for Terminating

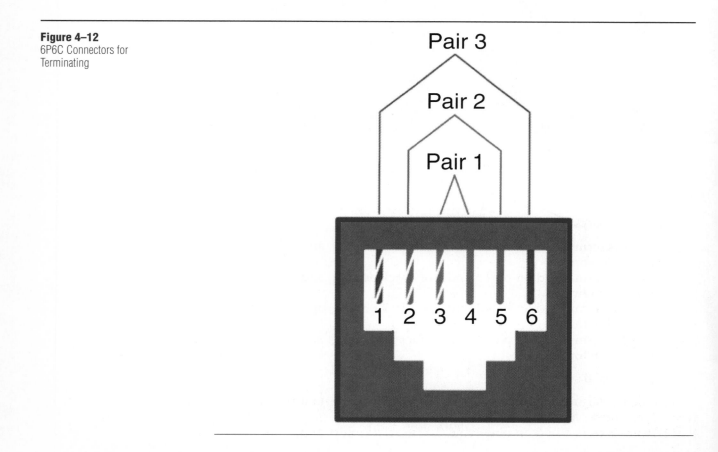

A jack or plug with this configuration is commonly referred to as an *RJ-11*.

RJ-11

NOTE Telephone handset cords use 4P4C (four-position four-contact) connectors.

8P8C (RJ-45) Jacks and Plugs

Data communication lines and patch cords use 8P8C (eight-position eight-contact) connectors. These are referred to as *RJ-45* connectors. The number of positions (8P) refers to the connector's width. The number of contacts (8C) installed in the available positions refers to the maximum number of conductors the connector can terminate. Use an 8P8C for making the connection to a CAT 5 UTP cable at the telecommunications outlet.

RJ-45

The individual CAT 5 wires are punched down into the slots according to color. A firm punch-down is required to make an electrical connection. The other side of the jack is a female plug, which looks like a standard phone jack, except that the 8P8C jack is larger and has eight pins. The 8P8C plug is wider than the 6P6C plug.

RJ-31x Jacks and Plugs

RJ-31x jacks are used for the phone lines and alarm system. The RJ-31x jack, as shown in Figure 4–13, can be configured so the security system will be able to "seize the line." When the alarm system is tripped, the RJ-31x jack automatically disconnects the phones, and the alarm system dials the appropriate emergency response unit. Additionally, RJ-31x allows a user to dial into the home network from a remote telephone and turn the alarm on and off or control the temperature inside the home.

RJ-31x

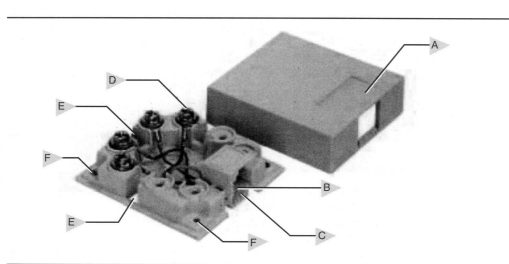

Figure 4–13
RJ-31x Jack

RJ-48 Jacks and Plugs

RJ-48 jacks, as shown in Figure 4–14, are used as a telephone company network interface for T-1 circuits. *T-1 circuits* are digital high-capacity circuits that transmit data at 1.544 Mbps through the telephone-switching network.

RJ-48

T-1 circuits

The RJ-48 has four pairs of wires like the RJ-45, but the wiring scheme for the RJ-48 is different.

Figure 4–14
RJ-48 Jack

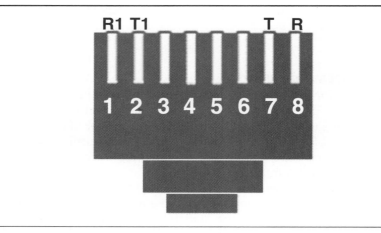

TROUBLESHOOTING AND DOCUMENTATION

Understanding exactly what the homeowner wants to accomplish and creating a detailed design document can show the HTI what the job will look like. This helps pinpoint any potential problems that can be addressed before cabling is installed. Once installed, properly labeled diagrams allow the HTI to troubleshoot more efficiently.

Troubleshoot Problems

Making sure that the installation is completed according to the design does not guarantee that problems will not happen. Sometimes, problems in the network are found immediately, but problems can also occur after the network has been working for a time. One common problem is encountered by homeowners who have DSL. They may complain that there is a low hum on the phone line. This hum is eliminated by adding a low-pass filter to the RJ-11 plug.

Regardless of when the problem occurs, you can take specific steps to troubleshoot the network. The troubleshooting process starts with eliminating the obvious causes, including loose connections on the device and to the outlet.

The following are steps to troubleshoot problems:

1. Observe the symptoms. What is happening? Where is it happening? Does it affect one telephone or every telephone?
2. Isolate the cause. Is this the first time it has happened? Does it occur regularly? Was anything changed recently?
3. Check the connection. Are the wires securely fastened to the devices? Are the wires cut or twisted?
4. Test hardware. Does the telephone work at another outlet?
5. Test software. Consult the user's manual that was supplied with the equipment, and conduct tests to determine the cause of the problem.

Floor Plans

Floor plans, such as the one shown in Figure 4–15, should be accurate and drawn to scale in order to determine obstructions. These obstructions, such as walls and doorways, will help determine whether special installation considerations are needed. The floor plans will identify outlet and jack locations. After the telephone network is installed, indicate on the floor plan where each outlet is and then label the outlet.

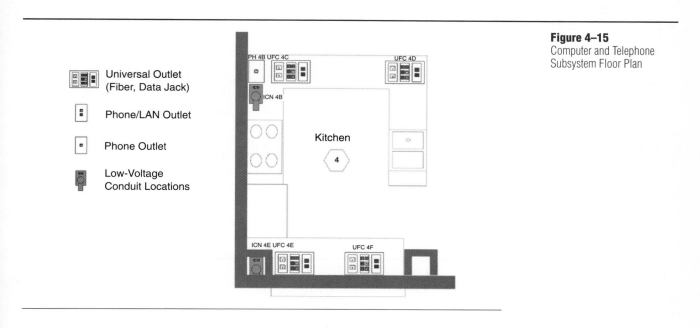

Figure 4–15
Computer and Telephone Subsystem Floor Plan

Schematic Diagrams

Schematic drawings diagram the devices to be installed and show how the devices interconnect with each other. They are not typically drawn to scale. Detailed in the drawing are the cables and where they run to specific components in the home network. It does not usually show the termination points or specific outlets and jacks. An example of a schematic diagram that shows a PBX connected to the telephone unit of the structured media center is shown in Figure 4–16.

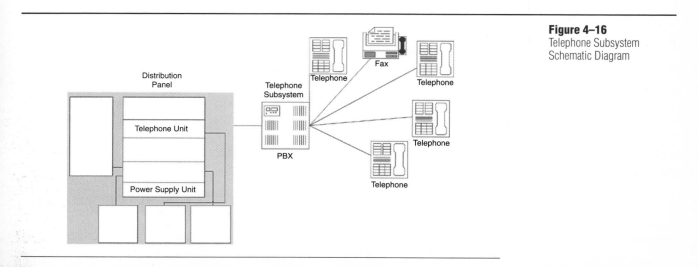

Figure 4–16
Telephone Subsystem Schematic Diagram

SUMMARY

In this chapter, you learned the following:

- The telephone subsystem can act as a conduit for signal distribution of other subsystems, such as the security system and the computer network.
- A telephone ring is the result of a 90-volt AC wave at 20 Hz being passed down the line.
- The POTS system, when used for Internet connection, is restricted to about 56K.
- HDSL needs two phone lines.
- The private branch exchange (PBX) allows each phone in the home to have a different extension.
- In VoIP, the voice signal is segmented into frames by a Digital Signal Processor (DSP).
- The point at which telephone service enters the home is known as the demarcation point, or demarc.
- Termination blocks are used to terminate telephone wires on the home network.

- Key system units perform a switching function to connect different incoming phone lines to the various room extensions in the home.
- The communications outlet should be located at least 16 inches away from an AC outlet to prevent signal interference.
- CAT 5 wire is recommended by ANSI/TIA/EIA-570-A for all home-cabling installations, although CAT 3 is the minimum accepted cabling.
- For a single phone line, only the green (tip) and red (ring) pair is normally used. The black (tip) and yellow (ring) pair is normally a spare and available to install a second phone line.
- An RJ-11 jack and plug is used to terminate phone lines.
- The USOC wiring scheme causes crosstalk whereas the ANSI/TIA/EIA wiring scheme keeps pairs close together to avoid crosstalk.
- RJ-31x jacks seize the line when the alarm system is tripped to inform the alarm monitoring company.

GLOSSARY

Analog transmission   Signal transmission over wires or through the air in which information is conveyed through the variation of some combination of signal amplitude, frequency, and phase.

CAT 3   Cabling used in 10BaseT networks that can transmit data at speeds up to 10 Mbps. Commonly used for telephone wiring.

CAT 5 or 5e   A category of performance for unshielded twisted-pair copper cabling that can transmit data at speeds up to 100 Mbps.

Centrex (Central Office Exchange Service)   The telephone company owns and manages all the communications equipment necessary to implement the PBX and then sells various services to the customer.

Crosstalk   Unwanted coupling of signals from one or more circuits to other circuits.

Datagram   A unit of data, or packet, that is transmitted in a TCP/IP network such as the Internet. Each datagram contains the source and destination addresses and data.

Demarc   Point where the telephone company's responsibility ends and the homeowner's begins.

DSP   Digital Signal Processor. The DSP segments the voice signal into frames that are then transmitted over data networks such as the Internet.

PBX   Private Branch Exchange. A central telephone switch that connects incoming calls to the appropriate extension or room.

Plain Old Telephone Service   The standard telephone service.

Public Switched Telephone Network   The international telephone system.

RJ-11   A 6P6C (six-position six-contact) connector for terminating. The number of positions (6P) refers to the connector's width. The number of contacts (6C) refers to the maximum number of conductors the connector can terminate.

RJ-31x   Used for the security system. It provides a feature known as "seize the line." When the alarm system is tripped, the RJ-31x jack automatically disconnects the phones, and the alarm system dials the appropriate emergency response unit.

RJ-45   An 8P8C (eight-position eight-contact) connector for terminating data communication lines. The number of positions (8P) refers to the connector's width. The number of contacts (8C) refers to the maximum number of conductors the connector can terminate.

RJ-48   Used as a telephone company network interface for T-1 circuits.

T-1 Circuits   Digital high-capacity circuits that transmit data at 1.544 Mbps through the telephone-switching network.

T568A   ANSI/TIA/EIA developed wiring scheme that specifies the order that pairs should be mounted in modular plugs and jacks.

T568B   ANSI/TIA/EIA developed wiring scheme that specifies the order that pairs should be mounted in modular plugs and jacks.

USOC   Universal Service Order Code. A code developed by Bell Telephone that defines the way wires are organized into a modular plug.

Voice over IP (VoIP)   The capability to carry normal telephony-style voice over an IP-based network.

CHECK YOUR UNDERSTANDING

1. The RJ-45 connector, which can be either a jack or plug, is used for terminating Category 3 cable.
   a. True
   b. False

2. PBX systems are the same as key systems.
   a. True
   b. False

3. PBX stands for:
   a. private branch exchange.
   b. phone bridge exchange.
   c. phone branch exchange.
   d. phone bridge extension.

4. A Centrex (Central Office Exchange Service) is a variation of the PBX. The telephone company owns and manages all the communications equipment necessary to implement the PBX and then sells various services to the company.
   a. True
   b. False

5. Demarc provides the point at which outdoor cabling interfaces with the intra-building backbone cabling.
   a. True
   b. False

6. The Home PBX unit offers some typical business PBX system features. What are those features?
   a. Call transfer
   b. Conference calling
   c. Music-on-hold
   d. All of the above

7. POTS stands for:
   a. Plain Old Telephone Service.
   b. Phone Operator Telephone Signal.
   c. Phone Operator Telephone Service.
   d. Plain Old Telephone Signal.

8. What type of network is constructed by using public wires to connect nodes?
   a. LAN
   b. VLAN
   c. VPN
   d. VoIP

9. Handicapped phones should be how many inches above the floor?
   a. 24
   b. 36
   c. 48
   d. 52

10. With call conferencing, as the number of participants on the home telephone system increases, volume:
    a. degrades.
    b. increases.
    c. stays the same.
    d. None of the above

11. ISDN supports data-transfer rates up to 64 Kbps.
    a. True
    b. False

12. UPS is the acronym for:
    a. uninterruptible power surge.
    b. uninterruptible power supply.
    c. uniformed power surge.
    d. uniformed power supply.

13. Schematic drawings are not to scale.
    a. True
    b. False

14. Telephone and data cabling is currently included in Division 15.
    a. True
    b. False

15. 6P6C (six-position six-contact) connectors are used for what type of cable?
    a. CAT 3
    b. CAT 5
    c. CAT 5e
    d. CAT 6

16. A 6P6C connector is also referred to as an RJ-45 connector.
    a. True
    b. False

17. 8P8C (eight-position eight-contact) connectors are used for data communication lines and patch cords and are referred to as RJ-45 connectors.
    a. True
    b. False

18. Describe the troubleshooting steps.

19. What is the difference between the T568A standard and the T568B standard?

20. What type of RJ jack is used with a security system?

5

Audio/Video Subsystem

OBJECTIVES

Upon completion of this chapter, the HTI will be able to complete the following tasks:

- Design a whole-house audio and video distribution system

- Understand externally provided services

- Identify audio and video subsystem equipment

- Select the appropriate low-voltage wiring for a project

- Install audio and video components

- Troubleshoot and test the audio/video subsystem

- Understand content-protection technology

INTRODUCTION

Whole-house (whole-home) audio and video distribution is becoming more popular with many homeowners today. It distributes signals from the media room to other rooms throughout the house. The goal of whole-house audio/video design is to provide the homeowner with the best choices for today and the ability to upgrade in the future.

WHOLE-HOUSE AUDIO/VIDEO DISTRIBUTION

Whole-house audio/video distribution depends on the wiring infrastructure. Whole-house audio/video wiring must be designed and installed to cover the majority of current and near-future technologies. This will assure that both the HTI and the homeowner have the widest choice possible in equipment selection, while keeping optional future upgrades open. The home floor plan in Figure 5–1 illustrates a well-conceived placement of audio and video components.

Figure 5–1
Floor Plan Showing Location
of Audio and Video
Components

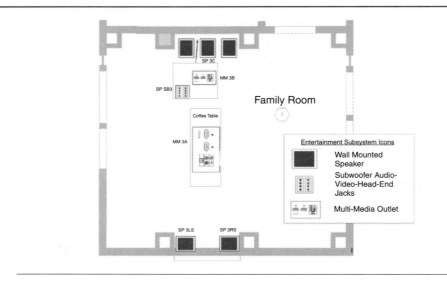

The floor plan reflects a design and installation that covers the current and future audio/video needs of the homeowner. Each room is labeled with the location of components, including the distribution device, input devices, speakers, outlets, and so on.

Whole-House Audio Distribution

The whole-house audio distribution system wires the entire house for audio. The family can listen to music from every desired room in the house. The audio can be turned off for different rooms or the volume adjusted so as to not disturb sleeping children or telecommuters. Multisource systems allow for specific audio signals to be selected in different rooms (zones).

The residential structured cabling for audio described here is wired in three runs. Each run can contain multiple cable runs, customized for the residence. The following is an example of how to do this:

1.   Amplifier locations to the Distribution Device (DD). In new constructions, wiring connecting the amplifier and/or source equipment location to the DD should include six conductors of speaker wire and one Category 5e cable. Four of the speaker wire runs feed speaker signals to the DD, whereas the other two can be used to deliver low-voltage (DC) current to the DD for such purposes as switching or controls. The included Category 5e will support other communications that relate to audio distribution: remote controls (infrared [IR]), data communications, and so on.

2.   DD location to the volume control locations. Normally, each audio zone (room or designated area) should have a minimum of one volume control. Some systems can have more than one volume control. Each volume control location should be wired to the DD using four speaker wire conductors and one Category 5e cable.

3.   Volume control locations to the speakers. At a minimum, each speaker will require two speaker wire conductors from the speaker to its associated volume control location. It is recommended to install one Category 5e cable along with the speaker wire. The Category 5e cable will serve for future upgrades such as installation of amplified speakers or IR applications.

The home theater is a dedicated A/V (Audio/Visual) system and requires special considerations, as you will learn later in the chapter.

> **NOTE** Multiple sets of Category 5 cable and speaker wires will need to be run if the A/V system design includes a cabinet, closet, or shelf location in which all input devices will be housed (DVD, VCR, CD player, stereo, satellite, and so on).

Whole-House Video Distribution

As with whole-house audio distribution, at the center of the design for whole-house video distribution is the distribution device. The design considerations for the type of A/V distribution unit to install include the infrastructure and any existing devices (in retrofits) and their suitability for the project. The HTI takes these considerations and combines them with what the homeowner wants, budget constraints, and type of features needed to determine which audio/video unit to install for the home.

With the whole-house video distribution system, every television in the home—as well as any video camera, VCR, and DVD player in the house—will be able to view cable/antenna/satellite channels simply by changing channels.

A coaxial wall plate with two connectors is installed in each room to accommodate the twin coax cable system. The top connector carries the output signal and is connected to the antenna/cable input of the TV. The bottom connector takes signals generated in that room from a VCR or video camera, and feeds them back to an input on the distribution device so that it becomes available to all other televisions in the house.

The A/V floor plan will also show the location of the twin coax connector wall plates. Twin coax connectors make it easy to add to the system later or move components around, therefore creating a scalable solution. The distribution system can be customized by adding a selection of amplifiers, modulators, tilt compensators, and splitter/combiners.

Zoning

Zoning is an important element of a whole-home A/V distribution. It is a concept that more specifically applies to audio distribution. Zoning refers to a system where you decide to run audio, for example, from the media room (or central location) to the other rooms in the house. Each room receiving the audio signal is called a zone. For example, a bedroom is a zone and so is a living room. A zone can be further defined as an area that is not separated by a wall. A kitchen that opens into the family room cannot be treated as a separate zone in terms of audio distribution. It's more convenient for both spaces to be designed into a single zone with multiple speakers because of the difficulty of listening to different sounds coming out of the kitchen and the family room speakers, all filling the same space.

Some simple issues to consider in audio zoning for a home are the number of rooms in the house, hallways, exterior (patio, front entrance, and so on) and the nature of the source—either single source or multiple sources.

In a single-source system, all the zones are driven with the same audio signal. The example in Figure 5–2 shows a system with a single receiver and CD player that drives all the speakers around the house.

Figure 5–2
Single-Source Audio Network
Hookup

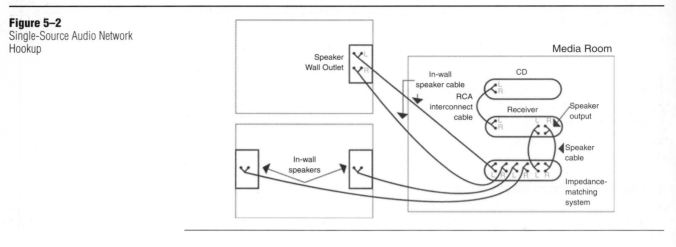

Although the single-source system is an uncomplicated system, it has two flaws:

1. Audio selection is limited to only one audio source for the entire house.
2. Some form of whole-house control system must be installed before adjustments to source equipment can be made.

In contrast, a multi-source system allows each zone to select a different audio signal. The example in Figure 5–3 shows the multi-zone amp/receiver. It's a more flexible audio-distribution system, which makes it possible to share expensive media-room source equipment such as CD jukeboxes. One simple technique for installing multi-source systems is to install a bank of identical audio receivers, each driving speakers in a different zone.

Figure 5–3
Multi-Source Audio Network
Hookup

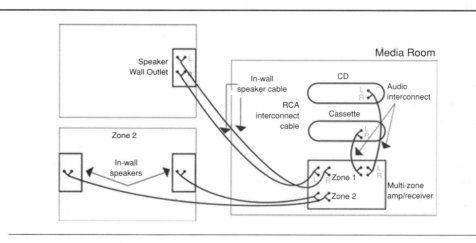

Remote Access

Remote access in residential A/V systems can be viewed from two perspectives: distant remote access and local remote access.

Distant remote access of the audio/video system refers to any type of access that is external (out/in), usually dial-up or web-based. With the availability of high-speed Internet services, it is possible to access most residential systems when these are integrated with the computer network. Web access is made possible by use of a browser and Internet protocol (IP) addressing, discussed in Chapter 3, "The Computer Network." One common web-related way of distant remote access is via a home control portal, which is a personalized website that can be set up for the homeowner by the HTI or the customer's Internet service provider (ISP).

Local remote access refers to the in-house use of an infrared (IR) target connected to the audio/video system (volume controller or TV) for controls. There are many remote access systems that can be used to control whole-house audio/video.

- **Infrared-repeater-control system:** Available as wired-IR-repeaters or wireless-IR-repeaters. These systems allow the use of the original remote control or a universal-type of IR remote in the zones around the house, as though the equipment were in front of you. Wired-IR-repeaters typically have an infrared sensor eye in each zone in the house, wires connecting the sensors to the media room, and some additional electronics that reconstruct an infrared beam to control the source devices whereby the infrared beam is invisible to the human eye. Conversely, the wireless-IR-repeaters have an infrared sensor eye located in a small radio transmitter, which is installed in each of the zones. Wireless-IR-repeaters work by using the IR code pattern to key the small radio transmitter. A radio receiver device located in the media room (DD) picks up and reconstructs the signal and then sends it to an IR emitter array.
- **RF control system:** These systems use remote controls, which transmit radio frequency (RF) commands to a controller/receiver. These systems are more efficient because there is no IR-to-RF repeating being performed, as is the case with wireless-IR-repeaters. The way this system works is simple: The controller unit will learn the infrared control codes of your audio/video equipment, and internally store them. When a button is pressed on the RF remote control requesting an IR code, the unit will use the previously learned command to access the audio/video device.
- **Infrared learning keypads:** These are wall-mounted keypads that can learn the infrared control codes of audio components. They can be mounted in J-boxes around the house, and wired back to the media room or distribution device. They can make use of the extra Category 5e cable run with the audio residential structured wiring. Special electronics in the media room route the keypad IR commands to the destination A/V component, using IR emitters or an IR-blaster.

There are a large variety of manufacturer product types, and each comes with the specifications for installation.

Dedicated versus Distributed

When designing a residential audio/video system, two aspects to take into consideration are dedicated and distributed signals.

A *dedicated signal* is one that is used for a specific purpose. A *distributed signal* can be channeled for multiple purposes. Both types of signals are used to design the system with the appropriate components. Some components will take care of dedicated signals only, whereas others can make the A/V signals from one source or multiple sources available to all the rooms or designated areas of the home.

Video signals are typically modulated and all the channels made available to all the rooms by using a central distributor that includes a modulator and an amplifier (for long cable runs). Conversely, audio volume control is more often dedicated because volume control devices are usually designed to control specific zones (rooms) in the house. When you have listen-only speakers (e.g., guest room, hallway, and so on), volume control may be distributed.

Dedicated signal

Distributed signal

Whichever is the case, the HTI should understand the difference between dedicated and distributed signal delivery to be able to design a functional system, including the appropriate devices.

EXTERNALLY PROVIDED SERVICES

Television programs and music are brought into the house by a variety of means. An antenna brings in local channels. A satellite dish mounted on a roof or a cable service can capture signals and deliver programming for television and radio. The decision between cable and satellite will depend on several factors, including the requirements of the homeowner. In rural areas where cable is not available, a satellite will provide this important element of a home networking strategy.

Cable Television

Cable television is widely available, and the number and types of services that cable companies can provide today keeps growing. Currently, in addition to digital television programming, high-speed Internet access and telephone services are available.

There is a slight difference between the channel system used by broadcast television and cable television systems. The frequency of channel 36, for example, on cable television is slightly different from that of channel 36 in the broadcast television environment.

To receive cable television, homeowners need one of the following:

- **Converter (set-top box):** A converter box is required when the service is analog and a cable-ready television is not used. This box enables you to take cable television signals and feed them to your television using only one broadcast channel, and to surf other stations through the box. For digital services, a converter is required, even when a cable-ready TV set is used.
- **Cable-ready TV set:** Allows you to select a cable mode and then to use its internal television tuner to decode and display the TV signals. When digital cable is used, a converter is required.

The advantages of digital cable service:

- **Digital quality:** Cable companies offer a regular service and a digital service.
- **Audio channels:** Can plug into the home theater (or stereo) for video-only programming.
- **Easy-to-use graphical interfaces:** Provide on-screen channel guides.
- **Cable line to provider's site:** Allows pay-per-view programming on the upgraded digital service.
- **Multiplicity of channels:** You can get hundreds of channels, depending on which service and the level of service you choose to subscribe to.

Satellite Television

Satellite television requires a dish (which is often roof-mounted) to capture the signals. Satellite dishes provide high-quality video programming. For a small additional fee, local channels are available, and subscribers can order the local channels for their community and those of other locations. The most common example is Direct Broadcast Satellite TV (*DBS*).

DBS

The advantages of DBS:

- **Digital quality:** Most DBS systems use a computer digital video standard known as MPEG (Motion Pictures Expert Group), which ensures higher quality and less susceptibility to interference.
- **Audio channels:** Can plug into the home theater (or stereo) for video-only programming.

- **Easy-to-use graphical interfaces:** Facilitates programming and on-screen channel guides.
- **Telephone line to provider's site:** Allows pay-per-view programming.
- **Multiplicity of channels:** You can get hundreds of channels, depending on which service and the level of service you chose to buy.

Even with satellite TV available, users normally don't have access to local channels. This means that when satellite is installed, consideration should be given to watching the local channels.

Just as with cable, a receiver is required to have access to the service. The receiver acts like a cable converter box to capture and convert the satellite digital signals for your television. You need as many receivers as the number of televisions in your home because a receiver puts out only one channel at a time. Additionally, you need a telephone line to plug to the receiver in order to allow it to communicate with the provider for billing, pay-per-view ordering, and service provisioning.

The future trend is toward combined television, audio, phone, and data to the home (digital delivery). This will be the future gateway for quality video to the home. Achieving this goal is contingent upon the effective integration of voice, video, and data.

No matter which television service you bring to the home, service comes in through the roof or side of the house to the distribution panel. Service providers see the demarcation point as the legal border.

Broadcast Television

Broadcast television is the option for homeowners who do not subscribe to cable or satellite service. In the United States, broadcast channels include ABC, NBC, CBS, PBS, and local channels licensed to broadcast in a specific area.

Despite the widespread use of cable and satellite television, broadcast television is still popular. Broadcast television users still have to employ antennas; however, antennas have now been greatly improved and it no longer presents such a hassle to set up and use them. They are usually mounted on the roof of the house, in the attic (in some cases), or on top of the television set (for the more advanced types).

Video Formats

Currently, there are three major video formats for television broadcast and signal transmissions in use:

- National Television Standards Committee (*NTSC*)
- Phase Alternation Line (*PAL*)
- Sequential Couleur avec Memoire (*SECAM*)

NTSC

PAL

SECAM

The National Television Standards Committee developed the NTSC set of standards for TV broadcast transmission and reception in 1953 in the United States. This set of standards has changed very little. The only major addition has been the new parameters for color signals. NTSC is used in the U.S. and other parts of North America, whereas PAL and SECAM are used in other parts of the world. The picture frame with PAL has 625 lines of resolution. This differs from the NTSC of 525 lines per frame. All PAL variations transmitted in Europe are very much the same, with the only main differences being the luminance bandwidth and the sound subcarrier frequency. SECAM, for picture scanning and format, is the same as PAL. However, SECAM is a totally different system from PAL and NTSC when color transmission is concerned. Most SECAM TV studios use PAL equipment, and the signal is converted to SECAM before it goes on the air.

> **NOTE**   Video quality is rated in terms of the number of lines of resolution across the picture horizontally. This is a direct measure of the detail and resolution of the picture available. (This shouldn't be confused with the vertical lines of resolution, which is fixed at 525 lines.) Laserdisc and satellite receivers can resolve up to 450 lines, whereas broadcast television signals and standard VCRs can contain up to only about 260 lines of resolution. TV monitors normally can reproduce anywhere between 300 to 700 (or more) lines of resolution. The resolution for HDTV is 1050.

When integrating residential systems, keep in mind that the NTSC signals are not directly compatible with computer systems. An NTSC TV image, for example, has 525 horizontal lines of resolution per frame (i.e., a complete screen image). But these lines are scanned from left to right and from top to bottom, with every other line skipped. It therefore takes two screen scans to complete a frame, one scan for the even-numbered horizontal lines and another scan for the odd-numbered lines. Each half-frame screen scan takes approximately 1/60 of a second, and a complete frame is scanned every 1/30 second. This alternate-line scanning system is known as "interlacing."

There are adapters that can convert NTSC signals to digital video that a computer can understand, and there are also devices that can convert computer video to NTSC signals, allowing a TV receiver or monitor to be used as a computer display. Unfortunately these devices do not necessarily work well for all applications because a conventional TV receiver has a lower resolution than a typical computer monitor, even if the TV screen is very large.

In recent years, there has been a growing movement towards the adoption of a new set of television standards. At the forefront is a standard known as high-definition television (HDTV). The good news is that although there are still some engineering problems associated with it, the proposed HDTV standard is ultimately expected to be directly compatible with computer systems.

High-Definition Television (HDTV)

HDTV has higher resolution than a standard television: 1050 lines of resolution as opposed to 525 with NTSC or 625 with PAL. In a digital system, images and sound are captured using the same digital code found on your home computer. HDTV's advantages include the ability to receive high-definition television broadcasts, multicasting in standard-definition television, and data transmissions in enhanced digital broadcasts. HDTV is able to receive and display all 18 digital TV formats approved for broadcast in the United States. In addition, HDTV has a widescreen aspect ratio and can display pictures at a higher resolution. Digital-ready TVs (sometimes called HDTV-ready or HD-compatible) do not receive digital signals—they have no built-in decoder. Instead, they are equipped with inputs that allow you to add a separate decoder box later on to enjoy the equivalent of HDTV.

Plasma Television Screens   Plasma screens are flat panel displays, and they represent the latest in television technology. The standard screen size for a plasma TV is 42 inches with a depth of two or three inches. Not only is its picture better, the plasma screen takes up less space and can be hung on a wall. The resolution meets the HDTV standards with an aspect ratio of 16 : 9. The common displays for television and computers are cathode ray tubes (CRTs), which cannot provide the clarity provided by plasma TV screens.

Analog and Digital Audio

The two ways that audio signals are transmitted are analog and digital. Broadcast and telephone transmission, for example, use conventional analog technology. Digital audio produces a greatly improved sound and is becoming the preferred method for receiving music and other audio files.

There is a fundamental difference between the ways a digital sound signal is transmitted as compared to analog.

- In analog, the strength of the amplitude of the signal is varied or the frequency is varied in order to add information to the signal. Sound in an analog signal is transmitted as movement of individual molecules of air. An example is a microphone that turns movement of air into a changing voltage, which represents the air movement.
- The digital signal is a discontinuous signal that changes from one state to another in a limited number of discrete steps or voltage levels. A simple digital audio signal has two states or signal levels—"on" translates to digit "1", and "off" or zero level corresponds to digit "0." Both levels can also be represented as a "positive" or "negative" voltage. The digital audio message or sound is made up of a series of these digital pulses (also known as bits), transmitted at regular and defined intervals.

When making any connection for analog or digital services in the home network, both analog and digital devices have to be connected to the network to receive the respective signals. Otherwise, consider including an analog-to-digital (A-to-D) converter to ensure the functionality of certain devices. The A-to-D device simply converts analog signals to digital. Alternatively, there is a D-to-A converter that converts digital signals to analog.

Internet for Music, Radio, and Movies

Internet music, radio, and movies are common examples of external services that can be utilized in the home with the audio/video component of a residential network. These are examples of *streaming media* that can be listened to or viewed as live feeds from different external sources.

Streaming media

A wide variety of web companies can encode (for a small monthly fee or even completely free) in various formats and deliver these media to the home. Music can be downloaded periodically and stored on dedicated media home servers and shared on the home network. Immediate access is available with on-demand programming, such as music and videos from the Internet.

One exciting reason why some external services such as Internet radio have become a way of life in many homes is the fact that people can listen to their favorite hometown FM radio station while on the road or living in another part of the country. For example, suppose a person born and raised in New York is now a homeowner who lives and works in Phoenix. If the home is networked and connected to the Internet, it is possible for her to get connected to, listen to, or even download and store a favorite show at any time from a local FM radio station in New York, simply with a few clicks of the mouse.

Note that most of these external services, however, would not be a reality without a broadband connection to the Internet.

Streaming Audio and Video

Streaming audio is sound that is played as it arrives in real time. The alternative to this is a sound recording (such as a WAV file) that doesn't start playing until the entire file is downloaded and saved to a hard drive. (WAV is a digital audio format in Windows.) To take advantage of streaming media, the home A/V system would require some sort of plug-in player. Streaming media sound can be accessed via the Internet using a computer browser. When the home entertainment (A/V or home theater) is integrated with the computer network, streaming audio/video from external sources can be distributed to all the rooms or zones using the audio/video network.

Leading providers of streaming sound include Progressive Networks' RealAudio and Macromedia's Shockwave for Director (which also includes an animation player). For faster transmission, audio and video files may be compressed using a complex algorithm. Audio files can be sent in short standalone segments (for example, as files in the WAV file format). The most popular audio file format today is MP3.

In order for people in a home to receive and distribute sound in real time for a multimedia effect, including listening to music or video conferencing, sound must be delivered as streaming sound.

AUDIO/VIDEO EQUIPMENT

The A/V system, a part of the overall home entertainment system, involves several products including receivers, speakers, television sets, CD players, DVDs, VCRs, and various control devices. These products can be divided into two main groups: source devices and distribution equipment.

Source Devices

Source devices produce audio/video signals. In a residential audio/video system, typical source devices include security/surveillance cameras, DVDs, VCRs, CD players, stereo sound systems, and so on. Usually, the signals from source devices need to be modulated in specific channels by modulators. Sometimes, there is a need to amplify signals before they actually get to the display or speakers.

There is a multitude of A/V source devices in the market today, and the choice of what to use or recommend is largely dependent on the needs of the homeowner and the requirements of the specific project.

There are several newer source devices that customers may have or may want to purchase for the system: a personal video recorder (PVR), a media server, a plasma television screen, or a high-definition television (HDTV) set.

Personal Video Recorder (PVR)   The newest product for digital television is the *personal video recorder* (*PVR*). TiVo and Replay Networks (creator of ReplayTV) were the first companies to release the first generation of PVRs.

A PVR records, plays, and pauses TV shows and movies. It uses a large hard disk, along with sophisticated video compression and decompression hardware, to record television streams. The PVR software is optimized to allow recording and playback at the same time. This allows the user to pause live television, do instant replay, and quick-jump from place to place. It provides the ability to quickly move to any location on the recording media. Coupled with an easy-to-use electronic program guide and a high-speed online service, this device can pause live television, recommend shows based on your tastes, and record your favorite shows automatically. For instance, a viewer can pause a movie, take a break, come back, and resume viewing, even though the movie in reality has continued past that point.

For all intents and purposes, a PVR is a stripped-down PC with a large hard disk that reserves storage space for downloading program listings every day. A remote control is used to select what you want to record so the PVR can capture and store these programs on a hard disk for future viewing.

A PVR system typically includes the following components:

- Hard drive: The average size of a hard drive in a PVR is 10 GB and is suitable for recording approximately 20 hours of standard quality broadcasts. There is also an option to increase storage capacity by purchasing additional disks to increase the recordable storage time. The presence of a hard disk allows you to pause or replay a broadcast in progress while recording

Personal video
recorder (PVR)

continues. All of the drives are qualified to meet the thermal, acoustical, and mechanical requirements of a consumer electronics product.

- **RAM chips:** Similar to a PC or set-top box, the RAM chipsets serve as the operational memory for the PVR.
- **MPEG2 decoder and encoder:** Similar to a digital set-top box, these chips are used to compress and decompress video signals.
- **Modem:** Used to download TV schedules and system upgrades every night.
- **CPU:** Runs the real-time operating system.
- **Boot ROM:** Similar to set-top and cable modem network devices, the boot ROM contains the software that allows the PVR to start up as soon as it is powered on.
- **NTSC decoders and encoders:** Because these gateway devices have been designed for the North American market, the PVRs currently available use these chips to process NTSC (National Television Standards Committee) signals. The decoder is used to digitize analog television signals, and the encoder reverses this action (i.e., it restores the decoder's compressed digital output stream into an analog TV signal).
- **Input/output jacks:** Used for interconnecting the PVR to a TV or satellite receiver.
- **Tuner:** These gateway devices contain only one tuner, so you cannot watch one broadcast channel while recording another. A way around this restriction is to install a splitter and connect the VCR.
- **Processing chips:** The types and models of the processing chips are PVR-specific. However, most PVRs contain a video processor for manipulating and processing multimedia content.

Media Servers   Music servers are used to download and store music and/or movies from the Internet and then stream the music/movie to multiple speakers/video displays throughout the home. Purchasing these types of media services for a home network can effectively eliminate the need to purchase CDs at retail music stores.

Media servers are the foundation of use of streaming media in modern networked homes. Media servers are specialty media dedicated for audio/video storage.

Distribution Devices

Sharing external services (Internet radio, music, movies, television/on-demand programming, and so on) that come to the home via external sources such as the Internet, satellite, or otherwise is the advantage of a home network. The distributor for an A/V system is commonly a central receiver that distributes the incoming audio or video signal to the appropriate devices around the house. Secondly, audio or video images from various source devices within the home are modulated (put in various channels) and sent to the distribution device, which then makes the signals available to all the rooms in the house. Figure 5–4 shows A/V distribution modules.

Figure 5–4
Audio/Video Distribution
Modules

The A/V distribution device distributes the signals to all desired rooms or designated zones of the house. The appropriate unit is an important design consideration in a residential A/V system. The type of A/V unit to be used is determined by the choice of entertainment. It also depends on whether the A/V devices are spread throughout the home or if they are concentrated in a few rooms within the home. Another consideration is whether the system is single-source or multi-source. Finally, the choice depends on the distances of the devices from the media room or wiring closet. Other considerations are capacity, configuration, and interoperability of the components. Figure 5–5 shows an A/V distribution system. The decora media system hub (part of the A/V distribution system) is shown in Figure 5–6.

Figure 5–5
Audio/Video Distribution
System

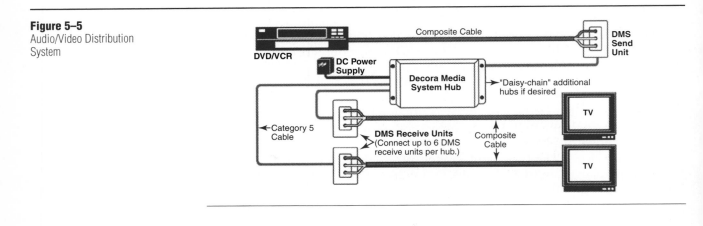

Figure 5–6
Decora Media System Hub

Other A/V distribution products include switchers that distribute audio/video signals to multiple devices. Products connect up to five audio/video sources to limited-input receivers and televisions, automatically select the active video input without manual switching, and automatically convert between optical and coaxial digital inputs. The distribution device also includes components such as video *splitters.* The notch filter filters out undesirable signals from external TV channels and makes those channels available for other custom uses within the home such as surveillance camera broadcasts.

Splitters

NOTE An X-10 control switch can be programmed to disconnect power at any time to any device such as a TV that it is connected to. Consider installing an X-10 control switch in any room where the homeowner needs to monitor the content or the hours the television is watched.

Keypads and Touch Screens

Keypads, touch pads, and touch screens are all input devices that can be used to control many systems in the home, including the audio/video system. They are available as wall-mounted pads, desk-mounted touch screens, or wireless pads. The Personal Digital Assistant (PDA) is an example of current technology that uses touch screens. The PDA uses a small stylus on the touch pad to input data for the calendar, phone book, or calculator; and it can access the Internet and work as a cell phone.

Speakers

Speakers are simply any electromechanical devices that play the role of converting electronic audio signals to sound that the human ear can hear. Criteria for selecting speakers for a home A/V system include the following:

- **Dispersion:** Make sure that the chosen speakers can be installed to give the customer maximum stereo sound coverage around the home. In most applications, ceiling speakers may work best, but flush or surface-mounted wall speakers can work equally well in areas with low or inaccessible ceilings.
- **Room size and sensitivity:** Balance is key to sound quality. Additional speakers may be installed in larger areas to provide better coverage and balanced sound levels. Sound-quality problems come mainly from areas such as the living room (where soft, sound-absorbing furniture and fittings can pose acoustical problems).
- **Aesthetics:** Aesthetics refers to how the speakers look. Speakers should be selected to fit the overall décor of the room, especially for open areas such as the living room. Outdoor speakers should be selected to blend with the landscaping. For outdoor purposes, you can even purchase speakers that look like rocks.

The location of speakers and volume controls must be carefully designed. Speakers should be placed so that each speaker (or set of speakers) covers its designated zone. The usual locations are ceilings or walls (surface or flush-mounted). Those with volume controls should be located at a convenient height. An example of a rocker-switch wall unit is shown in Figure 5–7. It is recommended to mount these at the same height as the wall light switches where possible. Speakers are best placed in a room so that they provide stereo left and right imaging when facing the main feature of the room, such as a large window or fireplace. In the bedroom, the ideal location of speakers is over the foot of the bed. Ceiling-mounted speakers provide the best sound dispersion in any space. In general, the speakers should be placed facing the typical sitting location of the listeners in a room or space.

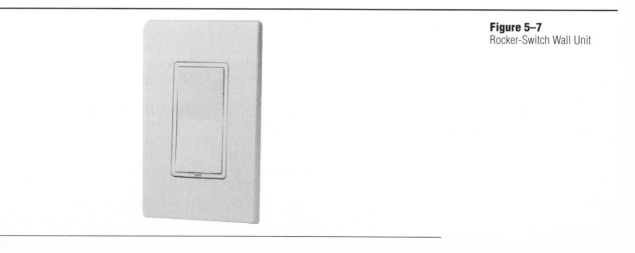

Figure 5–7
Rocker-Switch Wall Unit

In a multi-speaker stereo zone, make sure the speaker locations are roughly an equal distance from each other and from the room corners or boundaries. However, note that placing speakers too close to room boundaries can cause an undesired booming effect. It is usually recommended to keep speakers at least 24 inches from any room boundary.

Home Theaters

Home theater surround-sound systems are used to create a three-dimensional listening environment (surround-sound), allowing the listener to hear such sound effects as an airplane flying overhead from front to back. There are many vendor products that provide theater quality in ceiling and wall speakers. A surround-sound system may be composed of five channels, each connected to a speaker, as follows:

- Left front
- Right front
- Center front
- Right rear
- Left rear

A subwoofer is sometimes added to the home theater speaker system to emphasize such effects as dinosaur footsteps. A subwoofer may be powered or not, and can be placed anywhere in the room.

In a theater-quality speaker system setup, all front-channel speakers must be mounted at the same height from the floor. Other more-advanced systems even have mid- and rear surround-sound speakers. The Dolby system offers the most common standard for surround sound today.

To determine which system is appropriate to use, the size of a room must first be measured in cubic footage and then compared to the specifications for the speaker manufacturer's product. Home theaters are increasingly becoming a popular feature in many new homes, and this presents unique design and installation challenges to the HTI. Special training is required for many of the systems in order to provide a proper installation. The HTI may find it useful to complete this additional training or work with experienced installers.

Intercoms

Internal broadcasting with an intercom system can become an important issue in larger homes, especially those with children. Internal broadcasting is the use of internal speakers to broadcast a message to reach all areas of the house by using the intercom. A broadcast channel needs to be configured to carry these intercom signals.

Receivers and Amplifiers

Receivers, as mentioned in the previous section, constitute the distribution devices in the residential audio/video system. Video receivers include the television, satellite receiver box, and cable box. Audio receivers include any stereo receiver with FM or AM capabilities. For Internet radio, the home computer is the receiver.

Amplifier

Because audio/video signals are usually transmitted over long distances, the sound can fade out before it reaches its destination device (TV/video display or audio speakers). An audio *amplifier*, as shown in Figure 5–8, can be used to boost the signals back to the required threshold. The choice of what to use depends on the size of the home and the distribution system. In whole-house audio systems, make sure that the amplifiers you select are multi-zone amps. This means they have the capability to convert audio into multiple channels to feed the speakers in different zones (rooms) of the house as required.

Figure 5–8
Audio Amplifier

A television connected to outlets that are too close to the amplifier may receive signals that are stronger than the threshold for normal functionality. If this is the case, the use of an attenuator to reduce the signal strength may be required in order to reduce the signal to a tolerable limit.

Volume setting and equalization is an important part of the audio/video system configuration. Great care needs to be taken to avoid damaging the amplifier. Follow these steps to test volume and equalization:

1. Using an ohmmeter, test the resistance with the speakers installed, but not the amplifier. An ohmmeter is used to test the resistance of the speakers installed.
2. After the test, install the amplifier and then fine-tune the volume using the equalizer.

It is important to follow these steps to avoid any damage to the amplifier.

Multi-Zone Amplifiers

The distribution of zoned audio throughout the home requires a stereo (two-channel) power amplifier for each of the zones. When using a multi-zone controller that provides volume/bass/treble control internally, the zone amps will be fixed-gain (no volume control) or screwdriver adjustable-only power amps.

Instead of having a separate stereo amp per zone, the common practice is to deploy a multi-channel amplifier: two channels for left and right that are used for each zone. This means that a six-channel amplifier will service three zones. The type and quality of amplifier can be customized to the requirements of each zone; for example, a less-expensive amp for the kitchen as opposed to the living room.

Modulators and Filters

Video broadcasting is another important aspect of internal broadcasting. It is the capability of a video source device, such as a camera or DVD, to broadcast images to all rooms or predesignated areas of the house. This is made possible by the use of a *modulator*, making the images available on unused channels that can be browsed in different rooms using video monitors or televisions. Figure 5–9 shows a basic modulator.

Modulator

Figure 5–9
Basic Modulator

Figure 5–10 shows the front and rear views, respectively, of a stereo modulator. Note that the stereo modulator provides input and output for both the left and right channels.

Figure 5–10
Stereo Modulator: Front and
Rear Views

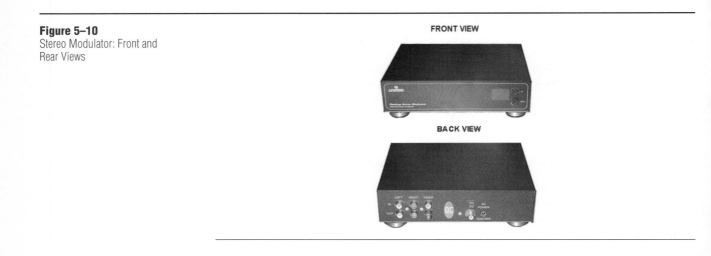

FRONT VIEW

BACK VIEW

Filter

Typically, a modulator is configured to use any vacant or unused channel for signal distribution. If there are no vacant signals available, a notch *filter*, as shown in Figure 5-11, can be used to create a vacant signal. This is accomplished by filtering undesired signals from existing channels and creating an available channel for use by the modulator.

Figure 5–11
RF Notch Filter Used to Create
a Vacant Video Signal

In a home in which the family is uninterested in certain foreign language channels, for example, an RF notch filter and modulator can be configured to filter and convert such channels for video-signal distribution from a home security camera or a DVD.

A multi-zone stereo system must include a multi-zone amplifier and should have sufficient capacity to connect all the zones designed into the whole-house audio distribution. Similarly, the video distribution must have sufficient capacity to carry all the channels that are sent to it by the various modulators. It should also be able to carry a vacant channel for any add-ons or upgrades. The addition of notch filters can free up channels from external TV sources carrying undesired programming, thereby creating more capacity for the video-distribution device.

Many manufacturers are offering modular distribution devices designed for small, medium, and large residential environments.

Equipment Layout

When planning the equipment layout for the audio/video system, criteria for consideration include the following:

- **Heat considerations:** Some audio/video equipment (e.g., VCRs, power amps, power adapters, and so on) can generate considerable amounts of heat. Avoid placing heat-generating equipment on top of or too close to heat-sensitive devices. This can negatively impact their performance.
- **Signal leaking (bleeding):** It is good practice to provide adequate separation between devices to avoid signals leaking from one device and negatively impacting the performance of other devices. This is particularly important for any device that is prone to generate electromagnetic interference (EMI) or radio frequency interference (RFI). Be on the lookout for any electric motor-driven device or control devices that employ radio receivers.
- **Accessibility:** The capability to get to devices for repairs and maintenance should be considered in the design phase.

A/V Rack

An A/V rack is the shelving located in the space designated for the equipment used for the home network. The location of the A/V rack is important because this is the point where most if not all of the externally provided services enter the home. It should be located near the point of entry for the CATV cable, satellite dish cable, TV antenna cable, and Internet access equipment.

Some homeowners want to display and have access to the A/V equipment in a living room, family room, or den. An entertainment center can be used in this case, with the cables from the externally provided services being fed from the demarc to the entertainment center. A typical entertainment center has provision for a television, an audio receiver, an amplifier, a digital cable box, a CD player, a DVD player, a VCR, and so on.

When a separate media room exists, a false wall can be built to separate it from the entertainment center. In this case, the audio/video source equipment located in the entertainment center can easily be wired back to the media room for signal distribution. For homeowners who want everything hidden, place the A/V equipment in the wiring closet or some other appropriate space.

LOW-VOLTAGE WIRING

Low-voltage wiring assures the connectivity of the A/V system within a home. It transmits signals using only small (and in some cases, trace) voltages for data transmission. Normal residential power lines carrying the nominal 120/240 volts supply most residential appliances. Low-voltage media is covered in more detail in Chapter 8, "Low-Voltage Media." This chapter focuses on the specialty media for residential A/V components connectivity.

Wiring Standards and Guidelines

The details for standards-compliant audio cabling are specified in *ANSI/TIA/EIA-570-A-3*. Audio- and video-cabling systems are generally designed using the following different types of cables:

ANSI/TIA/EIA-570-A-3

- **75-ohm coaxial Series 6 (RG-6) cables:** Used for video signals (both baseband video signals and RF-modulated antenna and cable-TV signals) and digital audio. Equipment patch cords also include coaxial Series 59 (RG-59) cables. Figure 5–12 shows RG-6 coaxial cable.
- **Shielded twisted-pair wiring (STP):** Used for balanced analog and digital audio signals in professional audio systems.

- **Shielded single-wire cables** (similar to coax in construction): Used for unbalanced analog audio signals that are used in consumer audio systems.
- **14- or 16-gauge speaker wire:** Used for wiring audio speakers. Chapter 8 describes the power budget table, which will help you choose the size wire for speaker installation.
- **Unshielded twisted-pair (UTP):** Used in some A/V system installations for transporting certain control signals between equipment. CAT 5 and 5e are examples of UTP.

Figure 5–12
RG-6 Coaxial Cable

ANSI/TIA/EIA-570-A-3 also defines the star topology for audio- and video-cabling subsystems. This allows the distribution device to be the center with speakers and other outlets running out like a star.

The coax cable transmits video and the Category 5 cable transmits data or control signals for volume control. A minimum of one coaxial Series 6 (quad-shielded) cable and one Category 5 cable are run to each room. Ideally, two coaxial Series 6 cables and two Category 5/5e cables should be run to accommodate future upgrades.

Even though RG-6 is the most acceptable standard for carrying video signals, in some large homes, a normal RG-11 cable is used to carry video signals for long distances.

Speaker wire is run in pairs to the left and right speakers to the appropriate locations in different rooms. The speaker wire should be terminated in one of two ways:

1. For stand-alone speakers, connect the ends of the speaker wires to a speaker-connector wall plate (the type with banana jacks that is discussed later in this chapter). The five-way binding post is preferable for this application.
2. For in-wall speakers, connect the ends of the speaker wires directly to each speaker.

In general, when specifying the cable type for audio/video installation, always verify the specifications on such characteristics as attenuation, temperature ratings, voltage and current capacity, and type of outer jacket used. The detail specifications for residential audio wiring are described in ANSI/TIA/EIA-570-A. Various relevant articles of the NEC (e.g., *NEC 2002 Edition, Article 820*) provide additional information on the specifics of this type of wiring.

Bundled Systems

Bundled cable systems are commonly used for audio/video and entertainment wiring. Typically, four or more cables are tied together with cable-ties or speed-wrap and pulled throughout the house. This minimizes the overall cost of installation, especially in retrofits. Some examples of bundled cable configurations include the following:

- Two Category 5 or 5e cables and two speaker (14- or 16-gauge) wires
- Two coaxial (RG-58 or RG-6) cables and one Category 3 or 5 cable
- Two RG-6 cables and two Category 5 cables

- Two RG-6 cables, two Category 5 cables, and two pairs of 14-gauge speaker cable
- Two RG-6 cables, two Category 5 or 5e cables, and two fiber-optic cables

Video Outlets and Connectors

There are many different types of video connections, each with different effects on picture quality. In order of delivering video quality from lowest to highest, they include the following:

- **Composite:** A single RCA connector combining the black-and-white and color portions of the video signal. This connection type depends on the comb filter in your video display. A comb filter separates the signals before displaying the image. Like the teeth in a comb, the filter alternately blocks off sets of frequencies to accept color information only and then brightness information only. This effectively splits up a composite signal into a chrominance component and a luminance component. Composite connections are the most common and should be connected with mid- to high-quality cable for the best results. These connections are commonly found on entry-level receivers.
- **F (or RF):** Uses an "F"-type connector that is commonly used in cable television installations. In this connection type, audio and video are combined into one signal, which is then separated by the video display when the image is displayed. It can be used only for connecting video components that are not primary home theater sources. Although available on older laserdisc players, some recent DVD players, and virtually all consumer-grade VCRs, F connections have two main disadvantages: They generally result in a poor picture quality, and only a mono audio signal will be received if additional connections are not made for audio.
- **A/V:** A/V jacks have three RCA connectors, and have better sound and video quality than F jacks. The yellow connector is for video, the red connector is for the right channel, and the white connector is for the left channel.
- **SuperVideo (S-Video):** Better video quality than all previous connectors. S-Video, or super-video (developed by Sony), uses a round multi-pin connector.
- **Component:** Component connections provide the best video quality, but they are rare (found mostly on newer consumer monitors and DVD players). Employs three separate RCA connectors.

Although Component connections provide the best video quality, S-Video connections, as shown in Figure 5–13, are preferred by most consumers because of the rarity of equipment with Component video connectors.

Figure 5–13
S-Video Connection

Audio Outlets and Connectors

There are several different types of audio connections, including the following:

- **Stereo Audio:** The most common connections for stereos and surround systems. There are two connectors: The red connector is for the right channel and the white connector is for the left channel.
- **Optical Digital:** Superior in sound quality to the Stereo Audio connections, Optical Digital connections provide connection to 5.1 surround-sound channels made up of right front, center front, left front, right rear, and left rear.
- **Banana:** Used to connect speaker wire to the audio amplifier.

Wireless Solutions

Wireless solutions are used not only for computer subsystems but also for audio/video and home entertainment subsystems. More audio and video products are being made that can be connected via IR or RF technology.

An audio system can be distributed around the home without any wires by utilizing RF technology. Wireless speakers receive signals from the wireless line-level distribution system. Wireless line-level distribution systems connect to audio source devices and send a signal to the receiver that is in turn connected to the amplifier.

Just as audio components can be wireless, so can video components. A wireless video network commonly consists of source devices such as VCRs, DVD players, or DBSs. A wireless transmitter and a wireless receiver are required. The transmitter connects to the video source via standard RCA-type connectors, whereas the receiver has an RCA and an RF output interface that connects to the TV. These wireless baseband video distributors transmit only one signal at a time. Make sure that the system can receive IR signals so that the user can use remote controls to pause, rewind, and control the volume of the video devices.

NOTE   The most important aspect of a wireless video solution is the frequency spectrum on which the signals are distributed. Some systems distribute over 900 MHz, and others use the 2.9 GHz band. The higher the frequency, the better the signals pass through physical obstacles and the less susceptible they are to interference. In addition, high frequency offers more bandwidth than lower frequency solutions.

INSTALLATION AND TROUBLESHOOTING

The installation process requires planning to ensure that all equipment and components are connected according to the design plan. A well-defined plan will ensure that all individuals involved can follow the step-by-step procedure. This section outlines examples of stepwise procedures for installing various A/V components.

Connecting Audio Components

Audio components should be located according to the design plan. The source devices (receivers, CD players, and so on) serve as the starting point when connecting to the speakers. Figure 5–14 shows how the audio components should be connected.

Figure 5–14
Connecting Audio
Components

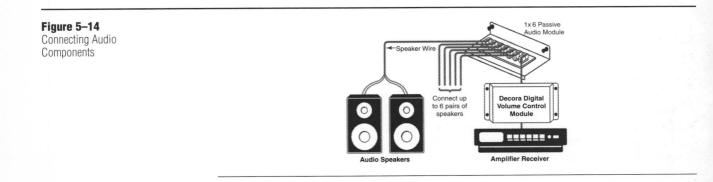

Follow these steps to install an audio network:

1. Run the wire from the distribution devices to the speakers. In the rough-in phase, the speaker cable should be run to every room in the house to ensure scalability.
2. Connect devices to control amplifier (= preamplifier).
3. Connect source devices to receiver (= integrated amplifier).
4. Connect the audio output interfaces of the receiver to a connecting block or speaker selector device to distribute the audio signal to multiple channels. A speaker connection module is shown in Figure 5–15.

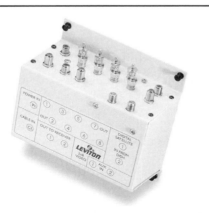

Figure 5–15
Speaker Connection Module

5. Connect the connecting block/speaker selector device to the speaker wiring.
6. In each room, terminate the speaker wire at the volume control device.
7. In each room, connect the volume control device to the speakers.

Connecting Video Components

To terminate video components it is necessary to connect devices in the rooms to the coaxial network and then connect the coaxial network to the distribution panel to enable the distribution of the video signals.

Figure 5–16 shows a 3 × 8 bi-directional video module.

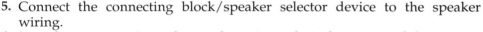

Figure 5–16
3 × 8 Bi-Directional Video Module

The 3×8 bi-directional video module allows for three devices into the module and eight connections out. The example in Figure 5–17 shows a cable or antenna along with satellite #1 and satellite #2 coming in and connections for up to eight televisions. Examples of modulated units include DVD, VCR, and security cameras. Two or more modulated signals can be run to modulated input with the addition of a combiner.

Figure 5–17
Diagram of Bi-Directional
Video Amplifier Setup

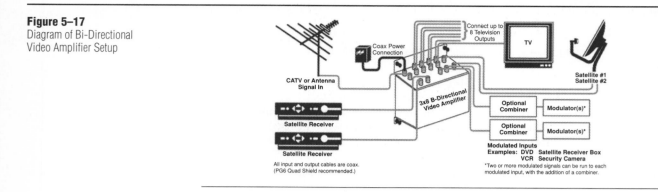

Figure 5–18 illustrates a video network. The input signal is run through the amplifier to the DD. From the DD, coaxial cable (specifically RG-6) is run to the output devices. The wall outlet with two female F connectors is used to connect to the modulator.

Figure 5–18
Diagram of Video Network

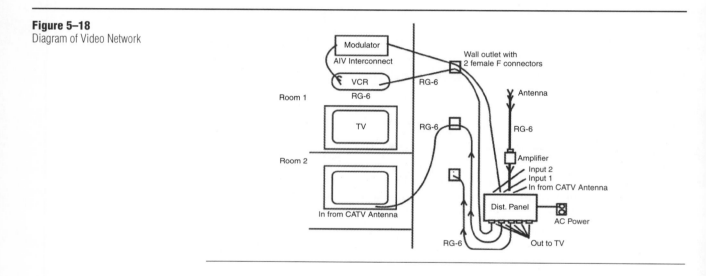

The following steps are used to connect devices:

1. Run the coax from the distribution devices to the rooms.
2. If the design incorporates modulators and RF notch filters, connect source devices to the modulator (to distribute video signal around the house by using unused television channels), and install RF notch filters.
3. In each room, connect the modulator to the coaxial wall jack that sends to the distribution panel.
4. In each room, connect the television to the coaxial wall jack that receives from the distribution device.

To connect the coaxial Series 6 cable at the distribution device:

1. Make sure that all prewired coaxial Series 6 cables end at the distribution device.
2. Label all cables (source, length).
3. Terminate each receiving coaxial Series 6 cable to an output with amplification at the distribution device.
4. Terminate each sending coaxial Series 6 cable to an input modulator at the distribution device.

Connectivity Documentation

As previously discussed, connectivity documentation is crucial for helping to figure out various aspects of the installation, including the equipment layout, device locations, and possible adjustments. With the audio/video system, useful connectivity documentation to be developed or secured include the following:

- Schematic drawing of audio/video subsystem. The illustration in Figure 5–19 shows the DD with the subsystem devices.

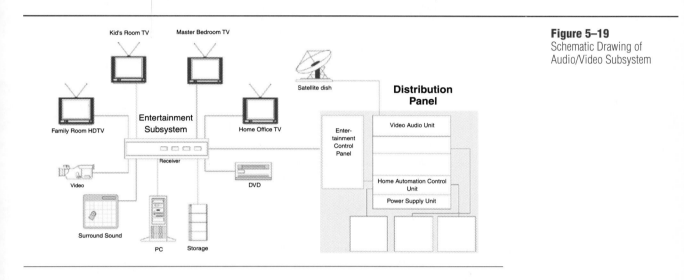

Figure 5–19
Schematic Drawing of Audio/Video Subsystem

- Audio/video signal flow diagrams. Figure 5–20 illustrates how the signal would flow through the various receivers and amplifiers.

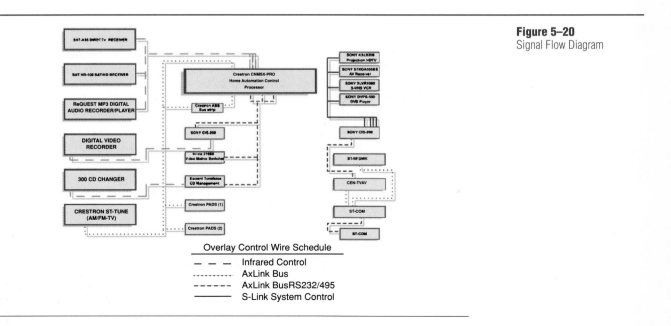

Figure 5–20
Signal Flow Diagram

- Floor plan of audio/video components. The floor plan, as illustrated in Figure 5–21, shows how the house is laid out and where the components would be installed.

Figure 5–21
Floor Plan of Audio/Video
Components

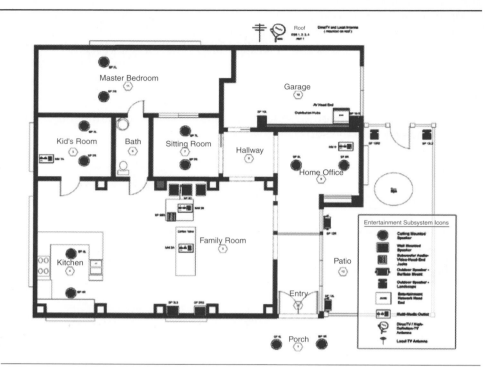

Figure 5–21
Floor Plan of Audio/Video Components

Connectivity documentation provides the HTI with the information needed to design, install, and maintain the A/V system. These documents are also useful for future upgrades and troubleshooting.

Customer Service

Although user manuals come with just about all the devices ordered and installed, it gives the homeowner, especially one not very well-versed in electronics, a better understanding if the HTI explains the basics of how the equipment works.

Basic use and maintenance of the installed devices is important in training and orientation for the homeowner. It is helpful to provide a demonstration of how all the equipment works and then to be sure the homeowner can operate all equipment. Review all safety procedures, especially the danger to small children. Some equipment with heat-producing components (for example, power amplifiers and some receivers) can become hot enough to cause injury. Review issues related to cooling and the general climate control around them to protect the users and the equipment.

Another point of training and orientation has to do with basic equipment configuration. This is useful for temporary disconnections, including equipment testing and troubleshooting. The homeowner should be able to provide information for warranty claims regarding equipment or general installations—such as the media room locations, equipment locations, distribution device location, and warranty information.

Although the whole-house audio/video system can be easy to understand, more specialized installations such as the home theater may at times need professional maintenance and repair of problems that arise. One way to assure your customer of your continued availability for servicing is to offer a service and maintenance contract.

As mentioned in previous chapters, a maintenance contract is simply a signed commitment on some standard form between your company and the homeowner (maybe at a small monthly fee, or per service call) to send someone out to take care of any problems with their equipment.

Warranty papers are important for such services, especially if some equipment needs to be replaced. Extended warranties are also available for more expensive equipment. Equipment manufacturers and even the product suppliers/retailers, in some cases, have all kinds of such plans.

As always, with the proper information, homeowners can make the decision about which plan best suits their needs and their budgets.

CONTENT PROTECTION AND DIGITAL WATERMARKING

An important aspect of home integration that can easily be overlooked by the HTI is copyright protection. Internet radio, music, movies, and streaming media in general are a large part of home entertainment. The need to access these products is usually cited as one of the driving forces behind home networking. A large majority, however, of these products are copyrighted materials.

When you buy a DVD recorder and connect it to the home network, the DVD is capable of storing a very large amount of data, about a three-hour movie. There are fears that piracy could destroy this emerging industry and consequently hurt one of the strong rationales for home networking. A number of groups have been established to develop various methods for safeguarding copyrighted material. This section describes these organizations and the emerging technologies that are being used to protect digital content.

> **NOTE**   It is important to share knowledge of copyright protection with the homeowner during the customer-support phase.

Content and Copyright Threats

Original copyrighted content is delivered to the home network from a number of sources. It may be transmitted via satellite, terrestrial, or cable systems; or recorded on various formats of digital media such as CD-ROMs, DVDs, and hard drives. To protect this content from unauthorized copying, a number of different technologies are being developed. The development of these protection technologies is driven by the movie industry, which fears that the proliferation of digital technologies will encourage people to illegally copy content. For instance, it is relatively easy to download a new movie from your cable operator's broadband network and use the home network to make several copies of this film. The level of content protection that is used in a home networking environment largely depends on the type of person running the system. Table 5–1 shows the different levels of danger associated with people who actively copy digital content.

Table 5–1
Content Protection and
Copyright Threats

Groups of People	Threat to Content Revenue Streams
Casual copier	**Low.** The individual may record a film on their VCR for personal viewing.
Hobbyist	**Relatively low.** The individual may purchase or develop a device for storing Internet content.
Small-scale hacker	**High.** Operates a bank of recording devices (e.g., DVD players, VCRs, and CD-ROM drivers).
Professional pirate	**Very high.** Well-funded and very knowledgeable of the digital security systems.

Much of the copy-protection technology has been aimed primarily at the first group identified in the table: casual copiers. All the technologies that have been proposed or developed so far have been breakable, and the breaches have become accessible to the casual copier as upgrades to consumer audio/video (A/V) products.

An important part of the strategy behind copy protection is legal prosecution. Legal means are used against the latter two categories. It is impractical to initiate legal action against average consumers. The use of copy-protection technology is partly to stop the casual copier from making and selling illegal copies. Because the distribution of pirate technology is illegal, this gives the industry a means to prosecute those in the latter two groups who would make a business of distributing this technology.

For most people, an in-home copy-protection system must provide a number of features in order to be acceptable:

- It must not degrade the quality of the signal that is used by the in-home devices.
- The complex protection system running on the network should be seamless.
- It should not add significantly to the cost of devices.
- All content-protection systems must be designed to work with other computing and A/V devices.

DVD players with an IEEE 1394 interface port manufactured today are required by law to comply with an industry-wide digital-protection scheme. These factors have acted as a catalyst for developing new standards that address the copyright protection of video, audio, and data in a digital format.

Digital Watermarking

CSS

Current read-only DVD systems prevent unauthorized copying from prerecorded DVD disks by using licensed technologies such as the Content Scrambling System (*CSS*). The CSS is a form of data encryption used to discourage reading media files directly from DVD disks. But with the advent of digital recording equipment such as recordable DVDs, digital tape recorders, and personal computers with large disk storage capacities, additional copy-protection features are needed to prevent unauthorized copies of the copyrighted digital content. The solution has been the creation of invisible

Watermarks

electronic *watermarks* for digital movies and video. The use of watermarking technologies is becoming a popular method of protecting video content from unauthorized copying and/or playback.

These watermarks are embedded in the video content during the processing and digital compression of the content. For this method to be effective, the watermark must always be added to the content before it is distributed. Watermarking technologies make life challenging for counterfeiters because they are difficult to erase or alter. Although transparent to the user, electronic watermarks contain information that can be recognized by a detector chip in consumer digital set-top boxes, or by special detection software running on compliant PC systems as instructions for enabling or disabling its ability to make a copy. In addition to being transparent to users, electronic watermarks must survive through normal processes such as digital-to-analog conversion and repeated digital compression/decompression cycles while still remaining detectable by the digital recorder system. The addition of a watermark to video/audio material does not interfere with the quality of service for home networking users.

Protecting Digital Content

High-speed digital interfaces allow consumer A/V electronic devices to interoperate with PCs. Although this is very exciting for home network users, it poses some serious challenges to industries that are worried about the pirating of digital content. To confront these challenges, copy-protection systems are being developed to do the following:

- Protect content in transmission from one A/V device to another.
- Protect content that is stored on digital tapes, DVDs, CD-ROMs, and hard disks.

Currently, digital video sources, such as DVD players and PCs, are not permitted to exchange digital entertainment content with devices such as televisions and VCRs through an external digital bus. In an effort to overcome legal and technical restrictions associated with the transfer of digital data within homes, the consumer electronics industry has developed the following protection systems: DTCP and XCA.

Digital Transmission Content Protection (**DTCP**) is a cryptographic protocol that protects A/V entertainment content from illegal copying, interception, and tampering as the content traverses high-performance digital interfaces such as IEEE 1394.

DTCP

Extended Conditional Access (**XCA**) is a smart-card-based renewable encryption scheme that can identify content as "never-copy," "copy-once," and "free copy." These three states are represented using the Copy Generation Management System (CGMS) bits. The A/V devices on the home network should obey "playback control," "record control," and "one-generation control" rules. Table 5–2 summarizes these rules.

XCA

Table 5–2
Extended Conditional Access

Device Type/ Content Type	Never-Copy	Copy-Once	No-More-Copies	Free Copy
Player	Play	Play	Play	Play
Recorder	Do not record	Record and change content type to "no-more-copies" in the new copy	Do not record	Record

SUMMARY

The goal of a whole-house audio/visual installation is to provide the components that allow the homeowner to have music and video in the rooms they want and to ensure that it can be upgraded in the future. To show how to achieve this, this chapter discussed the following:

- Video distribution is achieved by using a twin coax cable system, running two connectors to each room with the top connector providing antenna/cable and the bottom connector providing VCR or video camera.
- Each room or area receiving audio signal is called a zone.
- There are three major video formats for television broadcast and signal transmissions: NTSC, PAL, and SECAM.
- Source devices such as DVDs, VCRs, CD players, and surveillance cameras produce audio/video signals.
- The A/V distribution device distributes the signals to all desired rooms or designated zones of the house.

- When planning the equipment layout for the audio/video system, consider heat sources, sources that cause signal leaking or bleeding, and accessibility to devices.
- The detail specifications for residential audio wiring are described in ANSI/TIA/EIA-570-A.
- An attenuator may be needed to reduce the signal strength when a television is located too close to an amplifier.
- Wireless line-level distribution systems connect to audio source devices and send a signal to the receiver that is in turn connected to the amplifier.
- The technology that has led to the sharing of copyrighted material throughout the home has led to fears that material will be copied and distributed outside the home.
- Watermarks, CSS, and XCA are examples of digital methods used to protect copyrighted material from being duplicated.

GLOSSARY

Amplifier   Increases a signal's strength.

ANSI/TIA/EIA-570-A-3   Details for standards-compliant audio cabling.

Component   Connector that produces high-quality video output. It is rare compared to the S-Video connector.

Composite   The most common type of video connector.

CSS   Content Scrambling System. A data-encryption system used to discourage the reading of media files directly from prerecorded DVD disks.

DBS   Digital Broadcast Satellite TV.

Dedicated signal   A signal that is used for a specific purpose.

Distributed signal   A signal that may be channeled for multiple purposes.

DTCP   Digital Transmission Content Protection. A cryptographic protocol that protects A/V entertainment content from illegal copying, interception, and tampering.

Filter   Eliminates specified frequencies.

Modulator   Device to connect multiple locations.

NTSC   National Television Standards Committee.

PAL   Phase Alternation Line. Video format used primarily outside the U.S.

Personal Video Recorder (PVR)   Device that uses a hard disk to record, play, and pause TV shows and movies.

RF connection   F-type connector commonly used in cable television installations. The RF connector combines audio and video into one signal.

RG-6   Cable type that meets the most acceptable standard for carrying video signals.

SECAM   Sequential Couleur Avec Memoire. Video format used primarily outside the U.S.

Splitter   Divides an incoming signal.

Streaming media   Media that can be listened to or viewed as live feeds from different external sources such as Internet music, radio, and movies.

Watermarks   Encryption system used on DVDs to prevent duplication.

XCA   Extended Conditional Access. A smart-card-based renewable encryption scheme.

CHECK YOUR UNDERSTANDING

1. Which technology is used to prevent unauthorized copying from prerecorded DVD disks?
   a. Content Scrambling System
   b. Content Protection System
   c. Copyright Protection System
   d. None of the above

2. Which of the following is not a remote access system that can be used to control whole-house audio/video?
   a. Infrared-repeater-control system
   b. RF control system
   c. Infrared learning keypads
   d. S-Video

3. Where are the details for standards-compliant audio cabling specified?
   a. ANSI/TIA/EIA-568-A-3
   b. ANSI/TIA/EIA-570-A-3
   c. IEEE 1394
   d. UL 13

4. Which type of cable is the most acceptable standard for carrying video signals?
   a. RG-6
   b. RG-59
   c. Category 5
   d. None of the above

5. What needs to be run if the A/V system design includes a cabinet, closet, or shelf location in which all input devices

will be housed (DVD, VCR, CD player, stereo, satellite, and so on)?
   a. One set of Category 5 and speaker wires
   b. Multiple sets of Category 5 cable and speaker wires
   c. Coaxial cable
   d. RG-59 cable

6. ANSI/TIA/EIA-570-A defines which type of cabling topology for an audio/video subsystem?
   a. Bus
   b. Daisy chain
   c. Ring
   d. Star

7. Streaming sound or video is played after the entire audio or video file has been downloaded.
   a. True
   b. False

8. What is the first step when connecting the audio system?
   a. Connect the stereo to the receiver
   b. Terminate the speaker wire at the volume control device
   c. Run the wire from the distribution equipment to the speakers
   d. Connect the volume control device to the speakers

9. Which type of connector combines audio and video into one signal, which are then separated by the video display when the image is displayed?
   a. F-type connector
   b. BNC

c. RCA connector

d. S-Video

10. A device that records, plays, and pauses TV shows and movies by using a large hard disk, along with sophisticated video compression and decompression hardware to record television streams, is called a(n):

a. RVP

b. PVR

c. VPR

d. TTV

11. Of the three major video formats for television broadcast and signal transmissions in use, which is used in the United States?

a. NTSC

b. PAL

c. SECAM

d. HDTV

12. In addition to right front and left front, a surround-sound system is composed of which channel(s)?

a. Center front

b. Center front and center rear

c. Center front, right rear, left rear

d. Center front, center rear, right rear, left rear

13. Which type of common video connector (preferred by most consumers) uses a round multipin connector?

a. RF

b. Composite

c. Component

d. S-Video

14. What is the resolution for HDTV?

a. 450

b. 1050

c. 1200

d. 1394

15. What is the commonly understood language that instructs devices about when and how to transmit signals across the network?

a. Protocol

b. Interface

c. Device language

d. Broadcast

16. What is the advantage of a whole-house video distribution system?

17. Explain the advantages of a multi-source system over a single-source system.

18. Explain the difference between dedicated and distributed signals. Provide an example of each.

19. What does the TIA Certification test measure?

20. Why is it important for the HTI to explain the basics of how installed equipment works to the homeowner?

Security and Surveillance

 **OBJECTIVES**

Upon completion of this chapter, the HTI will be able to complete the following tasks:

- Identify alarm system components

- Explain functions of the alarm system

- Connect an alarm system

- Instruct the customer on using the alarm system

INTRODUCTION

The purpose of a security system is to provide protection for the homeowner. This chapter details the two types of security systems: wired and wireless. The components of a security system are discussed including the installation and configuration.

Today's security subsystems are stand-alone systems that report on potential security problems via standard or cellular telephone services. These reports are sent to security monitoring services who can then notify fire and police agencies about potential fires or security breaches. A security subsystem that is integrated with the home network can provide the homeowner with constant monitoring from anywhere in the world.

There is a range of security systems offered, from a simple do-it-yourself kit that is self-monitoring to a professionally monitored system. The HTI should discuss with the homeowner whether the system must be monitored by a security system company. If this is the case, the company will probably require that their technicians install the hardware. The HTI can then work with the security system company to integrate the system into the home automation system.

DESIGN ASPECTS OF ALARM SYSTEMS

Residential security and monitoring systems provide intrusion and smoke/fire detection. A security system can be installed during new construction or added as a retrofit to an existing home. There are two options for a security and surveillance system installation: a wired or a wireless system.

Zones

Zones relating to audio/video were covered in the previous chapter. In a security system, zones refer to the layout of the sensors. The key element of a good security system design, installation, and configuration is the zoning and zone layout of the various sensors. Security sensors, such as door and window sensors or switches, are commonly installed in zones. The HTI programs multiple security zones to better track down where an alarm has been triggered. A good system design will collect the sensors into a logical group called a *zone*. For example, zone 1 in the graphic shows the master bedroom. Sensors on the windows and a motion detector could be installed in this zone. Any breach in security in zone 1 would mean that either the window is open or someone is moving in the room. Sample zones in a home are shown in Figure 6–1.

Zone

Figure 6–1
Zones in a Home

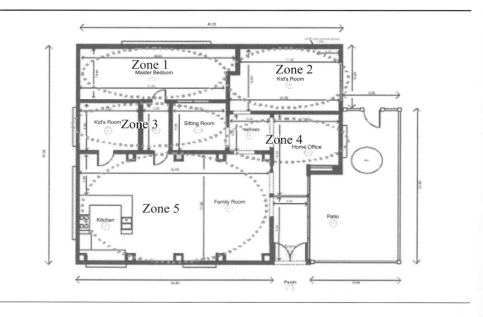

Multiple windows or doors can be managed as a single zone. For example, a zone may be composed of door sensors and window contacts for the master bedroom, master bathroom area, and sitting room. Similarly, another zone may include the dining room and living room windows; whereas the front door and a secondary entry door could constitute a separate zone with entry/exit delay. The building plan and homeowner requirements are major factors in the determination of the final system configuration.

Security zones are secure cells that are monitored independently by different types of sensors. Security breaches are detected by zone and then transmitted to a central controller to confirm an actual breach. The concept of supervision describes the monitoring of the security zones by a central controller and the capability to detect the status of the attached devices. Most systems control power failure, telephone-line problems, loss of internal time measurement, breaches in or attempts to tamper with the system, and low batteries. All disturbances send a signal to the central controller or panel. The major difference between wired and wireless security supervision is the type of media used to transmit signals.

Zones allow the central control panel of a security system to query the status of each device and also to interpret information furnished by each device monitored. A typical residential security control panel can monitor 6 to 48 different zones. It is important to understand the basics of the way zones are derived.

The control panel usually treats every circuit as an individual entity. The central control panel therefore tracks each circuit as a separate zone. To achieve this goal, each circuit is connected to only one security device (a sensor, for example). For example, if you have a room with two door sensors and two passive infrared receivers (PIR), they account for four security zones, not one. However, as mentioned earlier, the four zones are true zones only if all four devices are individual circuits.

Sometimes you may want to run more than one sensor on the same circuit, which creates what are known as *subzones*. Subzones have a major disadvantage because the security control panel "sees" all devices that connect to the same circuit as one zone. This makes it harder, if not impossible, to determine which of two doors or windows is causing the alarm to go off. It is recommended to install more wiring for extra sets of circuits and to avoid subzones.

Subzones

Wireless security systems also work by zones. This is achieved through port assignments to individual devices on the wireless control panel. Wireless security devices usually communicate via frequencies in the 300–900 MHz range.

Normally Closed (NC) and Normally Open (NO) Areas

Sensor switches for security zones are configured in two ways:

- *Normally closed* (*NC*) area. When the system is in protected mode (for example, a window shut or a door closed), the sensor switch contact is connected and the system is armed. Opening an NC area sets off an alarm.

  Normally closed (NC)

- *Normally open* (*NO*) area. Closing an NO area, which is usually in protected mode when open, sets off an alarm.

  Normally open (NO)

Most security alarm systems allow NO or NC zone configurations, or a hybrid of the NO/NC zone. Wiring for each of these configurations is different and will be discussed later.

Supervision

Supervision for an alarm system includes the monitoring of the security zones by a central controller and the capability to detect the status of any attached device. These concepts are an important part of a good home security system. Common things that security control panels supervise include low-battery conditions, power failures, phone-line problems, loss of the internal clock, and so on. The system supervises the devices that are attached to it; how often it does so may differ between products and manufacturers. Most alarm systems send in periodic signals to central monitoring stations about once a week.

Supervision

The importance of system supervision hinges on the fact that all sorts of things happen in the home that could throw part of its security system into chaos. Rerouting phone lines, installing a computer on the phone line, introducing new devices, or remodeling the home can change the dynamics of the home security system.

Monitoring

As previously discussed, one concept of supervision describes monitoring the alarm system by a central controller, which can be programmed to contact a monitoring service company. These companies monitor alarm systems for homes at a monthly cost. The monitoring company is connected to the security system with a phone line that calls them when there is an alarm. The monitoring company will then call the homeowner to find out what the breach is, or it will automatically contact the police or fire department depending on the level of service contracted. In most communities, the homeowner must register with the local police department when an alarm system is installed and a monitoring company is utilized. This requirement is so that the police can verify the situation when a monitoring company calls them for a disturbance at a residence. Some police departments also require a registration fee and have limits that might include fines for the number of false calls.

Homeowners can also be notified via their pager or cell phone that their alarm system has been tripped. Another option is to allow homeowners to monitor their own alarm systems via the Internet. Web-based monitoring allows the homeowner to view shots from video cameras, see which sensor was tripped, and program the alarm system from anywhere in the world.

Fire Systems

More-advanced security systems can integrate smoke detectors and carbon monoxide detectors. Fire systems are residential security systems designed to monitor fire-related incidents in a home. A fire system will detect fire, smoke, and heat. The components of a fire-security system include fire alarms, smoke detectors, and temperature (heat) sensors.

Smoke detectors are one type of fire system that works with a battery or electrical power. Smoke detectors, in general, work by sensing the rising smoke from a fire and sounding a piercing alarm. There are two types of smoke detectors:

- **Ionization chamber detector:** Uses a radioactive source to produce electrically charged molecules called ions in the air. Smoke entering the chamber attaches itself to the ions, which reduces the flow of electrical current, setting off an alarm.
- **Photoelectric smoke detector:** Uses an internal sensing chamber that includes a light source and a light-sensitive receiver. The chamber is designed so that ambient light cannot get in, but smoke can very easily flow through. Smoke entering the chamber scatters the very small amount of light that the detector normally sees from the device's internal light source. Light scattering causes more light to be seen by the receiver, setting off an alarm.

Smoke detectors also differ based on the source of power. A battery-powered detector can go for about a year without needing a battery replacement. When the battery gets low, the device will emit short beeps. Smoke detectors that use electricity depend on residential power and are also vulnerable to power failure. They operate as long as there is current in the circuit to which they are attached.

Over the years, smoke detectors have had other features added to them. These include an escape light and the capability to transmit their alarm to a central console (by radio signal) as part of an integrated emergency system.

Heat detectors are available, either as part of a smoke detector or as separate products. These devices use a special metal that melts or distorts when heat enters the air surrounding it, setting off an alarm. If built into a smoke detector, the metal distortion sets off the smoke detector's main alarm.

An understanding of the various fire systems and how they operate provides the HTI the ability to design these features into the overall security system.

NOTE Check the manufacturer of the system being installed to ensure that it is in compliance with the NEC and local fire codes. Also, check the detailed information for special installation instructions and any limitations regarding where it can be installed.

Video Surveillance

Video surveillance is both an in-house and an externally provided service. Closed-circuit TV (CCTV) cameras are mounted at various strategic locations to monitor the home. The images that are generated can be modulated and made available on a specified channel when connected to the video-distribution system. Any monitor within the home that browses those channels can also view the images. These images can also be recorded on a tape with a VCR, personal video recorder (PVR), or other storage system for viewing at a later time. The availability of a high-speed broadband Internet connection to the home makes it possible to use a web browser to connect to a residential security panel integrated with the home network. This makes it possible to remotely turn on, reconfigure, and even view images of various locations of the home being generated by real-time cameras. Figure 6–2 shows examples of outdoor and indoor video cameras.

Figure 6–2
Outdoor and Indoor Cameras

Environmental Monitoring

Environmental monitoring is another element of the residential security system design that includes video surveillance and security lighting. Environmental monitoring/surveillance employs various types of sensor devices that detect motion and lights to trigger alarms and send signals to a central monitoring station.

- **Photocells:** Commonly used to turn lights on at dusk. Photocells trigger events based on the amount of sunlight hitting the cell.
- **Motion detectors:** Used to turn on lights in a hallway when someone enters. Motion detectors trigger events when movement in a specified range is detected.

- **Motion-detector floodlights:** Can be used to trigger a chain of events such as turning on the porch light and turning on the hallway light. Motion-detector floodlights trigger events when movement occurs in the detection field.
- **PowerFlash modules:** Can be implemented in almost any stand-alone sensor. An event is triggered when the sensor is activated.
- **Closed-circuit television (CCTV):** A network of small closed-circuit cameras mounted near the front and rear doors and other strategic locations in the home to capture images of any disturbances.
- **Telephone autodialer:** Any programmed alarm system in the residence that dials one or more telephone numbers (the monitoring company, for example) on a wired or cellular phone line, and plays a prerecorded message when triggered.

Events that trigger the home-monitoring system alert the security company to take action. Monitoring services are available for a monthly fee. They are staffed 24 hours per day and 7 days per week. Some monitoring companies install their own equipment and contract with the homeowner for services. If a homeowner is using one of these services, their design documentation will help determine whether the system can be integrated into the home network.

Other companies that market and install monitoring equipment such as alarm systems do not have their own central monitoring station. They bundle another company's monitoring service—a business specializing in monitoring systems for alarm companies. The HTI should establish a framework for working with all the parties involved so that home security is the priority.

Emergency Response Systems

The security subsystem utilizes a monitoring service that is alerted when there is a security breach. The monitoring service, in turn, alerts the police and fire departments when necessary. The emergency response system should be able to call for appropriate help depending on the circumstances. The home control system (monitoring/surveillance alarm) calls the monitoring station when an emergency or other appropriately programmed situation arises. The central station in turn interprets the information coming from the alarm system and calls the appropriate parties.

The backbone of any emergency response system is its response time. Monitoring centers are high-security operations with backup power supplies, backup telephone lines, and even two-way radio contacts with some homes they monitor. A monitored alarm system in the home will use a digital communicator, which seizes a telephone line in the event of an intrusion or other breach, and signals a computer at the monitoring center or central control station. An operator receives the alarm, verifies it if necessary by calling the home or using a two-way radio contact, and notifies the appropriate authorities—police department, fire department, homeowner, and so on.

Deciding on a central monitoring station service is dependent upon the level of security the homeowner requires and the residential security system design. A security system that is contracted with a monitoring company provides a professional security response. Other security systems may not have external central monitoring. In these cases, the system's central control panel is designed and programmed to seize the home phone line and auto-dial the homeowner's mobile or office phone. Some systems can be programmed to page the homeowner or send a programmed message to a computer or personal digital assistant (PDA). This requires the homeowner to make the decision to call the police or fire department.

The options for emergency response systems are many. The type of system and the functions/features that are included should meet the needs of the homeowner while staying within the budget.

Wired Systems

Wired (wire-line) systems rely on low-voltage telecommunications media for functionality. A wired alarm system interconnects door, window, and other sensors to a main alarm panel using shielded or unshielded wiring. Resistors, electrical components that send the security control panel information regarding the status of sensors by regulating the flow of electricity, are located at the end of the wires along with the sensors.

The features of a wired system include the following:

- **Reliability:** Hidden wires provide direct communication with the monitoring company, and the battery backup system maintains the connection in the event of a power failure.
- **Efficiency:** Wired systems are constantly in touch with their monitoring companies to allow constant communication for emergencies and system updates.
- **Proven method:** Because the first security systems were wired and most security systems today are still wired, the technology has been proven.
- **Availability:** Because most security systems are wired, the choices of manufacturers, equipment, and decorative finishes are quite numerous.

Wireless Systems

Wireless systems provide the same security features as a wired system, but they are much easier to install, especially in an existing residence. Wireless systems require batteries that must be changed in order to keep the system working. Wireless systems provide the following:

- **Ease-of-use and installation:** The only tool usually needed to install a wireless security and monitoring system is a screwdriver. Complete installation instructions are included with the system, and an installation video is provided in some cases.
- **Portability:** A wireless system can be moved to another house simply by unplugging the control unit and moving the sensors.
- **No new wires:** A wireless system does not require the homeowners' walls to be drilled or ripped open.
- **Low installation costs:** Because wireless systems are easy to install and no construction work to walls is required, they have relatively low installation costs.

Wireless systems can be monitored by a security company or set up to call the homeowner when triggered, just like wired systems. When moving, the homeowner contacts the monitoring company to arrange for service at the new location.

Remote Access

Remote access is another aspect in a home security system's design and installation. A residential security system may be designed to allow the user to manage the system through the Internet with a broadband or dial-up connection. Some security companies offer advanced systems that use highly secured Internet access.

> **NOTE** The security subsystem devices communicate very little with other outside devices, and therefore present fewer opportunities for home integration. Ironically, this is what makes the security subsystem secure.

One goal of the residential network is the ability to use the Internet to access the control systems that manage the systems installed in the home. With a web browser interface and password, the residential security system can be securely used to control access to the home from anywhere and for many reasons:

- Delivery people can have access when the delivery is made and no one is home. When the delivery person rings the doorbell, the homeowner is notified through his/her personal digital assistant (PDA), cellular phone, or office workstation. The customer can deactivate the security system to allow access and monitor the delivery person's actions through a security camera. The homeowner can reactivate the security subsystem when the delivery person leaves.
- The housekeeper can have regular access to the home on specific days at specific times, and allow the kids access anytime, using different PIN codes for the system.
- Homeowners can access and reprogram security links via the Internet when away from home.
- Homeowners can program an individualized onscreen home page to receive updates on news, weather, mail, financials, or personal messages.
- Homeowners can browse the Internet, connect to the home network, access security cameras and reorient their view angle, monitor intercoms, program music for the kids, and so on.

Make sure the security-monitoring company allows the security subsystem to be connected to an Internet interface for remote control and access to the system. Currently, many monitoring services will not allow remote access because of the possibility of system hacking.

There are many things to consider with remote access. As the technology evolves, all the benefits of remote access can be realized. Always explore the homeowner's needs to ensure that the system has the capacity and capabilities to function as desired.

ALARM SYSTEM COMPONENTS

There are many components to a home security system. The most common device is a magnetic contact sensor that detects when a door or window has been opened. Motion detectors are also common. These use microwave or infrared waves inside or outside the house that trigger the alarm when an object moves past the sensor. Some security systems include audio and video monitoring of an entry.

The alarm system is the core of in-house security services. The various security sensors, contacts, or detectors are usually configured in zones using a combination of star and daisy chain topologies. Keypads for the user interface are connected to the wiring that runs from the central control panel. Other security devices include chimes and alarms, zoned security cameras for monitoring, and smoke detectors. The telephone may also be considered one of the devices, especially if a second line is connected to the security box as a backup.

The security subsystem may consist of most or all of the following components:

- High-speed Internet access
- Security control panel
- Operating system programming
- Security wiring that is usually laid by zones in a combination of star and daisy chain topologies
- Category 5/Category 5e wiring for connection to security cameras

- Phone wire for connection to the home network distribution panel and the demarcation point of the home (for the backup line)
- Security pads and contacts in each zone
- Keypad or other user interface device

Security Control Panel

The security control panel, as shown in Figure 6–3, is the distribution point of the residential security system. It is the command center for all the security programming. The security control, or main alarm, panel is the heart of the alarm system. All the sensors, keypads, and other warning devices are cabled to the location of the alarm panel to allow connectivity.

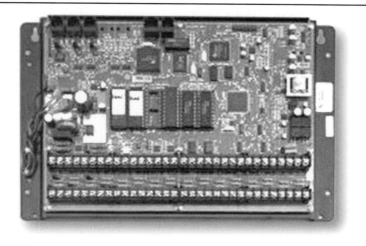

Figure 6–3
Security Control Panel

The security control panel may be a stand-alone product, a specialized module built in as part of the residential security system, or integrated into a security system keypad. Whether a stand-alone or integrated security control panel is selected, make sure that it can interface with the residential network distribution device for remote access and notification via the Internet.

The security control panel must have enough capacity and functionality to carry all the security devices that will be installed as part of the residential security system. It also can be connected to a second phone line so that a backup for the subsystem is available. The operating system in the security control panel can be programmed to function per the homeowner's requirements.

The alarm panel usually provides a 12 volts DC alarm output (at about one amp) that is used to drive one or more warning devices, including bells, sirens, buzzers and chimes, strobe lights, speakers, and even telephone auto-dialers.

Security panels are generally installed in the distribution panel. The security control panel has to be mounted close to or within the residential network distribution device. The security panel must have an outlet for telephone service so that the system can capture the phone line and connect to the monitoring service. In addition to a phone line, the control panel, which may be integrated to a system keypad, requires power and must be located within reach of a power outlet.

NOTE When selecting a security panel, make sure that it is approved by Underwriters Laboratories (UL) for the type of security service that you require.

Other considerations for locating the security panel to facilitate systems integration is secure Internet access and the potential need to connect the panel to a second phone line if used as a backup to provide off-network functionality to the system.

Sometimes, it is necessary to mount the security panel in a location isolated from the common wiring closet or equipment room for reasons of security. The final location comes down to the specific needs and circumstances of the homeowner, the installation design, the size of the installation, the features provided, and any special requirements.

Security Cameras

A closed-circuit television (CCTV) system is an important element of modern residential security systems. It includes video monitors and other components such as modulators, along with the security cameras. Figure 6–4 shows an indoor security camera, and Figure 6–5 shows the wiring for this type of camera.

Figure 6–4
Indoor Security Camera

Figure 6–5
Wiring of Indoor Security Camera

Resolution and sensitivity are factors to consider when selecting cameras for the CCTV installation. Resolution, which is the sharpness of a displayed image, is expressed as the number of lines onscreen. Although camera and monitor resolution can range from 200 to 500 lines, an adequate display can be achieved at a reasonable cost with a monitor that displays 350 lines.

> **NOTE**   Sensitivity of cameras to light is commonly measured in lux, which is a unit of measurement for the intensity of light. One lux is equal to the illumination of a surface that is one meter away from a single candle. A lower-lux camera is more sensitive in low-light levels than a high-lux camera. For a door application in which additional sensitivity at night is desired, infrared (invisible) light blasters are available that can be used to enhance the camera image.

Advances in technology have produced high-quality, low-cost cameras for home security uses. There are a wide variety of home security camera products on the market for indoor use. Others are specifically designed for outdoor use, such as the camera shown in Figure 6–6; these can be mounted anywhere to allow for the best view of the residence, including in the garage, on the front gate, and on trees. These outdoor cameras typically have a weatherproof housing. You can also find recessed designs that mount in a deep single or dual J-box. Some models have lenses small enough to hide behind a 1/16-inch hole. Using low-voltage wiring, these cameras can easily be networked and integrated with the other residential systems, including video distribution. If a webcam is used, it is connected to the computer subsystem distributor and will not have its own distribution unit.

Figure 6–6
Outdoor Security Camera

Wireless cameras can be small, portable, and very easy to mount. They come with a video receiver that is simply plugged into a TV and the wireless images sent by the camera from any location can be displayed.

A video camera can also be fitted with audio to be installed at the main doorway, in a nursery, overlooking the backyard, or in the children's room. This CCTV camera allows the homeowner to both watch and listen to children, nannies, baby sitters, and others from a different room in the house or from anywhere outside the house with remote access.

> **NOTE** A video camera and microphone can monitor any area that is considered private. Cameras and microphones that are located to monitor neighbors or outside a homeowner's property may be violating the law. Install exterior cameras so that only entrances or the customer's property are covered, not adjacent properties.

On the whole, the location of cameras is dictated by the purpose that the homeowner requires. Always discuss these issues with the customer before proceeding to design or recommend the type of system that would best meet their needs.

Some useful locations for monitors include the master bedroom, home office, and living room for easy access and viewing. When integrated with the computer network, it's also possible to route images to computer monitors such as the home office machine. The best monitor locations are those the homeowner considers most convenient. External monitors are located at the monitor service company when subscribed to by the homeowner.

Monitors and Switchers

Switchers

Sequencers

Video monitors are used to view the images taken by the security (CCTV) cameras. Some devices, known as *switchers* or *sequencers,* are installed with a multicamera security system to sequentially switch from camera to camera for single camera viewing on a video monitor. Figure 6–7 shows a Leviton switcher or sequencer. These inexpensive units usually allow a camera's picture to be sequenced, bypassed, or held. Only the actual camera image being viewed on the monitor can be recorded.

Figure 6–7
Switcher or Sequencer for Switching between Cameras on a Multicamera System

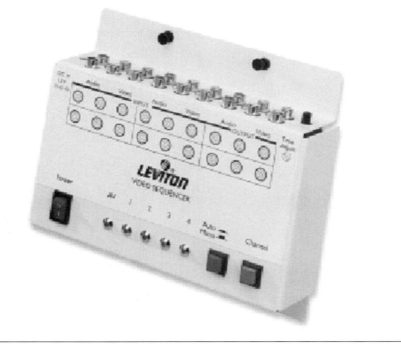

Splitters

Quads

Other devices known as *splitters* or *quads* can be used in multicamera systems to allow the viewing of all camera images on a single video monitor at the same time. Figure 6–8 shows a Leviton splitter. A quad, for instance, divides the monitor screen into four quadrants, each showing the image from a different camera or allowing a single camera view to be shown full-screen. When recording, only the camera images being viewed on the monitor are recorded. The quality of recorded images is greatly reduced in this mode.

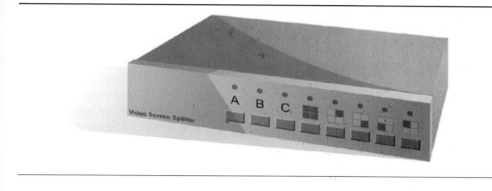

Figure 6–8
Splitter Divides the Monitor
Screen into Four Quadrants

Switchers and splitters come in a number of varieties. Each product has its own feature set and is shipped with a specifications sheet completely defining its characteristics and capabilities. Monitors are available in both black-and-white and color display. Some color monitors allow the system to adjust/hold/sequence images. Installing switchers and quads with a combination of features will provide functionality in a network security system.

Define capabilities and features required by the security system before proceeding with the installation of the monitor, switchers, and quads for optimal video displays.

Keypads

The keypad is the homeowner's primary interface with the security system. It allows the user to activate and deactivate the security system with a numeric code. Older units use key switches, which use a physical key. They are used less frequently in new installations. Security keypads, which resemble the telephone keypad, are used in the vast majority of new applications, due to their flexibility and the fact that they are programmable.

The system can be programmed from the keypad. Keypads usually have one or more function buttons programmed to execute specific functions. The most common functions are buttons that quickly activate a fire alarm, a medical emergency alarm, and an audible panic alarm.

Some security alarm systems even have a module, which can be programmed to turn any touch-tone phone into a functional keypad. This feature allows the homeowner to arm, disarm, program, and check the status of the security system from outside the home. This feature is also cost-saving when considering the installation of a security system: All the touch-tone phones on the premise can become fully functional keypads, which eliminates the need to distribute wired keypads all over the house.

Some keypads can use remote controls. The remotes come in keychains that are easy to carry around. With these products, one button alone can disarm the system, turn on the lights in the garage and around the house, and open the garage door.

Most keypads provide the means to identify which security zone in the home has been violated. In many instances, a number corresponding to the zone (an open door, open window, and so on) appears on the keypad display.

Keypads are typically mounted on the wall in close proximity to the doors used for entering and leaving the residence. This makes it easier for the homeowner to set the system or turn it off. This location also can act as a deterrent to would-be thieves because the keypads are visible to someone looking through a window. It may be necessary to place additional keypads in other locations, such as the master bedroom, bedroom hallway, and home office. These locations allow the homeowner to control the system more conveniently. Keep in mind that the more locations you choose, more wiring and cost are required to connect them to the main control panel. As mentioned earlier, there are also wireless hand-held or portable IR devices that can be carried around and enhance the use of keypads remotely.

Sensors

The most common sensors used in a base system include contact sensors, passive infrared sensors, and smoke detectors. Within these categories is a multiplicity of sensor types based on where they are located and their use. Some of the location possibilities based on the type of sensor are summarized as follows:

- **Contact sensors:** Mounted in windows and doors to detect when windows and doors are opened or closed.
- **Glass-break sensors:** Installed near windows to detect sounds related to shattered glass.
- **Motion sensors:** Mounted in bedroom hallways and stairways to detect movement by intruders. Uses infrared technology.
- **Flood sensors:** Mounted in places such as basements or other flood-prone spaces in and around the house and frequently used to trigger sump pumps.
- **Carbon monoxide (CO) detectors:** Mounted on interior walls of a home to detect hazardous levels of CO.
- **Rain sensors:** Mounted outdoors to trigger or cancel sprinkler systems.
- **Water disturbance sensors:** Installed in a swimming pool or a hot tub to determine when someone or something (for example, a crawling baby) has entered the water.

There are many more possibilities for sensors. The security design should allow for all the needs of the homeowner.

PIR

Passive infrared (**PIR**) detectors, also known as motion sensors and microwave detectors, are security system devices that can tell the homeowner when someone is moving around in an area that should be empty. These devices help keep the doors and windows secured in the home. Additionally, the motion detectors can be turned off separately from the perimeter sensors, allowing the family to move around freely in the house while keeping the outside secured. When installing motion detectors in large or odd-shaped rooms, direct the infrared eye to the entry points into the room.

Simple PIR detectors are not a good recommendation for a home with active pets because they can activate a false alarm. In such cases, dual-technology sensors using both microwave and PIR sensors to confirm an intrusion are more reliable.

 **CONNECTING THE SYSTEM**

The security control panel is the distribution point of the security subsystem. The security control panel is connected to the home network distribution panel for remote access and notification via the Internet. A computer is then connected to the subsystem for remote access via a web browser interface.

A reliable backup to the security system is an important part of the home. If a second phone line is used for system backup of the network, telephone wire or Category 5/Category 5e wiring is also used between the security control panel and the demarcation point of the home.

For wired systems, wires are terminated on one end at the security panel and on the other end at the other components of the security system such as

motion sensors, cameras, keypads, and contacts. The wiring is usually configured in zones using a *star topology* (also called a homerun topology).

Star topology

Low-Voltage Wiring

The wires for the security system are selected based on the control panel and
the devices that will have to be installed. It is also based on economics. In a
retrofit installation, a bundled structured wiring system that includes security
wires will cost less because all wires for the installation are provided.

The connection of security system devices in a home is achieved using
low-voltage media, wireless, or combination wireless and low-voltage wiring.
Although several wireless security systems are available on the market that
emit high-frequency coded signals to the base station, a phone line is still
needed to call out to the monitoring service in case of a security breach. Phone
lines are wired using a minimum of Category 3 twisted-pair copper wire, with
Category 5 recommended. A wired phone line is usually required for wireless
security systems because wireless systems are prone to interference from other
signals (for example, radio, television, other wireless phones, and so on).

Security wiring is usually laid in zones in a star topology with the devices
connected by a daisy chain. A *daisy chain* refers to the way devices are connected. Each device gets connected in a series, one after the other. Signals go to
the first device, then to the second, and so on. For new constructions, the security cabling is pulled as part of the bundled cable system, as is the case with
other subsystem wiring. Examples of such bundled systems were mentioned
earlier in this chapter.

Daisy chain

Types of security system wire include the following:

- **22 AWG two-conductor copper wire:** Used to connect inside sounding devices
  to control panel terminals and to connect external contacts to transmitters.
- **22 AWG four-conductor copper wire:** Used to connect consoles to control panel
  terminals. Used to connect the RJ-31X to the phone company's interface box.
- **18 AWG two-conductor copper wire:** Used to connect external power
  supply to control panel terminals, and outside sounding devices to control
  panel terminals.
- **14 AWG single-conductor copper wire:** Used to ground the security system.
- **CAT 5:** Used to connect the security camera. Category 5 cable is also used to
  connect the security system to the residential network and sometimes to
  supply power to low-voltage devices such as CCTV cameras. See Figure 6–9.
- **Coaxial Series 6 (RG-6) cable:** Used for video connectivity and tied to the
  video distribution system. See Figure 6–10.

Figure 6–9
CAT 5 Cable

Figure 6–10
RG-6 Coaxial Cable

The security system design can be as simple or as complex as needed. It is dependent upon the features required of the alarm system and the type of sensors to be installed. Most systems use 22- to 26-gauge unshielded twisted-pair (UTP) or shielded twisted-pair (STP). A preselection of all the components before wiring will allow you to follow the manufacturer's cable specifications. However, from the pre-wire standpoint, it is recommended that you lay wiring for all the sensors and warning devices you can foresee being connected to the system, even if some of it is initially unused. You can also plan for a generic wiring installation that could work with a variety of systems, if you are not sure about what type of equipment will be installed.

If the security system includes a fire alarm, special fire alarm system cable is required. Use either Fire Power Limited Plenum (FPLP), which is the type designed for use in ducts, or Fire Power Limited (FPL), the type designed for general use (see the NEC, section 760-61, for more on fire alarm system cable).

> **NOTE**   The type of transformer used in most security system control panels is a UL Listed Class 2 (it meets requirements set forth in UL 1585). With a UL Listed Class 2 transformer, Class 2 low-voltage wiring must be used.

Prewiring is usually planned together with the residential structured cabling. Security wiring is further discussed in Chapter 8, "Low-Voltage Media."

NO and NC Zone Contacts

The sensor switches for a normally closed (NC) zone are cabled in a series-connected loop, whereby a loop is formed by running one of the zone wires through each of the sensor switches (in series) and returning back from the last switch to complete the zone loop. In contrast, sensors having normally open (NO) contacts are wired in parallel or across to the zone wires. As mentioned earlier, most alarm panels typically allow the zones to be wired in either NC or NO configurations, or provide some of each.

There are two important considerations regarding the two connection methods:

- An NC loop (zone) may pick up less noise than an NO zone, depending on the alarm panel design.
- If an intruder cuts off the wires of an NO zone, provided both wires are not cut simultaneously, the alarm will not go off (activate). This is not the case with NC zones.

In most cases, zone wiring is not accessible to an intruder, although wire cutting may be a possibility to consider in some installations.

The security zone connectivity is carried out with cabling daisy-chained between the various monitoring devices within a zone (for example, glass-break detectors and window or door contacts). However, whenever possible, a homerun topology, also known as a star topology, is recommended because it is much more reliable, is easier to troubleshoot, and can be easily configured.

The main control panel of a good security system is a self-contained unit located somewhere other than the main distribution panel or central wiring closet. This allows for better security. A three-wire power cord from the panel plugs into a 120-volt power source (receptacle). After proper installation of the security control panel, the number of control lines established depends on the number of security devices installed. For example, conductors can be run to the different detectors (smoke, heat, and so on), window and door pads, keyboards, chime or alarm devices, and window foil strip. You also can run conductors to outside strobe lights and even to a telephone dialer to alert the security system provider and the police or fire departments.

Telephone Connection

Most wireless systems still require some parts to be wired. Although several wireless security sensors are available that emit high-frequency coded signals to the base station, a phone line is needed to call out to the monitoring service in case of a security breach. A wired phone line is usually required for security systems because wireless security systems are prone to interference from other signals (for example, radio, television, other wireless phones, and so on). The phone line is actually seized by the security system. The *RJ-31x* jack is configured so that the security system can seize the phone line if the alarm is tripped. The alarm company is then called by the alarm system to indicate that the alarm was tripped. The phone line actually runs through the security unit in the distribution panel. It is also necessary to consider dedicating a second phone line to the security system as backup. This could even be connected to the computer network for a backup dial-up service, which can activate in the event of failure of high-speed broadband Internet access to the home.

RJ-31x

Programming the Operating System

The security panel uses a specialized program, called an operating system, for all system programming. Operating systems are designed by their manufacturers for their specific needs. However, when integrated with the computer network, these programs can actually interface with computer operating systems, such as Microsoft Windows, on a desktop computer or a home server.

The operating system allows the homeowner to program passwords to activate or deactivate the alarm. Passwords will be discussed later in this chapter.

Follow the directions for installation and programming of the operating system software. Then test the software.

 **HOMEOWNER ORIENTATION**

Training and orientation are critical for the homeowner to properly operate the system. All individuals who need access to a residence should be trained in the use of the security system after the installation is complete. The training should not be limited to a walk-through, but should also include basic features programming and password uses. The homeowner should know how to read the "vital signs" of a failure or malfunction. When a false alarm is detected, the homeowner should notify the monitoring company immediately to avoid police notification and possible false alarm fines.

The homeowner should know where the RJ-31x termination point, which interfaces the system with the home phone line, is located. Knowing where to plug or unplug the system from the phone line is important for vital repairs, testing, or troubleshooting the phone line or security system.

Passwords

A password is a series of letters and numbers, usually four to eight characters in length, which protects a system from unauthorized use. It is a crucial part of programming and configuring the security system.

Each authorized user with access to a security system should know the password used to arm and disarm the system. Ideally, each authorized user should have his own password. Passwords on advanced security systems can be configured with different levels of access. For example, the homeowner would be assigned as the administrator with privileges that provide access to all functions of the system. Other users, such as service personnel who only need to arm and disarm the system, would be assigned specific access just for that purpose. Passwords can then be changed when the homeowner no longer wants that person to have access to the home.

The most important aspect of passwords is keeping them secure. Hackers are sometimes able to compromise a security system by using programs designed to "break" passwords. To help guard against such threats, recommend to your customers that they do the following:

- Change passwords regularly.
- Use a combination of letters and numbers.
- Do not use familiar names and dates such as middle names or birthdays. It is too easy for unauthorized users to figure them out and gain entry.
- Do not use words that can be found in a dictionary. There are programs designed to crack this kind of password.

Maintenance Contract

The goal for a security system is that it functions at all times and according to the way it is programmed to work. Providing a maintenance contract gives the homeowner peace of mind so that any problem that occurs, including any form of failure or malfunction, can be corrected quickly. This ensures that downtime is kept to a minimum.

To keep the security system up and running all the time, a maintenance or standing contract can be signed with the installer, usually for a monthly fee, to carry out periodic maintenance, testing, and repairs. However, this is usually an issue mostly for do-it-yourself systems. With most of the more advanced systems, the company installs everything for a specified down payment and then maintains the network and monitoring services using its own specialists for a monthly fee.

SUMMARY

This chapter discussed the types of security systems available, their components, and how to establish connectivity. You should have a firm understanding of the following topics:

- Wired or wireless systems are the two options for a security and surveillance system installation.
- Sensor switches are configured in two ways: In a normally closed (NC) area, an alarm sounds when the zone is open; in a normally open (NO) area, closing a zone sounds the alarm.
- Closed-circuit TV (CCTV) cameras generate images that can be modulated and made available on a specified channel when connected to the video distribution system.
- Wired (wire-line) systems rely on low-voltage telecommunications media that interconnect door, window, and other sensors to a main alarm panel using shielded or unshielded wiring.
- Wireless systems are much easier to install, especially in an existing residence.
- With remote access, a web browser interface and password can be used to monitor and program the security system from any computer with an Internet connection.
- Switchers are installed with a multicamera security system to sequentially switch from camera to camera for single-camera viewing on a video monitor.
- Quads or splitters are used in multiple-camera systems to allow the viewing of all camera images on a single video monitor at the same time.

- The security control panel is the distribution point of the security subsystem.
- Security system components are usually configured in zones using a star topology.

- The RJ-31x jack seizes the phone line in order for the alarm system to contact the monitoring company to report that an alarm has been tripped.
- Passwords protect a system from unauthorized use.

GLOSSARY

Daisy-chain   A wiring scheme in which each device is connected as a series. Signals are sent to the first device, then to the second, and so on.

Normally closed (NC)   Configuration of a sensor so that the alarm will sound when the zone is open.

Normally open (NO)   Configuration of a sensor so that the alarm will sound when a zone is closed.

PIR   Passive Infrared Receivers, also known as motion sensors and microwave detectors. Security system devices that indicate movement in an area.

Quad   Device that allows images from four cameras to be viewed at the same time. Also known as a splitter.

RJ-31x   A jack that seizes the phone line and makes it available to the alarm system.

Sequencer   Device that allows the homeowner to view images from multiple cameras sequentially. Also known as a switcher.

Splitter   Device that allows images from multiple cameras to be viewed at the same time. Also known as a quad.

Star topology   A wiring scheme in which each device on the network or system has a single, dedicated cable running from the main panel. Also known as a homerun topology.

Subzone   Running more than one sensor on the same circuit.

Supervision   (1) Monitoring of security zones by a central controller. (2) Capability to detect the status of any attached device in an alarm system.

Switcher   Device that allows the homeowner to view images from multiple cameras sequentially. Also known as a sequencer.

Zones   Physical layout of the security system sensors.

CHECK YOUR UNDERSTANDING

1. What is the term for a logical grouping of security system sensors?
   a. Domain
   b. Area
   c. Zone
   d. Perimeter

2. A CAT 5 or coaxial cable is used to connect which device to the control panel?
   a. Power supply
   b. Security camera
   c. Motion detector
   d. Keypad

3. Which type of wiring topology is most common for security system installations?
   a. Daisy-chain
   b. Bus
   c. Star or homerun
   d. Ring

4. What is the term for the monitoring of security zones by a central controller?
   a. Zoning
   b. Video surveillance
   c. Supervision
   d. Remote access

5. Homeowners with a high-speed Internet connection can use a _____ from anywhere in the world to connect to a residential security panel integrated with the home network.
   a. security camera
   b. web browser
   c. switcher
   d. splitter

6. The main advantage of a wired system is:
   a. reliability.
   b. cost.
   c. aesthetics.
   d. no new wires.

7. Which type of jack is used to connect the telephone wire to the security system in order for the alarm system to seize the telephone line in an emergency?
   a. RJ-11
   b. RJ-31x
   c. RJ-45
   d. RJ-48x

8. What is the device that allows the homeowner to monitor views from different cameras in sequence?
   a. Quad
   b. Splitter
   c. Switcher
   d. Router

9. Passive infrared (PIR) detectors are also known as:
   a. smoke detectors.
   b. glass break sensors.
   c. contact sensors.
   d. motion sensors.

10. A 14 AWG single-conductor copper wire is used for which function?
   a. To ground the security system
   b. To connect inside sounding devices
   c. To connect an external power supply
   d. To connect a security camera

11. The distribution point for the residential security system is the:
   a. keypad.
   b. security panel.
   c. wired panel.
   d. security device.

12. CCTV stands for:
   a. closed-camera television.
   b. closed-caption television.
   c. closed-circuit television.
   d. circuit-component television.

13. What is lux?
   a. Unit of measurement for the intensity of light
   b. Unit of measurement for the intensity of sound
   c. Unit of measurement for the intensity of video
   d. Unit of measurement for the intensity of a camera

14. The major difference between a wired and a wireless system is:
   a. the type of signal sent.
   b. the way it is programmed.
   c. the media used.
   d. the number of zones supported.

15. Smoke detectors that use an internal sensing chamber are called:
   a. photoelectric.
   b. ionization.
   c. battery powered.
   d. receivers.

16. What is the backbone of all emergency response systems?

17. List two ways a homeowner might use remote access.

18. Passwords should not be words that can be found in the dictionary. Why?

19. What are the benefits of providing a maintenance contract?

20. Why should the homeowner know where the RJ-31x termination point is?

Lighting, HVAC, Water, and Access

OBJECTIVES

Upon completion of this chapter, the HTI will be able to complete the following tasks:

- Select an automation system
- Install lighting controls
- Integrate the heating, ventilation, and air conditioning (HVAC) system with other systems
- Set up controls for gates and doors
- Automate miscellaneous systems

INTRODUCTION

In this chapter, you learn about the automation products that control environmental systems such as lighting, heating and cooling, access and security, and water. You learn the basic issues involved in system design, component selection, and implementation. You identify major issues in component placement, commonly implemented standards, and methods for automation system installation.

Lighting, HVAC, water, building access, and accessory systems must be designed and installed with user safety, security, and convenience in mind. Some systems, such as lighting and HVAC, have to conform to national and local standards and building codes upon installation. Others have less stringent formal requirements, but they must still be designed for the unique needs of the homeowners and their families.

There are a wide variety of products available to automate home systems. They range in appearance from the most traditional residential to high-tech industrial, and may affect one device or an entire building. The HTI should work with the homeowner to choose devices that offer the necessary functionality while adhering to codes and standards.

163

AUTOMATION SYSTEMS

Automation systems are used to control many subsystems in a residence. One of the most common automation systems communicates commands from controllers to the controlled devices at electrical outlets using existing electrical wiring. This system is known as *power-line technology*. Automation can provide complex control without additional wiring by sending modulated commands over standard household AC wires. Another option that does not involve new wiring is wireless automation systems, which use either infrared or radio frequency signals to communicate between devices.

The most common types of automation systems are X-10, CEBus, ANSI/TIA/EIA-485, and ANSI/EIA/TIA-232. X-10 and CEBus are most commonly used for controlling lighting systems. X-10, ANSI/TIA/EIA-485, and ANSI/EIA/TIA-232 are the most frequently used control architectures for HVAC automation.

There are several items to consider when selecting an automation system:

- Wiring architecture
- Control-feature set that provides the required automation functions
- Appearance of the wall controller
- Nature of controlled outlets, whether built-in or plugged in

These considerations are particularly important if other residential systems are also being automated because the chosen architecture must support all required systems on the **Structured Media Center (SMC).**

Structured Media Center (SMC)

X-10

X-10

X-10, which is the most common way to connect lighting automation devices in North America, uses a set of commands that are transmitted by modulating the 60 Hz AC electricity that flows to household AC power outlets. A set of commands are sent when the alternating current passes through the zero voltage state. The X-10 architecture is popular because it doesn't require new wiring to retrofit an existing home. X-10 is largely software-based architecture. The lighting and appliance modules may require that addresses be set with dual in-line package (DIP) switches, which are small switches on some devices used to set up or adjust the equipment or a pair of dials. Other types of systems can use a controller to assign and address the module.

Addressing takes the form of a house number and a unit number. Multiple house numbers can be used, which is convenient for controlling different floors or sections of a house. House numbers are usually alphabetic, whereas unit numbers are usually numeric. "B12" is an example of an address. The B represents the house portion of the address, and the 12 represents the unit number.

Originally, X-10 provided only one-way communication, but communication is a two-way street in the X-10 architecture with modern systems. Sensors and switches can report their status back to the control program. This capability makes complex conditional control sequences possible. It also enables users to receive detailed status reports.

Power line carrier systems such as X-10 can sometimes interfere with similar systems in a neighboring house. If two neighboring houses use the same address scheme, controllers in one house may activate lights in another house. Special AC filters are available to prevent X-10 signals from leaving the house and to prevent signals from adjacent residences from entering.

CEBus

Consumer Electronic Bus (*CEBus*) also uses modulated household AC for communications between the controller and the controlled lights and appliances, but it is designed to be used with appliances and lighting devices that have CEBus capability built in. This is why CEBus is sometimes referred to as a "plug-and-play" home automation standard. As with X-10, connecting devices using CEBus is largely a software issue. The HTI must ensure that each device has registered a unique address and that the control software is communicating properly with the devices.

The following items can interfere with the power line signal:

- Wireless intercoms
- Electric razors and hair dryers
- Food processors, blenders, mixers, and grinders
- Vacuum cleaners
- Wireless doorbells
- Electronic ballasts
- Television receivers
- Laser printers
- Non-FCC Class B computers
- Low power factor (LPF) fluorescent lamps

ANSI/EIA/TIA-232

The *ANSI/EIA/TIA-232* standard defines a type of serial communication; the ANSI/EIA/TIA-232 specification describes a number of signal names and voltage levels. The way they are implemented depends upon the particular manufacturer. The two most frequently encountered ANSI/EIA/TIA-232 variants use either 25-pin or 9-pin D-series connectors. These variants were first made popular on two early models of the IBM PC. The first variant is based on a 25-pin connector, of which only three or four pins are typically used: transmit on pin 2, receive on pin 3, and ground on pin 7 for the three-pin version; transmit on pin 2, receive on pin 3, ground on pin 7, and carrier on pin 8 for the four-pin version. A later version of ANSI/EIA/TIA-232, popularized on the IBM PC-AT, reduced the pin count from 25 to 9, but still used only pins 3 and 4 for most applications. (The pins for transmit, receive, ground, and carrier are the same on both connectors.) Figure 7–1 illustrates RS232 pin out.

CEBus

ANSI/EIA/TIA-232

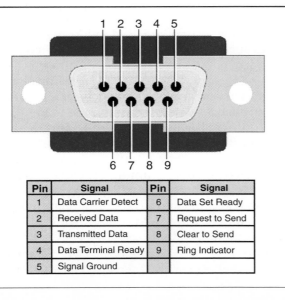

Figure 7–1
RS232 Pin Out

Pin	Signal	Pin	Signal
1	Data Carrier Detect	6	Data Set Ready
2	Received Data	7	Request to Send
3	Transmitted Data	8	Clear to Send
4	Data Terminal Ready	9	Ring Indicator
5	Signal Ground		

ANSI/TIA/EIA-422

ANSI/TIA/EIA-422

ANSI/TIA/EIA-422 is a standard used to extend serial communications up to 4000 feet between two devices. Each channel of communication is converted to a differential signal and carried on a twisted-wire pair so that when the differential signal is received, common mode noise is rejected. ANSI/TIA/EIA-422 generally has at least two channels: Receive and Transmit. They are carried on four-wire systems, and a signal common/ground provides a reference for the receiver and transmitter.

ANSI/TIA/EIA-485

The dedicated ANSI/TIA/EIA-485 cable may provide greater protection from accidental interference than the relatively noisier channel provided by X-10 (using household electrical wiring).

ANSI/TIA/EIA-485

The *ANSI/TIA/EIA-485* communications standard is an upgrade of the ANSI/TIA/EIA-422 standard. It uses either a two-wire or a four-wire system, both of which also need a ground wire. In a two-wire system, all devices, including the transmitter of the main controller, are normally in the receive mode. When one device transmits, all the other devices receive the transmitted signal and the response from the transmitter. Each device must wait until the transmission is finished before beginning a new signal of its own. In a four-wire system, all devices are connected to the transmitter and receiver of the main controller, and each device must respond only to commands addressed to it. The receiving device can respond to any request immediately, even while receiving the request.

ANSI/TIA/EIA-485 can support up to 32 devices at a distance of up to 4000 feet. Repeaters can be used to increase the number of devices supported and extend the signal. Each repeater can add 31 more devices, and the signal can travel an additional 4000 feet.

> **NOTE**   The signal ground is important to maintain data integrity. Data sent over power lines can be lost or damaged if the signal grounds of each device do not conform to one common ground. For example, ANSI/TIA/EIA-485 systems can fluctuate between -7 and $+12$ volts, but any deviation from this range can cause data to be damaged or lost, or can cause damage to the device's port.

European Automation Systems

Two standard automation systems are used in Europe. It is important to be familiar with them because at least one has already entered the market in North America.

Conson

The first system, *Conson*, was developed by a Danish company and uses a European-standard DIN rail for its commands. DIN rail modules are used to control lights and appliances directly through the home's breaker panel. The DIN rail-connection system is specified in two widths to handle different current levels. The slots within the DIN rail are sized to make both electrical and mechanical connections from devices to the signal rail. DIN rails are not yet common in North America, but are rapidly growing in popularity in Europe. The HTI can expect to see more DIN rails, especially as customers ask for higher-end European equipment and systems in their homes.

Dynalite

The second system, *Dynalite*, which is growing in popularity in North America, was developed in Britain. It uses ANSI/TIA/EIA-485 serial cable to carry commands between modules. ANSI/TIA/EIA-485 is a two-wire system capable of reliably carrying signals over distances approaching a mile in length.

Some Dynalite systems use ANSI/TIA/EIA-422 in conjunction with ANSI/TIA/EIA-485 as part of a total communications architecture. ANSI/TIA/EIA-422 is a four-wire standard that does not support as many devices in a circuit as does ANSI/TIA/EIA-485, but the former is useful when connecting two devices that must pass large amounts of control data with high reliability and speed. The additional cabling adds control flexibility, but makes these systems easier to install in new construction rather than as building upgrades.

NOTE   Another automation system, C-Bus, was developed in Australia and is becoming popular in North America. C-Bus uses Category 5 network cable as the medium for command and control.

Wireless

Wireless is another option for an automation system. For example, a homeowner pulling into the driveway might use a transmitter similar to a garage-door opener to turn on lights in the house.

Wireless communications systems use two media: infrared (IR) and radio frequency (RF). Of the two wireless-transmission methods, IR is more directional than RF. With an IR remote control device, the beam must be pointed toward the receiver for a command to be accepted. Additionally, IR does not have the capability to pass through walls, whereas RF often has that capability.

With both types of remote control, the wireless command must not be blocked when the user is in the most likely locations for control. For example, in a family room, an IR receiver should be clearly accessible from the sofa or chairs where family members are most likely to sit. RF receivers should not be placed where large metal structural members can act as a screen between the receivers and the user.

An in-home IR system can use a system of targets and emitters to transmit and relay signals to devices in different rooms. Targets are normally wired to emitters through the structured cabling system within the home. With this type of system, a device such as a DVD player in one room can be controlled by an IR remote in a different room.

Common wireless RF systems within the home include Bluetooth and 802.11B. *Bluetooth*, an RF technology operating in the frequency range of 2.4 GHz, is normally used in limited-distance applications. Common applications include cordless phones, wireless keyboards, and personal digital assistants (PDAs). The 802.11B system and its descendants are most commonly used in computer networking.

Bluetooth

LIGHTING SYSTEM CONTROLS

Lighting control is an important part of a home automation system. Unlike standard light switches and faders that normally control lights in the room in which the control device is located, each lighting automation station or control location can control lights anywhere in the house.

When selecting lighting automation controllers, function and architecture should be considered before appearance because most vendors produce controllers that can be covered by a broad range of faceplates.

Components of a Lighting Control System

The components of a lighting control system can be broken down into six categories:

- Central control unit
- Receivers
- Motion sensors
- User controls (switches and dimmers)
- Scene controls
- Lights

Each category is discussed in the following sections to ensure that the appropriate components are selected for each lighting control system installed.

Central Control Unit

The central control unit is a device that generates commands to one or more receivers (Figure 7–2 shows an example of a central control unit). A central control unit can control individual lights or can create lighting scenes that control lights as a group. In either case, the controller sends out commands that provide automatic on, off, dim, and brighten control for all lighting devices in the home. By adding the appropriate sensors, the lights can also be programmed to turn on or off when detectors sense dusk, dawn, or motion. In addition, the lights can be manually controlled from the central control unit.

Figure 7–2
Central Control Unit

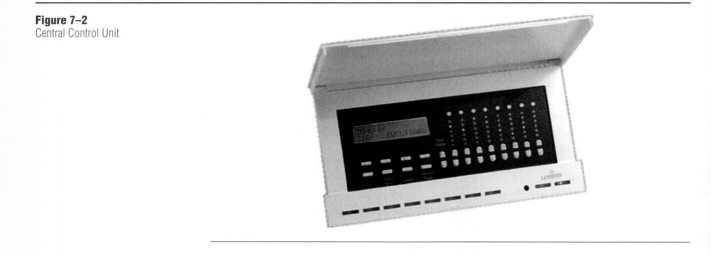

The central control unit may have the capability to be connected to a PC via a dedicated ANSI/EIA/TIA-232 port if complex lighting sequences need to be programmed.

When using X-10, the central control unit may be either a dedicated home automation terminal or a personal computer (PC) running special home control software.

NOTE   Use a surge protector to protect the central control unit from spikes or surges (see Figure 7–3).

Figure 7–3
Surge Protector

Receivers

Controllers send unique coded command signals to receivers (dimmers, switches, and receptacles) to perform various functions such as on, off, brighten, dim, all lights on, and all lights off. Infrared receivers are attached to the lighting devices and can be wall-mounted or ceiling-mounted. Figure 7–4 and Figure 7–5 illustrate receivers for dimmers and switches, respectively.

Figure 7–4
Dimmer Receiver

Figure 7–5
Switch Receiver

Additional devices may be needed between the central control unit or user controls and the receivers. These devices include the following:

- **Repeater:** Used to control devices that are on more than one circuit.

- **Amplifier:** Used if the signal strength decreases to a point that the receiver is having trouble recognizing the signal (see Figure 7–6).
- **Noise attenuator:** Used to prevent interference generated by lights (see Figure 7–7).
- **Noise filter:** Used to prevent interference generated by other appliances (see Figure 7–8).

Figure 7–6
Amplifier

Figure 7–7
Noise Attenuator

Figure 7–8
Noise Filter

The following items can weaken or distort the control signal on the power line:

- Distance
- Improperly installed home control devices

- Wireless intercoms
- Compact fluorescent lamps
- Halogen lamps
- Dimmers
- Some fax machines, VCRs, and televisions
- Printers

Motion Sensors

Motion-sensing detectors are used frequently in lighting systems. Motion detectors use either ultrasonic sound transceivers or infrared change sensors to determine whether an individual has entered the controlled area.

NOTE   An infrared system is often called a passive infrared (PIR) system because an infrared detector essentially monitors the low-frequency light spectrum and can detect body heat. An ultrasonic transducer, on the other hand, sends out pips of high-frequency audio and listens for changes in the echo, which can indicate movement.

Determining where to place motion detectors will mean surveying rooms and passageways for expected resident movement patterns and for potential obstacles to the reflected ultrasonic waves. Care must also be taken when installing a series of motion detectors so that they do not interfere with each other. Avoid facing the transmitter of one sensor directly into the receiver of another.

Almost any interference source can generate problems. The infrared beam of a remote control can cause problems, for instance, or the beam of an IR relay used to control appliances in a different room. And interference can be external as well as internal. For example, one of the most powerful sources of interference is the sun. Windows that admit direct sunlight must be considered and avoided for successful implementation of an IR system.

Pets pose a unique problem for motion sensors. A dog entering a room could cause the light to go on—or worse—trip a burglar alarm. Special motion sensors are available that ignore motion from small animals, and they are often rated in pounds. A 40-pound motion sensor ignores motion from an animal smaller than 40 pounds. Other very sophisticated systems avoid pet-generated false signals by using a combination of technologies to detect both IR and motion.

User Controls

User controls for lighting automation usually take the form of either switches or dimmers. Traditional lighting controls are wired to the AC mains and are controlled by touching a button, toggle, or knob on the control's front. Modern control systems allow users to control the switches or dimmers by sending them special signals. In the case of X-10 or CEBus, these signals travel over the power lines, although it is possible to have a control device that accepts IR or RF signals directly.

Many of these devices plug into an outlet and sit in the circuit between the outlet and the device to be controlled. Many homeowners prefer controllers that mount in the same outlet box as a standard switch or dimmer, which often gives a more finished look to the installation.

Some touch panels fit into standard wall plate openings, as seen in Figure 7–9. Single-unit touch panels can control the lighting instruments on a single electrical circuit or affect the lighting levels throughout the entire residence. Multiple lighting circuits can also be controlled with large multiswitch wall plates that have a significantly more industrial look, as shown in Figure 7–10.

Figure 7–9
Switch and Wall Plate

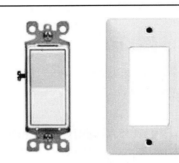

Figure 7–10
Multiswitch Wall Plate

Choosing between single-opening or multiopening wall plates is largely a matter of visual style and the homeowner's preference. Beyond the aesthetic issues, however, are the questions of which automation architecture the controls support. X-10, CEBus, C-Bus, Conson, and Dynalite are just five of several standards available for linking lighting controls to one another and to other home automation systems. It is not common for these systems to interoperate with each other.

Another common user control is a dimmer switch, which enables the homeowner to set different intensity levels of light. Dimmers can be used to manually adjust the intensity or set up with preset adjustments. They can also have X-10 controls built in. (Figure 7–11 shows dimmer controllers.)

Fluorescent lights may require the use of special switches and dimmers. Due to the ballast associated with fluorescent lights, however, dimmers and X-10 controllable switches may not function with fluorescent lights.

Figure 7–11
Dimmer Controllers

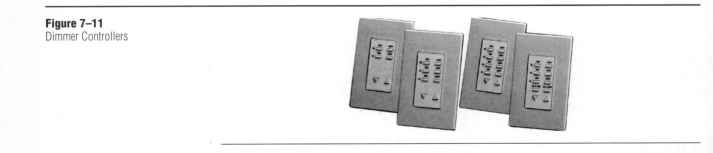

Scene Controls

Scene

A *scene* refers to the lighting throughout a designated area of the home. Each switch, button, or slider controls one or more scenes—preset combinations of lights and light levels that may include devices on many different circuits. A scene controller is used to program the various scenes in the house, as shown in Figure 7–12.

Figure 7–12
Scene Controller

Depending on the manufacture and control standard, a scene controller can send commands for many scenes, with several scene-capable dimmers per circuit. In the case of X-10, six scenes with 20 controllers per scene are possible. Multiple scene controllers can be used for multi-point scene access.

A scene dimmer, shown in Figure 7–13, can respond to lighting-level commands sent by the scene controller. Commands sent by a scene controller usually follow the X-10 protocol. Users can also locally adjust the scene with the dimmer without affecting programmed settings for scene lighting.

Figure 7–13
Scene Dimmer

Figure 7–14 is a scene lighting control diagram that shows how scene-capable dimmers with programmable scene controllers are used to control scenes. Each scene-controllable dimmer should have an address assigned to it. X-10 is a popular protocol for scene-controllable dimmers.

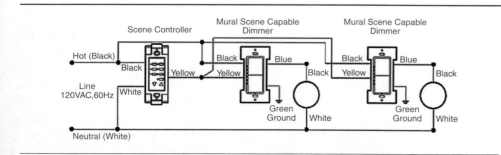

Figure 7–14
Scene Lighting Control Diagram

Lights

There are three types of lighting available:

- **Incandescent:** A lamp that provides light when a filament is heated to incandescence by an electrical current. Halogen is one example of an incandescent lamp.
- **Fluorescent:** An efficient lamp whose ultraviolet light is converted to white light by internal white phosphor powder.
- **Low voltage:** Commonly used for outdoor lighting applications.

System Layout

The layout of the various components can vary, depending on the type of communication system used.

Power line systems, such as X-10, offer a great deal of flexibility for positioning components throughout the residence. No control-specific wiring or cabling is required for an X-10 system. Each lighting control module is given a unique address; commands are then directed to individual addresses, either by direct input switches or under automated software control. User input can be delivered through wall switches or hand-held remote control units. Each input device is programmed by the HTI to send commands to a specific address or group of addresses. A complex series of commands to groups of lights and other electrical devices can be initiated by the central command unit under software control.

The HTI will often choose to place the central control computer in a wiring closet or some other out-of-the way area to avoid unintentional input or tampering with the unit. Individual lighting control modules can be wired into the fixture that holds the light bulb; lamps can be plugged into the wall outlet, with the lamp plugged into the power module.

Lighting-control systems that depend upon a dedicated cable for passing messages must be designed around the wiring plan for both electrical service and control. Whether the control mechanism is ANSI/TIA/EIA-485 or another architecture, most HTIs choose to *home run* the command wiring to a central wiring closet. Home run wiring means that cables from each lighting control module are run directly back to the central computer's location—there is no daisy-chain wiring or wiring from one lighting control module to another. Each lighting instrument is on its own dedicated circuit.

Home run

Locating User Controls

In most cases, controls for lighting automation are placed at the same locations used for nonautomated lighting controls—at the entrances to rooms and to the home. In addition to single-room or single-location controls, however, lighting-automation systems generally have one or more master panels from which lights in many rooms can be controlled. These master panels are typically located in rooms that are household activity centers, such as family rooms or kitchens, or in rooms with special security considerations, such as the master bedroom. Family activity patterns determine whether individual lighting controllers in other rooms activate the lights in a single room or in a group of rooms.

It's important to remember that lighting controllers are subject to local building codes. Make sure that all applicable codes are met when positioning residential lighting controls.

The location of a lighting control is first and primarily governed by national and local building and electrical codes. In general, these codes tend to specify that the overhead lighting for a room be controlled by a wall-mounted switch located near the entrance to the room at least. This important safety consideration cannot be overlooked in the push to automate the lighting features within a home.

HVAC CONTROLS

Heating, ventilation and air-conditioning (HVAC) controls are similar to lighting systems in the variety of controllers and architectures available. All major home automation architectures include control commands and remote activation modules that interface with furnaces, air conditioners, and heat pumps. In addition, HVAC controls can work on different zones in the home. A *zone* refers to the way a house is divided to efficiently cool and heat the various living spaces. Zones can be heated or cooled by multiple HVAC units, or else one larger unit can serve the entire dwelling. It is not uncommon for modern homes to have two furnaces and two air conditioners. Major HVAC system vendors have released control panels and sensors that meet popular home automation standards. HVAC is an area in which homeowners have become comfortable with automation through the energy-saving potential of programmable thermostats.

Dampers

In addition to controlling the temperature and timing of the heating and cooling unit, most automated systems can control *dampers*, which are mechanisms that open to allow air into a room and close to stop the flow of air. When selecting control systems for dampers, the HTI must know whether the dampers are normally closed and must have power applied to open; or are normally open and must have power applied to close. There are also some systems, such as the one developed by Carrier, that open and close dampers gradually in regular increments instead of the simple open/close choice provided by most systems. Most home automation architectures are capable of providing the commands to control either sort of air-flow system, but the HTI must be certain of the system type so that unintended control consequences can be avoided.

Damper motors are generally operated on 110 volts, but are controlled through low-voltage wiring and relays. Other actuators include solenoids, which are pistons that are electrically activated.

Stage and Relay Requirements

There are two types of heating and cooling systems:

- **Single-stage:** A single-stage furnace tends to be on or off in its operation. Its output provides a set number of British thermal units (Btus), the standard measurement for amount of heat generated.
- **Dual-stage:** A dual-stage furnace can fire at a set number of Btus with its first stage and can add additional Btus if necessary. Staging in furnaces can be done by lighting part of the burner as the first stage and lighting the rest for the second stage. Either dual gas valves or a variable gas valve can be used to control the flow of gas.

Similar to a two-stage furnace, two-stage air conditioners generally have two compressors. The first stage is used most often, and the second is turned on when extra cooling capacity is needed. Virtually all new construction is dual-stage because it offers greater energy efficiency. Most newer systems in retrofit situations are also dual-stage systems.

Each stage of the heating or cooling system requires a control relay, meaning that the controlling system might need up to four control relays for both heating and cooling. It is crucial to check with the heating and cooling system manufacturer to understand the relay and stage requirements of the HVAC system. HVAC compressors can be damaged if cycled between heating and cooling too quickly, or if cycled on and off repeatedly.

Dampers

Sensors and User Controls

HVAC temperature sensors in nonautomated residential settings tend to be located in central air flow locations such as hallways. Unfortunately, these are locations in which people tend to spend the least amount of time. Automated HVAC systems combine zoned air flow with programmable controllers to allow users to regulate the temperature in the rooms where people are located. Figure 7–15 shows an HVAC control.

Figure 7–15
HVAC Control

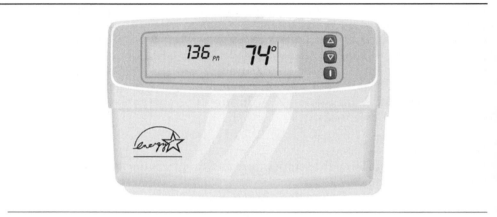

HVAC controls must be placed where they can accurately regulate the temperature of rooms and spaces where the homeowner spends time. Modern home automation systems, combined with zoned HVAC systems, allow temperature and humidity sensors to be placed in each major room of a house, with air flow directed to each room in order to maintain the desired temperature.

As with traditional thermostatic controllers, the key consideration for placing home automation HVAC controllers is making sure that they are located away from factors that affect their temperature readings. For example, controllers should not be placed in the air stream from a register because they would then register temperature changes faster than the rest of the room. Keeping them away from windows, fireplaces, and skylights is also crucial, as is making sure that they are several feet from large electronic devices such as projection televisions, computer monitors, and stereo amplifiers. When mounting a thermostatic controller, place it approximately five feet off the ground or at a height that is specified by local code. If the homeowner is physically challenged, mount the controller at a height that is accessible to that homeowner.

Central Control Unit

HVAC controllers typically communicate with whole-house automation controllers via one of the standard serial protocols: ANSI/TIA/EIA-485, RS-488, or ANSI/EIA/TIA-232. Of these, ANSI/EIA/TIA-232 is most commonly used for sending information between controllers or between a controller and a PC.

Ethernet-enabled controllers are increasing in popularity. As home networking expands, Ethernet-controlled thermostats will become just another node on the home network.

To ease installation and testing and to simplify system integration, the HVAC central automation control unit should be placed in the Structured Media Center with other home automation system-control units.

Wiring

HVAC systems communicate between thermostat, controller, furnace, air conditioner, or heat pump with a variety of four-, five-, and six-wire architectures that have been developed by HVAC manufacturers. Although the specifics of wiring color and function differ from manufacturer to manufacturer, the essential functions of the connections are the same, regardless of the vendor. Terminals on equipment are well labeled and easily understood on virtually all HVAC systems.

HVAC component configuration is similar to that used for lighting control; all thermostats and heat sensors are provided with a home run cable back to the main control unit. Low-voltage cables are run to the main HVAC control unit, which then interfaces to the heating and cooling unit controls, to air-handling controls, and to any additional interfaces needed to integrate the HVAC system into the whole-house automation plan.

The latest trend is for thermostats to be Ethernet-enabled so they can interface with the computer network and be controlled via a web browser.

WATER CONTROLS

Although automated lighting, HVAC, access, and entertainment systems are typically sold and installed on the basis of user convenience and enjoyment, water system automation is aimed primarily at energy savings and safety. Safety is enhanced by water system automation because the temperature maintained by the water heater can be held to 120 degrees, thus reducing the chances for accidental scalding. But when hotter water is required for sanitizing dishes, for example, commands can be sent to the water heater to raise the water temperature for the duration of the dishwashing cycle. Convenience is added to safety when cold, warm, or hot water—each temperature level predetermined and thermostatically controlled—is delivered instantly to each faucet or appliance with the touch of a single button.

Another advantage of water automation is its capability to detect and control leaks. In many areas, cold winter weather can cause pipes to leak or burst. Left unchecked, this flow can result in flooding and other water damage. The automation answer to this is to install leak monitors that warn the homeowner in the event of unusual water flows. Water control also can be used to purge certain sensitive systems, such as evaporative coolers or sprinkler systems, if the temperature falls below a certain point. Such systems often open a drain valve at their lowest elevation so that gravity allows water to drain.

Flow and Temperature

Water control systems are simpler in some respects than those used for lighting or HVAC in that there tend to be fewer components involved in the whole-house automation of water flow and water-based features. Water flow control modules tend to be incorporated into the valves that actuate the flow. This integration saves parts and allows the HTI to deal with integrating flow control units and (in some cases) water-heating appliances into a single system.

The most important configuration for the HTI is the output water temperature at the faucet. Output water temperature is a settable configuration item for virtually all automated control faucets, and the HTI should take care to keep the water temperature at the hot water heater below 120 degrees in any household in which there are young children or elderly family members. Bathroom water temperatures in such residences should be limited as well. One option for dish sterilization is to link dishwasher water demand to a hot water heater temperature. Under automated control, dishwashers can be set to run between midnight and 5:00 a.m., when small children are unlikely to be awake. At this time, the hot water heater can be programmed for a higher temperature of 140–150 degrees for dish sterilization and then reset to the safer temperatures for the remainder of the day.

CEBus and X-10 each has vendors who work to integrate flow and temperature into a single system. Both CEBus and X-10 are power-line systems, so no command-specific wiring is required.

Types of Controls

Water control comes in at least three types from a variety of vendors. The first type includes the controls used to regulate temperature and flow at sinks, lavatories, and tubs. These control units, shaped and located to duplicate nonautomated units, provide water at predetermined temperatures, protecting users and allowing hot water to be produced on demand. These control units can communicate with the hot water heater in order to ensure an adequate flow of hot water, but they are rarely tied to other automated systems.

A second type of water control, for spas and whirlpools, deals with water level, water temperature, and water and air flow within the tub. The control at the tub can communicate with the hot water heater to ensure that sufficient hot water is available to fill the tub. Unlike the standard lavatory and tub controls, whirlpool controls can be linked to overall home automation systems so that the spa can be brought to the proper temperature by remote control or placed into operation as lighting levels are changed as part of an overall scene. Figure 7–16 illustrates a typical pool and spa control.

Figure 7–16
Pool and Spa Control

A third category of water controls, fountain controls (such as those for spas and whirlpools) can be tied to larger household automation systems for the creation of overall scenes. They do not need to communicate with hot water heaters, however. Fountain controls are also frequently under the control of timer devices so that they are on when residents and guests are awake and in the home, and they are still at other times.

Location of Controls

Control placement for water systems is much more limited than that for lighting or environmental controls because water controls tend to be placed at plumbing locations. One notable exception to this is that controls for residential irrigation systems can be located away from external hose bibs, faucets, and irrigation outlets.

Spas and whirlpool tubs are another exception to the rule of placing water controls only near the point of water application. In order to save energy, systems can be designed to hold water in a spa or whirlpool tub at temperatures below ideal use levels. When a user wants to relax in the spa, systems raising the temperature of the water can be activated from hand-held wireless controls or from control panels located away from the tub or spa. Alternatively, water temperature controls can be placed on timer activation so that water temperatures are lowered during times when all residents are sleeping or away from the home and raised when spa use is more likely. In all these cases, remote spa and whirlpool controls are located in areas where family members are likely to be.

Control panels for spas and whirlpools must be located safely away from the water because water-filled tubs tend to be excellent conductors of electricity. Allowing a user to touch a wired control while in the water, or even standing on a wet deck, is asking for trouble. Similar standards require that wired telephones be placed several meters from the water's edge.

A typical location for dedicated whirlpool and spa control is near the lighting control for the room in which the spa or whirlpool is found. One of the advantages of whole-house automation systems is that control for spas and whirlpools can be incorporated into the panels used for lighting, access, and accessory control. With these panels, whether located in walls or as part of a wireless hand-held remote control, the user can bring spa water up to temperature and begin the water and air flow when in another room.

Decorative fountains, both inside and outside the house, are also subject to automated water control. Fountains differ from most other residential water features in requiring a pump for water motion, rather than relying on the pressure of the water entering the house. The automated control of fountains tends to focus on electric pump control rather than valve actuation. Fountain controls are best located within sight of the fountain so that the operator can be sure that there is sufficient water in the fountain for circulation and that proper pump operation has begun. Fountain operation can also be included in household scene controls in which lighting, window covers, fireplace flame height, and other details are controlled in sequence to create a total home atmosphere with the push of a single button.

NOTE External water controls must be located in consideration of local weather, with freeze conditions taken into account.

One necessary addition found in standard plumbing is the electrical service required for the control of electronics and valve actuation. The HTI should check local electrical codes for requirements of electrical service in close proximity to running water and areas of high humidity, and plan for *ground fault interrupters (GFIs)* as needed. A GFI is an electrical device that prevents electrical shock by stopping the flow of current when a flow of current to ground is sensed, rather than tripping when current flow exceeds a certain amperage, as is the case with conventional circuit breakers. Electrical codes such as the National Electrical Code (NEC) usually mandate installation of GFI outlets and breakers in areas where water is normally present (typically spas, swimming pools, bathrooms, and kitchens). Section 680 of the NEC requires a GFI to be installed on all spas.

Ground fault interrupters (GFIs)

Automation Systems

Water systems typically communicate between controller and controlled devices through simple two- or four-wire systems. Whether the water feature is a faucet, a whirlpool, or a lawn sprinkler, the possible control features are relatively simple compared to systems such as building access.

Wireless remote control is a popular option for water features as it is for many other home automation systems. Wireless remote control systems have an advantage in convenience for users and a safety advantage by working on low voltages that are harmless to humans even if the remote comes into contact with water.

The safety advantage is a reason why many manufacturers of electronically controlled faucets use battery power for electronics and valve activation. The HTI should review manufacturer literature for anticipated battery life on these systems and include this information in system documentation left with customers.

When water features are brought into the whole-house automation system, ANSI/EIA/TIA-232 is the most common mechanism for communication. Serial communication, in general, and ANSI/EIA/TIA-232 allow for inexpensive components and the capability to communicate over the distances frequently encountered in residential automation.

 **ACCESS CONTROLS**

The physical aspects of access systems are primarily concerned with allowing entry. Access systems require entry-activation devices such as electrically operated strike plates for gates, doors, and overhead garage door openers. These entry-control devices can be activated by standard keys, touch pads, magnetic identification keys, or remote control units attached to the user's key chain.

Control Stations

Control stations for building access can be grouped into three categories of activity: those that control locking mechanisms, those that control and display monitoring information, and those that control multiple functions. The three types of controls are typically located in three different locations. *Locking controls* are usually located at the points of entrance to the building or in vehicles that users drive to the buildings. *Monitoring controls* are generally located in rooms in which users like to get information, such as bedrooms, kitchens, and home offices. *Function controls* are usually located at the distribution panel, where the system is integrated with other home automation systems.

Systems use several types of sensors to detect entry. *Infrared motion detectors* activate if a change in infrared energy is detected. *Contact sensors* generally have a time delay after a sentry contact is opened.

Access control panels have three different location types, each fulfilling a special need of the homeowner. All must allow control of and access to information from the major systems. First, *point-of-entry controls* enable the access system to be armed or disarmed when residents are leaving or returning to the home. Second, *mobile system controls* enable residents to arm and disarm the system from vehicles. Third, *bedroom and living area controls* are positioned to allow the homeowner to check out visitors, control the status of the various access subsystems, and arm the perimeter system before retiring for the night.

Configuring Controls

Configuring access controls is important for the safety of the customer. Prime considerations revolve around the usage patterns of the various members of the family at different times of the day and night. Young parents with infant children almost certainly have households with different traffic and usage patterns than a family with several teenage children or a family with elderly grandparents in residence.

Gates and doors can be programmed to open, close, or lock at certain times. They can also be opened or closed if an approaching object is detected.

Typically, programming allows a gate to close when exiting, but requires that a code be entered on a keypad before it opens. A homeowner can enter the code remotely, from a car for instance, by sending a signal to open the gate.

Wired Options

The majority of traditional access components communicate with the control system through simple two-wire cables. Two wires are sufficient when the only communication necessary is "open" or "closed" for most input components. (This same communication can take place via radio frequency wireless links. These RF links have become popular, especially in retrofit or renovation installations. RF is discussed in the next section.)

An exception to the two-wire communications rule comes with the closed-circuit television (CCTV) cameras that are an active part of many access systems. The following wired transport mechanisms are common for CCTV signals:

- Coaxial video cable
- Category 5 networking cable

Coaxial video cable (coax) is the same cable type used to carry cable or satellite television signals from a set-top box to a television. The most common types of coax used in residential applications are RG-6 and RG-59 (also known as Series 6 and Series 59 cable, respectively). Of these two types, RG-6 has less loss of signal and should be used in most situations to help ensure better picture quality. Most manufacturers now recommend the use of a quad-shield RG-6, frequently abbreviated as RG6Q. The additional shielding prevents signal egress and noise entry.

Category 5 (CAT 5) networking cable is the same as the unshielded twisted-pair cable used for Ethernet computer networking. It is used for most installations in which the camera is a network device controlled by a computer. A converter may be required to change the signal into a digital state for better transmission and improved noise resistance.

Wireless Option

A third transport mechanism for access control is RF networking, often making use of the 802.11b wireless Ethernet standard (although lower priced systems may attempt to use an analog scheme). Wireless networking has the same advantages in this application as in others; it doesn't require cables to be pulled in retrofit situations or from difficult locations. On the other hand, there is a significant concern that HTIs and their customers should keep in mind: There is currently no law against receiving wireless network signals from someone else's network. This means that a hacker or neighborhood snoop can go "war driving"—traveling through the neighborhood with a laptop computer and wireless networking adapter card—and see the image from the camera. This is likely to be of little concern if the camera is used for front door external security. If it is used to monitor children's rooms, the concern is much greater. HTIs should take special care to make sure that the network is secure against unwanted interception and intrusion.

 **MISCELLANEOUS SYSTEM CONTROLS**

In addition to lighting, water, and home access systems, convenience and safety systems such as television lifts, fireplace ignition, and ceiling fans can be placed under automated control.

If separate control wiring is required for these systems, home run cables to a central wiring closet in which control computers and systems are located are preferable. Some systems, such as the X-10, do not require any additional wiring, but it is imperative to avoid electrical circuits with heavy motors and lighting ballast loads if using X-10.

Television Lift Systems

Television lift systems are popular in bedroom locations because television sets can be placed out of view in furniture and brought into view when turned on. Controls for lift systems are typically tied to wireless remote controls for the television and not activated through stationary, wired controls. When there are stationary controls, consider installing them in locations built into bedside tables for convenience when users want to access the television from bed.

Fireplace Controls

Fireplace controls have gained in popularity as gas-fueled ventless and direct-vent fireplaces have brought decorative fireplaces into bedrooms, bathrooms, home offices, and other rooms throughout the house. Hand-held remote control units allow users to control flame intensity without leaving the comfort of their favorite chair. Fireplace controls can also be made part of an overall home automation panel and set as part of overall "scenes" defined by the HTI as combinations of lighting levels, fireplace intensity, water fountain activity, and shade extension. When considered as part of overall room control, it makes sense to incorporate fireplace controls into full-house control panels located at room entrances or (more often) in wireless remote control units.

Controls for fireplaces are popular additions to living rooms, bedrooms, bathrooms, and other living areas in which fireplaces are now installed. Control units are typically hand-held remotes similar in size to television remotes, and similar in complexity to those used for ceiling fans.

Ceiling Fan Controls

Ceiling fan controls can be either wall-mounted or hand-held. Wall-mounted controls are generally sized to be mounted in standard outlet boxes and covered with wall plates matching the room's décor. Hand-held fan remote units tend to be simple, with few controls beyond those for the fan speed and any lighting that can be built into the fan.

Window Covering Controls

Window shade and skylight cover controls are popular additions to home theaters, media rooms, and houses, particularly in cities that receive a lot of annual sunshine. Window shades can be controlled individually or as part of lighting scenes in preparation for watching movies or other programming on television. Shades are also used to help maintain a cool temperature in the house by shutting out the hot sun. Many newer homes include windows high in the ceiling for light and use electronic control for operation.

Window shade actuators that operate on low voltage can be installed and wired by the HTI or low-voltage wiring contractor. Window coverings that are mounted outdoors often have wind-speed sensors associated with them. If winds rise to a certain level, shades are withdrawn to protect them from damage.

Types of Controls

The miscellaneous systems within a total home automation system—fireplace, window, and fan controls, among others—are usually controlled through a combination of wall-mounted switches and keypads, and hand-held remote control devices. For example, controls for fans can begin with simple wall switches for fan speed and light activation.

Individual remote control units and wall switches can work well if there is a single fan or if the homeowner has not decided on full-house automation. For more involved or complete home automation situations, the sheer number of controls can become unwieldy if they are not combined into a single automation system. Although no single manufacturer provides all the miscellaneous systems, there are architectures such as X-10 that provide mechanisms for controlling the disparate products that go into making the complete home environment.

Central Automation System

All controls for miscellaneous systems in the home can be controlled through a central home automation system. The central home automation system has the advantages of eliminating the many hand-held remotes that can accumulate when separate systems are employed for each control, and allowing the systems to be combined into a unified whole for creating scenes of distinct atmosphere for different user needs.

Configuration of Systems

Configuration of miscellaneous systems in the automated home often means coordinating their activities with lighting levels, water activity, or access control. Proper sequencing of controlled actions is crucial to avoid unintended consequences from combinations such as a fireplace radiating varying levels of infrared energy into a room monitored by an infrared security sensor or a ceiling fan beginning to revolve in a room with an ultrasonic motion detector.

Much of the configuration work on these systems is integration work to make sure that the command set required by one device will be understood by another or that the total automation system—X-10, for example—is properly interpreted and acted upon by all components.

As the number of systems placed under total control grows, it becomes more important for the HTI to ensure that one component is not affected by unintentional interference from another. Although CEBus and X-10 are standards, there still may be differences in the way standards are implemented by various vendors, with the consequence that the status report of one module is interpreted as a command by another. Care must also be taken when electric motors or lights with ballasts—such as fluorescent or halogen lamps—are placed in a circuit with controlled devices. These motors and lights can generate electrical interference on a circuit, thereby masking legitimate control signals.

Wiring

Most residential accessory systems use two-wire or four-wire serial cables as the communication media between controller and controlled devices. Most accessory system vendors also make wireless remote control devices available for their systems. In virtually every case, these accessory systems are not candidates for home run cable installations in which the controller is located in a central wiring closet. Accessory systems tend to locate controllers alongside or in close proximity to the device that is being controlled.

Two- or four-wire serial cables are the most common control transport media. For example, there are systems for controlling motorized window shades that use both two-wire serial and RJ-11 four-wire modular cable for communications transports. Despite this, many accessory systems also include ANSI/EIA/TIA-232 or ANSI/TIA/EIA-485 serial ports for communications with a home automation control computer. The HTI should keep in mind that ANSI/EIA/TIA-232 describes signal names and voltage levels. Most implementations make use of D-shell connectors and 9 or 25 pins, whereas some do not. Only a few device automation systems used in residential construction use more than three or four signal pins, regardless of the connection media implemented.

TESTING AND DOCUMENTATION

After the controls have been installed, the system should be tested. The HTI will need to troubleshoot any problems encountered. After testing is complete, the HTI should document the entire system with its subsystems. This section of the chapter explains the different types of diagrams needed for this documentation. At this stage of the implementation, the HTI should also design user training to ensure that no important functions or corrective actions are missed. Finally, a complete documentation package should be presented to the owner during user training.

Test the System

After the controls have been installed, test each individual component. Before applying power to systems, double-check the wiring. Improperly wired components can be damaged if power is applied. Next, test each subsystem (access, water, fireplace, and so on). Finally, when all the subsystems are in working order, perform tests on the entire system. If any of the tests fail, troubleshoot the problem. After the problem is solved, test the systems and components again.

Each HTI develops a methodology that is appropriate for the components used on a project, but there are a number of considerations to be taken into account for a testing process.

First, make a list of every component that is part of the system. Include a check box for each test that is to be performed on the component, beginning with the simplest: Have the necessary connections been properly made? Some tests can be made at the time of component installation; other tests must wait until complete subsystems or the entire system is in place. Making the list and checking items off as they are completed help to ensure that no step is missed. The completed list, presented to the homeowner at the end of the installation as part of the documentation package, will provide a sense of confidence in the installation.

Test every function of each component. In the case of lights, make sure that every level, from off to fully powered, is tested. Access system components should be tested to make sure that each sensor and contact provides the expected input in the expected manner when it is activated. This level of testing is especially important for HVAC controls, for which incorrect control input can cause damage to compressors and dampers; and for access systems, in which incorrect input can threaten the safety of the residents or result in monetary damages from false alarms to public safety agencies.

Test subsystems before testing the entire system. Make sure that all lights and all light combinations are functioning properly before beginning the test of complete home automation scenes involving multiple subsystems.

Test components, subsystems, and total systems for the results of both incorrect and correct input. What happens to the system if a toddler begins to play with a touchpad? What happens to lighting scenes if someone uses a wired control while someone else is using a wireless remote control? Remember to test for the unusual circumstance to ensure confidence that the system will continue to function properly, even when the users don't use it correctly.

> **NOTE** All symbols used on plans should be defined and labeled, and all device locations should be carefully noted along with any installation instructions and special needs.

Troubleshoot the System

Troubleshooting home automation is an important skill. The HTI can save time and money by looking for common and simple problems first. For example, many problems are caused by plugs not being fully inserted into receptacles. This is why it is important to test each component first. You can narrow down the problem immediately if you use this method.

Common things to check:

- Verify that lamp plugs are firmly inserted into receptacles.
- Verify that lamps have bulbs installed, are turned on, and are in working order.
- Verify that loads fall within 60–1000 W range.
- Verify that the transmitter and receiver are set to the same address code.
- Verify that an incandescent dimming module is not controlling a fixture containing a built-in dimmer.
- Verify that the system amplifier power lamp is on.
- Verify that no faulty connections exist between aluminum and copper conductors.
- Verify that a neighbor who shares a utility transformer does not have a DHC or other automation system that is sending command signals on the same house code.
- Verify that a neighbor who shares a utility transformer is not generating electronic noise that can interfere with DHC.

When troubleshooting equipment that uses AC power to control devices, the following are common reasons for problems:

- Improper installation
- Improper or insufficient signal coupling
- Misapplication
- Electrical noise on AC lines
- Improper grounding
- Electrical system alterations
- Faulty hardware
- Use of incompatible surge protection

When troubleshooting, keep a careful record of each change made. Record the reason for the change, the person making the change, the time and date of the change, and the complete details of the change. These records, although tedious to make, provide valuable information for maintenance, troubleshooting, and upgrading.

Documentation

Complete documentation is an important part of any complete home automation installation. One of the key pieces of a complete documentation kit is a schematic diagram of the system components and their relationship to each other. A schematic diagram provides a visual representation of how all the components are connected (see Figure 7–17).

Figure 7–17
Schematic Diagram

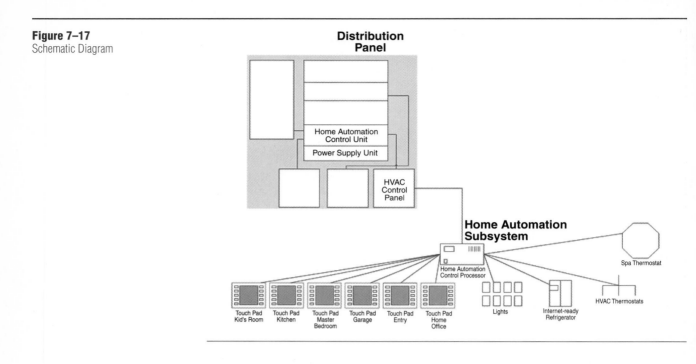

Figure 7–18 shows another necessary diagram: the signal flow diagram to document how the devices communicate.

Figure 7–18
Signal Flow Diagram

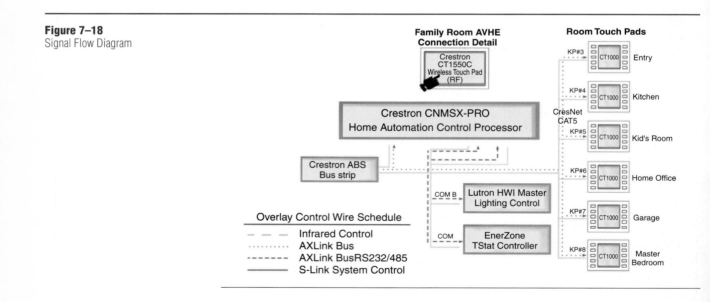

A complete plan for the layout of all equipment used in an automation system is required, whether the system encompasses one function, such as lighting, or total home automation. Figure 7–19 shows a typical floor plan. The latter is necessary for any local code-mandated inspections, and it serves as a guide for the HTI and other installers. In addition, the floor plan becomes a valuable piece of the system documentation for the homeowner.

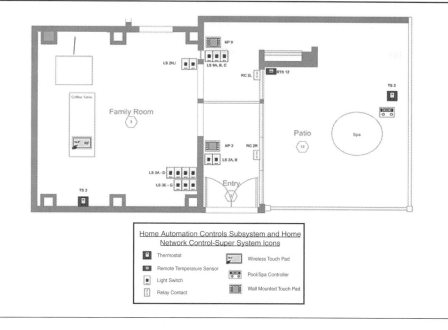

Figure 7–19
Floor Plan, Showing Devices

Any complete plan must begin with a description of all symbols used in the plan. The symbols should adhere to accepted architectural standards to the greatest extent possible. Where accepted architectural standards do not exist for a particular component, the symbol used should be clear and unique. Do not depend solely on color to make a symbol unique because copies in black and white are common.

Following the description of symbols, the plan must include accurate scale renderings of the involved space with the placement of each component noted. On the equipment placement pages, all relevant details must be shown, including distance from the floor, ceiling, or other architectural details; orientation of the device; and the nature of the connection required to wiring, plumbing, electrical service, or any other building subsystem.

The equipment placement diagram or blueprint is the place to note any special requirements for installing a piece of equipment, whether the requirement is structural, electrical, or something else. Remember that the goal of any plan is to allow the installation of the system by someone other than the HTI—if details are needed, make sure they're included.

NOTE   Although there are a number of software tools available for creating diagrams, Microsoft's Visio has become a *de facto* standard in the commercial networking industry.

Customer Training

The goal of the automated system is to make life easier for the homeowner by ensuring that the home will respond as designed. Training should cover the use and simple maintenance of the system to make the homeowner comfortable with its operation.

The HTI should cover the proper operation of the system. Improper operation can be covered as well. When a system is new, the input can be given out of sequence on one or more occasions. The user should know what to expect when this happens and how to correct the situation.

Look at worst-case scenarios. What happens when the power fails? Do systems automatically reboot or is user intervention required? What happens if one system becomes "confused"? Can the user force a reset of a malfunctioning system without causing additional malfunctions?

Provide complete documentation for the system to the homeowner. This includes an explanation of all commands and functions, all necessary user intervention to correct malfunctions, block diagrams, blueprints, manufacturer-supplied literature on individual components, and your business card. Presentation of the system documentation is an important milestone in the project, and an opportunity for the HTI to present a professional, positive impression to the customer. Providing the documentation in a professional binder, slipcase, portfolio, or other appropriate container will help build the user's confidence in the performance of the HTI.

SUMMARY

In this chapter, you learned about the automation products that control environmental systems such as lighting, heating and cooling, access and security, and water. The following are the major issues identified in this chapter:

- Lighting, HVAC, water, building access, and accessory systems must be designed and installed with user safety, security, and convenience in mind.
- Lighting and HVAC controls must conform to national and local building codes upon installation.
- X-10 and CEBus, the primary automation systems used in North America, use existing power lines to control devices.
- The central X-10 control unit may be either a dedicated home automation computer or a personal computer (PC) running special home control software.

- Home run wiring, in which cables from each control module are run directly back to the central computer's location, is the preferred wiring scheme because it allows for scalability and makes troubleshooting easier.
- Lighting scenes are created using scene controllers and scene dimmers.
- Dampers are used to control the flow of air in HVAC systems.
- Most water controls handle the flow and temperature of water.
- Gates, doors, and garage doors can be controlled both manually and automatically.
- Other home products—such as fireplaces, fans, and window shades—can be wired so that they can be controlled by the home automation system.

GLOSSARY

ANSI/EIA/TIA-232   A standard that supports up to 32 devices. Most serial ports follow this standard. This standard was previously known as RS-232, and is still commonly referenced by that designation. The current version, ANSI/EIA/TIA-232-F, was published in 1997.

ANSI/TIA/EIA-422   A four-wire standard that does not support as many devices in a circuit as does ANSI/TIA/EIA-485, but can pass large amounts of control data with high reliability and speed. This standard was previously known as RS-422, and is still commonly referenced by that designation. The current version, ANSI/TIA/EIA-422-B, was published in 2000.

ANSI/TIA/EIA-485   A two-wire system capable of reliably carrying signals over distances approaching a mile in length. This standard was previously known as RS-485, and is still commonly referenced by that designation. The current version, ANSI/TIA/EIA-485-A, was published in 1998.

Bluetooth   An RF technology operating in the frequency range of 2.4 GHz that is normally used in limited distance applications.

Common applications include cordless phones, wireless keyboards, and PDAs.

CEBus   Automation system that uses modulated household AC for communications between controller and controlled lights and appliances that have CEBus capability built in.

Conson   European automation system developed by a Danish company.

Damper   An adjustable plate that regulates the flow of air in the ducts.

Dynalite   Automation system developed in Britain that is growing in popularity in North America.

Ground fault interrupter (GFI)   An electrical device that is installed to prevent electrical shock to the users of electrical appliances.

Home run   Wiring running from a control module directly back to the central computer or control center; that is, a dedicated circuit.

Scene   Preset combination of lights and light levels.
Structured Media Center (SMC)   The central point of the integrated home network where all subsystems converge. Also known as a distribution device (DD) or distribution panel.

X-10   Automation system that uses a set of commands transmitted by modulating the 60 Hz alternating current in North American household electricity.

CHECK YOUR UNDERSTANDING

1. What is the combination of lights turned on in a space at a given time called?
   a. Program
   b. Scene
   c. Situation
   d. Mood

2. HVAC controls divide living spaces into areas known as:
   a. rooms.
   b. halls.
   c. zones.
   d. scenes.

3. For greatest safety, a hot water heater should normally be set to no more than what temperature?
   a. 150 degrees
   b. 100 degrees
   c. 212 degrees
   d. 120 degrees

4. Which automation protocol uses household AC wiring to send commands?
   a. X-10 and CEBus
   b. Dynalite
   c. ANSI/TIA/EIA-485
   d. Conson

5. Which automation protocol is referred to as plug-and-play?
   a. Dynalite
   b. ANSI/TIA/EIA-485
   c. Conson
   d. CEBus

6. Which standard is used to extend serial communications up to 4000 feet between two devices?
   a. ANSI/TIA/EIA-232
   b. ANSI/TIA/EIA-422
   c. ANSI/TIA/EIA-485
   d. ANSI/TIA/EIA-400

7. HVAC compressors can be damaged if switched too quickly between:
   a. on and off.
   b. night and day.
   c. heating and cooling.
   d. automatic and manual.

8. Which of the following should HVAC temperature sensors be placed away from?
   a. Fireplaces
   b. Heat registers
   c. Skylights
   d. All of the above

9. Water control devices that are used specifically for spas must have what safety device in the household AC circuit?
   a. Circuit breaker
   b. Ground fault interrupter
   c. Fuse
   d. Dead short

10. What is the term for routing all control cables to a single control location?
   a. Cable channeling
   b. Structural wiring
   c. Home run
   d. Daisy chain

11. Which of the following is a 2-wire serial communications standard?
   a. ANSI/EIA/TIA-232
   b. X-10
   c. ANSI/TIA/EIA-422
   d. ANSI/TIA/EIA-485

12. All symbols used on plans or blueprints must be:
   a. blue and white.
   b. ANSI standard.
   c. at least one inch in size.
   d. labeled and defined.

13. Which of the following is the standard used to connect a central control to a PC when complex lighting sequences are needed?
   a. ANSI/EIA/TIA-232
   b. X-10
   c. ANSI/TIA/EIA-422
   d. ANSI/TIA/EIA-485

14. What is used to control devices that are on more than one circuit?
   a. Amplifier
   b. Repeater
   c. Attenuator
   d. Noise filter

15. What is the first step in testing the system?
   a. Check sensor contacts
   b. Check power source
   c. Test the subsystems
   d. Make a list of every component

16. What is the difference between a single-stage and a dual-stage furnace?

17. What are the three different location types for access control panels?

18. What are the three transport mechanisms common for CCTV signals?

19. What are the items to consider when selecting an automation system?

20. What is the goal of an automated system?

Low-Voltage Media

 OBJECTIVES

Upon completion of this chapter, the HTI will be able to accomplish the following tasks:

- Identify types of signals
- Troubleshoot common signal problems
- Identify categories, gauge, and grades of wire
- Describe types of low-voltage wiring
- Use color-coding schemes
- Select wiring equipment
- Determine equipment layout
- Test and certify the installation

 **INTRODUCTION**

Low-voltage media is the cabling that supports such services as telephone, data (computer network), security systems, doorbells, entertainment systems, and other services within the home. This type of wiring conveys lower voltages than the higher voltages carried in AC power wiring, which supplies power to outlets and appliances. Low-voltage wiring is the core of residential structured cabling infrastructure. This chapter discusses the types of residential wiring that will be used as well as other aspects of low-voltage wiring that need to be taken into consideration in the design and installation of residential structured cabling.

SIGNAL TRANSMISSION FUNDAMENTALS

Transmission can be described as the movement of information in the form of signals from one point to another via a medium. The signals consist of electrical or optical patterns that are transmitted from one connected device to another. More specifically, the signals represent digital bit streams that move down the media as a series of voltages or as light pulses.

The three common methods of signal transmission are

- **Electrical signals:** Transmission is achieved by representing data as electrical pulses on copper wire.
- **Optical signals:** Transmission is achieved by converting the electrical signals into light pulses.
- **Wireless signals:** Transmission is achieved by using infrared, microwave, or radio waves through free space.

Today, the most commonly used media for signal transmission are copper conductors and glass optical fiber. However, the choice of a specific transmission medium is usually influenced by both economics and technical considerations that include the following:

- Distance over which signals must be transmitted
- Size of the network
- Type of services to be provided (voice, data, or video)
- Bandwidth requirements and the transmission path

It is important for the HTI to understand the fundamentals of electronic transmission in order to better appreciate the impact that an improper residential cabling infrastructure installation can have on media-transmission characteristics. The sections that follow further explore the various signal transmission methods.

Electrical Signals

One of the first things that must be understood about electronic transmission is how a current *propagates* or travels through a wire. Signals flow as the result of complex actions of very small particles called electrons and the charges that they carry. Most electronic devices send and receive information through electrical pulses. To make this possible, wires or cables that carry these signals provide pathways that interconnect the devices.

The HTI must be exceptionally careful when installing copper cable. A cable that is improperly terminated may not be able to transfer all of the energy from the wire to the next circuit. Also, a wire that is positioned too close to sources of electrical noise or radio noise may act as an antenna. This introduces stray signals that may impede the information going down the wire in the desired path. Improperly handling or terminating a cable can also affect its *impedance,* which is the measurement of how much resistance the flow of electrons encounters. *Impedance mismatches* prevent the transfer of energy from one piece of cable to another and cause reflections, where energy bounces back up the wire instead of being propagated downstream. When stray electrical signals enter the cable, they interfere with the desired signal by distorting the amplitude and/or frequency of the data, thus confusing the receiving devices. This interference can require the network to retransmit the data, which slows down the network.

Optical Signals

There are two ways of transmitting information using light as a signaling method:

- **Optical fiber:** Optical signals propagate down glass threads called fiber optics. Fiber optics is one of the most popular methods of data transmission. It is the core of long-distance data transmission.

- **Optical free-space:** Optical free-space communications can take the form of microwave or other point-to-point transmission systems.

> **NOTE**   A particular brand of optical fiber cabling, plastic optical fiber (POF), has recently become an alternative patch-cabling medium that is used for transmission of signals primarily between audio devices in the residential environment. Although it is a non-standard signal-transmission medium, POF has the same immunities from external interference as its glass counterparts EMI and RFI, as you will learn later in this chapter. This is an especially important factor in the typically noisy residential electrical environment.

Wireless Signals

Wireless is a term used to describe communications that do not require interconnecting wires of cables, in which electromagnetic waves carry signals. Wireless transmission works by sending high-frequency waves into free space. *Free space* in this context relates to transmission without cables in which high-frequency waves are sent into the air. Waves propagate through free space until they arrive at their intended destination and are converted back into electrical impulses so that the destination device can read the data. Wireless can be one or more links in a network composed of other types of media.

A common application of wireless data communication is for mobile use such as cellular telephones; satellites used for transmitting television programs; walkie-talkies used to dispatch emergency services; and telemetry signals from remote space probes, space shuttles, and space stations. A common application of wireless data communication in residential networking is wireless LANs. In some cases, such as retrofits, it is easier to set up a wireless system that serves an entire floor or portion of a floor and equip each user with an individual receiver and transmitter than it is to wire every room into the network. Wireless reaches areas of the home that are hard to reach by wired networking.

There are three distinct methods of wireless transmission:

- **Light wave:** *Infrared* is light waves that are lower in frequency than the unaided human eye can see. This is rarely used over long distances because it is not particularly reliable, and the two devices must be in line of sight of each other. In the home, infrared is used for remote controls for televisions, VCRs, DVD players, and stereo systems. There are also infrared applications in computer networking, although techniques involving radio waves are more popular.
- **Radio and microwave:** A very effective and practical system of wireless communication uses radio waves or microwaves for signal transmission. Common examples of wireless equipment and applications in a residential setting include the following:
    Cordless telephones that connect the handset to the base station via radio frequencies
    Home entertainment systems in which the VCR and TV sometimes signal each other on an unused TV channel
    Garage door openers that work with radio frequencies
    Baby monitors that use a transmitter and receiver for a limited range
    Cordless computer peripherals
    Wireless LANs used for business and residential networks
- **Acoustic (ultrasonic):** Some monitoring devices like intrusion alarms employ acoustic sound waves at frequencies above the range of human hearing. These are sometimes classified as wireless.

Signal Distortion and Degradation

A signal sent over a wire must reach the other end of the wire, while still resembling as closely as possible the original signal that entered the wire. If anything should happen to the signal along the way that could reduce its strength or change its pattern, the received signal may be unintelligible and the receiving device unable to interpret it. Several factors can cause the degradation of a signal. Some of these are the result of problems in the cable itself (physical properties that degrade signals). Other factors are caused by internal or external noise that interferes with the signal as it travels down the wire. These factors include the following:

- **Resistance:** The property of a wire that tends to oppose the flow of electrons or electrical signals. Resistance decreases the strength (energy loss) of a signal; when this happens, it is called attenuation.
- **Noise:** It is another cause of signal degradation and distortion. Noise problems in cabling are caused by electrical signals, radio or microwave waves, or signals on adjacent wires.
- **Inductive reactance, capacitive reactance, and impedance:** Other reasons for distortion and degradation are the result of the wire's shape, its position with respect to other wires and to ground, and the frequencies the wire may carry. These effects are called inductive reactance and capacitive reactance. *Inductive reactance* is the "reaction" of the inductor to the changing value of alternating current. The opposition that a capacitor offers to alternating current is inversely proportional to frequency and to capacitance. This opposition is called *capacitive reactance*. Together, these two factors contribute to a special form of resistance called impedance.

The HTI can do little to change the effects of resistance and reactance; they are characteristics of the wire itself. Other problems, however, are directly within the HTI's control. Only careful wire termination, layout, and routing will keep the wiring reliable, resistant to noise, and resistant to degradation.

Attenuation

Attenuation is a general term that refers to any reduction in the strength of a signal. Attenuation occurs with any type of signal, whether digital or analog. Sometimes called *loss*, attenuation is a natural consequence of signal transmission over long distances. Attenuation is usually measured in units called decibels (dB), and this measurement is proportional to the cable length. Figure 8–1 shows a cable with a kink and a cable with a snag.

Figure 8–1
A Cable that Has a Kink (Above) or Snag (Below) Can Cause Attenuation by as Much as 2.5 dB

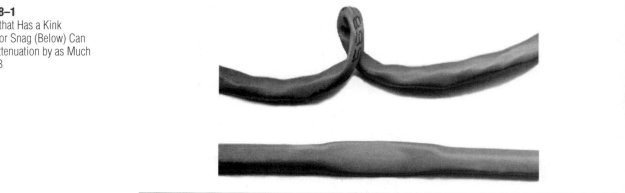

Attenuation can affect a network because it limits the length of network cabling over which signals can be sent. If the signal travels for a long distance, the bits can become indiscernible by the time they reach the destination. When it is necessary to transmit signals over long distances via cable, one or more repeaters can be inserted along the length of the cable. The repeaters boost the signal strength to overcome attenuation. This greatly increases the maximum attainable range of communication cable.

Attenuation also occurs in fiber-optic transmission. The light pulses that travel through optical fiber are subjected to energy loss. Attenuation can occur in fiber-optic cables as a result of the following:

- Impurities within the fiberglass that scatter some of the light energy as the light pulse travels down the fiber
- Macrobends (from poor installation practices) in the fiber strands
- Microbends (manufacture defects) in the fiber strands

The HTI must understand that the most unexpected attenuation in wiring results from improper installation practices. Maintaining adequate bend radius using the proper wire for applications and conducting the work according to industry standards and manufacturer guidelines are necessary to provide a fully functional and reliable transmission system for the home network.

Noise

Noise is unwanted electrical, electromagnetic, or radio frequency energy that can degrade and distort the quality of signals and communications of all types. Noise occurs in both digital and analog systems. In the case of analog signals, the signal becomes noisy and scratchy. For instance, a telephone conversation can be interrupted by noises in the background. In digital systems, bits can sometimes run into one another, becoming indistinguishable to the destination computer. This translates into network instability.

Signals that are external to the cables, such as emissions of radio transmitters and radars, or electrical fields that emanate from electric motors and fluorescent light fixtures, can interfere with signals that are traveling down cables. This noise is called *electromagnetic interference (EMI)* when it is caused by electrical sources; it is called *radio frequency interference (RFI)* when it is caused by radio, radar, or microwave sources. Noise can also be conducted from alternating current (AC) power lines and lightning. AC power line noise as well as EMI and RFI are discussed later in this chapter. Figure 8–2 shows a noise attenuator for a lighting management system.

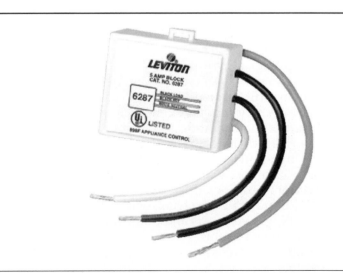

Figure 8–2
Noise Attenuator for Lighting Management Systems that Prevents Interference Generated by Fixture Components

Each wire in a cable can act like an antenna. When this happens, the wire actually absorbs electrical signals from other wires in the cable and from electrical sources outside the cable. If the resulting electrical noise reaches a high enough level, it can become difficult for the computer to discriminate the noise from the data signal, just as background noise disturbs a telephone conversation. Figure 8–3 shows a plug-in noise filter for lighting management systems that prevents interference generated by electrical appliances.

Figure 8–3
Plug-In Noise Filter for Lighting Management Systems Prevents Interference Generated by Electrical Appliances

The biggest source of signal distortion for copper wire occurs when signals inadvertently pass out of one wire within the cable and onto an adjacent one—a phenomenon called crosstalk.

Crosstalk

Crosstalk is the unwanted transfer of signal from one or more circuits to other circuits. Crosstalk typically occurs between conductor pairs in close proximity or between adjacent cables. During the manufacture of cables, crosstalk is decreased by wire-pair twists, cable lay, shielding, and physical separation. Although such measures are helpful in mitigating crosstalk in networking media, the most effective prevention methods still involve proper cable installation practices—ensuring careful terminations, maintaining the twisting of pairs, and respecting existing industry wiring standards.

EMI and RFI

There are external sources of electrical impulses that can interfere with the quality of electrical signals on wires. The main culprits here include lighting (in particular, fluorescent lighting), electrical motors, radio and television systems, cordless telephones, and so on. These types of interference are referred to as electromagnetic interference (EMI) and radio frequency interference (RFI).

In general, any device or system that generates an electromagnetic field has the potential to disrupt the operation of electronic components, devices, and systems within its vicinity. This is known as electromagnetic interference, or EMI. Moderate- or high-powered wireless transmitters can produce EMI fields strong enough to upset the operation of electronic equipment nearby. Ensuring that all electronic equipment is operated with a good electrical ground on the system can minimize problems with EMI. Specialized line filters can also be installed in power cords and interconnecting cables to reduce the EMI susceptibility of some systems.

Transmitters operating in the radio spectrum generate RFI interference. This includes radio transmitters, television stations, cordless telephones, radar television stations, and portable two-way radios or any radio-transmitting device.

There are a number of measures that the HTI can take to mitigate the impact of EMI and RFI on the residential cabling infrastructure. First, EMI and RFI can be minimized by the selection of appropriate cabling components and by using proper installation methods. Shielded cable, for example, may be more resistant to chronic, high-intensity exposure to interference than unshielded cable. Second, grounding is essential in reducing noise effects and interference in residential cabling installations. If the environment is particularly prone to electrical noise, optical fiber may be a better choice of media. Because fiber conveys light pulses rather than electrical signals, it is immune to EMI and RFI.

Additionally, copper cable manufacturers commonly use the technique of cancellation to mitigate the effects of EMI and RFI. Cancellation is achieved by twisting wire pairs together within the cable and by carefully controlling the manufacture of the cable to ensure precise physical tolerances. The way this technique works is simple: Signals flow in the twisted pairs in opposite fields, and the electromagnetic field generated by one wire of a pair is used to cancel out the field in the other wire of the pair, thereby shielding the wires from any external interference.

Finally, the HTI should select electronic and communications products for home integration that comply with FCC's electromagnetic compatibility (EMC) directive. The EMC is a set of guidelines set by the FCC for the manufacture of electronic and communications equipment. It establishes acceptable limits for electromagnetic emissions from electronic devices and also specifies the amount of EMI such devices should be able to tolerate.

AC Power Line and Reference Ground Noise

AC power line and reference ground noise can be the source of many problems in residential networking. The noise in AC power lines can create problems in homes, schools, and offices. Electricity is carried to home appliances and other networking devices by wires that are concealed in walls, floors, raceways, and ceilings. Consequently, AC power line noise is all around inside a building. If not properly addressed, it can cause all kinds of problems for a network.

A signal reference ground is the point tied to an earth ground that serves as the zero or no voltage point against which signals are compared. Ideally, the signal reference ground should be completely isolated from the electrical ground. Isolation keeps AC power line leakage and voltage spikes off the signal reference ground. The chassis of a computing device serves as the signal reference ground and as the AC power line ground, however. Because there is a link between the signal reference ground and the power line ground, problems with the power line ground can lead to interference with the data system. Such interference can be difficult to detect and trace. Usually, it occurs because the electrical contractors and installers did not install the appropriate length of the neutral and ground wires that lead to each electrical outlet. Unfortunately, when these wires are too long, they can act as antennas for electrical noise. It is this noise that interferes with the digital signals a computer must be able to recognize and process.

AC power line noise coming from a nearby video monitor or hard disk drive can be enough to create errors in a computer system. It does this by changing the shape and voltage level of the desired signals. This problem can be further compounded when a computer has a poor ground connection.

Other Losses Occurring in Signal Transmission

Other losses can occur in residential network cabling. These include the following:

- **Intrinsic loss or dispersion:** Occurs in fiber-optic cabling. Optical signals are susceptible to losses when small particles are trapped inside glass.

When the light pulses hit the particles, the light scatters; as a result, the signal can be lost. This is sometimes called intrinsic loss or dispersion. Dispersion can also occur because of misalignment of fiber-optic connectors.

- **Coupling losses:** A coupling is a connector for two wires. Because the signal has to pass from one wire to another, the signal can suffer if the coupling is not done properly. In most cases, poor connections cause reflected electromagnetic energy. Fiber-optic connectors can suffer a coupling loss when contaminants or improper bonding decreases the amount of light that can enter or leave a connection.
- **Wireless losses:** Wireless losses can be caused by water vapor, weather, or suspended particles in the air. The signal can scatter or be absorbed. In fact, signal loss of a laser beam is used in measuring the amount of pollutants in air. Overcoming these environmental losses is difficult to accomplish. Hence, the wireless system should be built with these losses in mind.

WIRE AND CABLE SPECIFICATIONS

In this section, you learn about the various types of wiring used for the integrated home network and each subsystem. Each wiring type has different specifications and purposes. Understanding the advantages and limitations of each will help you choose the best wiring type for each project. This section of the chapter focuses on wiring size, ANSI/EIA/TIA grades, UL ratings, Anixter levels, and EIA/TIA categories that apply to all wires and cables.

Wire Gauge (AWG)

Low-voltage wiring varies by type and (to a lesser extent by wire size, known as gauge). In the United States, wire gauges are rated using the American Wire Gauge (AWG) system. The higher the AWG number, the thinner the wire, as shown in Table 8–1. Thus, a 24-gauge wire is thinner than a 22-gauge wire. The TIA/EIA Standard 570-A proposes different grades of residential cabling, depending on the services the cable will be providing within a residence. Speaker wire, for example, is usually 14- or 16-gauge copper wire. Thermostat cable has four pairs of copper wire with a rating of 16 AWG. Computer cabling typically has wire gauges around 23 to 26 AWG. Understanding such ratings is important for picking the right media type for home networking and home control subsystems.

Table 8–1
The Diameter of the Wire Decreases as the AWG Value Increases

AWG	Diameter (mm)
10	2.59
12	2.05
14	1.63
16	1.29
18	1.02
20	0.81
22	0.64
24	0.51
26	0.40

UL Cable Ratings

Underwriters Laboratories (UL) certifies all twisted-pair cables for data networking, including unshielded twisted-pair (UTP) and shielded twisted-pair (STP). Network cable falls into one of the following UL safety standards:

- **UL 444:** Standard for Safety for Communications Cable
- **UL 13:** Standard for Safety for Power-Limited Circuit Cable

The UL LAN Certification Program addresses not only safety, but also performance. UL has adopted the ANSI/TIA/EIA-568-B.2 performance standard with minor variations:

- The UL program deals with both STP and UTP cable, whereas the ANSI/TIA/EIA-568-B.2 standard focuses on UTP cable.
- The UL markings range from Level I through Level V, whereas the ANSI/TIA/EIA-568-B.2 has Category 1 through Category 6.
- The UL level markings deal with performance and safety, so the products that merit UL level markings also meet the appropriate NEC as well as the TIA/EIA standard for a specific category.

As part of its certification process, UL evaluates samples of cable; after granting a UL listing, it then conducts follow-up tests and inspections. This independent testing and follow-through make UL markings valuable symbols to buyers. Companies whose cables earn these UL markings display them on the outer jacket of the cable as "UL Level I," "LVL I," or "LEV I," for example.

Plenum Grade Cable

The *plenum,* a separate space in a building that provides air circulation for the HVAC system, is typically provided in the space between the structural ceiling and a drop-down ceiling or under a raised floor. This prevents smoke from being sucked through the air circulation system and returned into rooms that are not on fire. In buildings with computer installations, the plenum is frequently used to house networking and telecommunication cables. Because ordinary cable can produce toxic fumes in the event of fire, special cable is required in the plenum areas.

Plenum cable is made of fire-resistant materials such as Teflon, so it is more expensive than ordinary cabling. In the event of fire, the outer material is slower to burn and produces less smoke than ordinary cabling. The smoke is also non-toxic. Typical plenum cable sizes in the United States are 22 and 24 AWG. Both twisted-pair and coaxial cable are made in plenum cable versions. Installers must choose plenum cable when running cable through heating and air-conditioning ducts.

Grades of Residential Cabling

The ANSI/EIA/TIA-570-A standard defines two grades of residential premises telecommunications cabling for a home: Grade 1 and Grade 2. The use of these grades is based on the nature of services to be supported within the home, and the grading is intended to assist in the selection of the appropriated cabling infrastructure for each home integration project. Grade 1 cabling meets the minimum requirements for telecommunications services, whereas Grade 2 is for more advanced services.

Grade 1 For each cabled location within the home, Grade 1 provides a generic cabling system that meets the minimum requirements for telecommunications services. Typical applications are telephone, fax, modem, and CATV services including cable modem and DSL services.

The minimum cabling requirements for Grade 1 are

- One (1) 100-ohm, four-pair UTP (unshielded twisted-pair) cable and connectors that meet or exceed the requirements for Category 3 (Category 5 recommended)
- One (1) 75-ohm, coaxial cable Series 6 (quad-shield recommended)

Grade 2 For each cabled location in the home, Grade 2 provides a generic cabling system that meets the requirements for basic, advanced, and multimedia telecommunications services. Typical applications consist of telephone, fax, modem, CATV, cable modem, and DSL services.

The minimum cabling requirements for Grade 2 cabling include the following:

- Two (2) 100-ohm, four-pair UTP cable and connectors that meet or exceed the requirements for Category 5 (Category 5e recommended)
- Two (2) 75-ohm coaxial cable Series 6 (quad-shield recommended)
- Two (2) multimode fiber-optical cables are optional

As seen in the preceding description, Grade 2 residential cabling specifies twisted-pair cable, coaxial cable, and (optionally) fiber-optic cable. Additionally, it provides for both current and developing telecommunications services.

Low-Voltage Cabling Recommendations Following are some useful recommendations for wiring for both Grade 1 and Grade 2 cabling:

- Multiple outlets should be installed in each of the living areas of the home to allow for a flexible furniture configuration and avoid the use of long patch cords.
- Each living area (room in which voice/data/video [VDV] applications may be used) should be cabled with a minimum of one outlet. Examples of living areas include kitchens, bedrooms, living/family rooms, and den/library/home offices.
- All the cabling should be home run directly from each outlet to the distribution device (DD).

Anixter Levels

Early attempts to improve the quality of wire resulted in a multitude of different specifications because there were no set standards. Because this caused compatibility problems, Anixter, a major cable distributor, established a system of levels that described the characteristics of various cable grades and what they could be used for, as follows:

- **Level 1:** Used for plain old telephone service (POTS)
- **Level 2:** Used for low-speed computer terminals and networks
- **Level 3:** Used for Ethernet operating at 10 Mbps and for Token Ring

This system served well for a while, but the EIA and TIA industry organizations established standards for building wiring in the early 1990s.

EIA/TIA Category Classification

One of the first moves by EIA and TIA in the early 1990s was to rename the Anixter system levels as Category 1, Category 2, and Category 3 cables. Category 4 and Category 5 standards were then developed. *Category 5*, commonly referred to in the industry as CAT 5, has been the standard cabling used in Ethernet networks. An improved version of Category 5, *Category 5e*, is the current standard. Category 5e contains more twists.

- **Category 1:** Once used for telephones, it is still found in older homes. Other uses include ISDN and doorbell wiring. This two-pair cable has a maximum

Category 5

Category 5e

data rate of 1 Mbps. Unlike Category 3 and 5 wiring, Category 1 wiring is not recognized as part of the ANSI/TIA/EIA-570-A standard; therefore, its transmission characteristics are not specified. The same is the case with Category 2 and Category 4, which have no further coverage in this course.

- **Category 3:** Generally used for telephone wiring, Category 3 has a maximum data rate of up to 16 Mbps. This designation applies to 100-ohm UTP cables that have four pairs of copper wire.
- **Category 5/5e:** Category 5 wiring, often referred to as CAT 5, is the most commonly installed media. This designation applies to 100-ohm UTP cables with four pairs of copper wire whose transmission characteristics are specified. Category 5 has a maximum data rate of up to 100 Mbps.

Category 3 twisted-pair is the minimum grade of residential cabling recommended in the ANSI/EIA/TIA-570-A standard for all new installations. It is also now the minimum specified by the FCC for new home constructions.

Newer cabling standards are being developed to meet the demands for more speed and bandwidth. More and more applications require high bandwidth, such as streaming video, video conferencing, and IP telephony (telephone conversations over network wiring using the Internet Protocol). As a result, Category 6 cabling has been ratified. Category 6 has even more twists than Category 5e as well as a plastic separator to keep the pairs of wires apart.

TYPES OF LOW-VOLTAGE MEDIA

When working with copper wire, an installer will deal with two basic types of cabling: twisted-pair and coaxial.

Twisted-pair cable consists of pairs of insulated copper wires that are twisted together and then housed in a protective sheath. Twisted-pair cable comes in a variety of types:

Twisted-pair cable

- Unshielded twisted-pair (UTP)
- Shielded twisted-pair (STP)
- Screened twisted-pair (ScTP)
- High-count twisted-pair (25 or more pairs)

Coaxial cable has a center conductor that is wrapped in insulation and covered with one or more layers of braid and foil. The cable is then encased in a durable outer jacket.

In this section, the structure, use, advantages, and disadvantages of each type of cabling are explored.

Unshielded Twisted-Pair (UTP)

Unshielded twisted-pair (UTP) cable is used in a variety of networks. The most common number is Category 3, 5, 5e, and 6 with four pairs of wires. This type of cable relies solely on the cancellation effect, produced by the twisted-wire pairs, that limits signal degradation caused by electromagnetic interference (EMI) and radio frequency interference (RFI). These terms were explained earlier in this chapter. Figure 8–4 illustrates UTP cable.

Figure 8–4
UTP Cable

UTP cable must follow precise specifications about the number of twists that are permitted. More twists result in fewer problems with signal degradation, but they can be costly. For instance, using more twists means that the cable uses more copper, making it more expensive. It also means that the path the data travels is longer. In addition, because the number of twists varies between pairs in the same cable, it can cause *delay skew*, which refers to signal distortion and errors due in part to the longer cable. That is, data bits traveling in different pairs that arrive at slightly different intervals cause signal distortion and errors.

The simplicity of UTP can be misleading. Unlike other cable packages that have strong central wires or shielding elements, UTP has relatively thin wires (22–24 AWG). Even tight twisting does not prevent the pairs from separating or from bunching together, which can lead to degraded performance.

Although there are several categories of UTP cables, Category 5e is the standard when installing networks in North America. European networking systems use less UTP and more shielded twisted-pair (STP) cabling. The other common type of UTP cabling encountered in the United States is Category 3, which was very popular for telephone wiring until recently.

Shielded Twisted-Pair (STP)

There are electrical environments in which EMI and RFI are so strong that shielding is needed to make communication possible. A cable that carries its own screening with extra shielding is needed. As shown in Figure 8–5, shielded twisted-pair (STP) is copper wire with a screening layer to protect the wires from outside interference. In STP, the individual pairs are wrapped in a shield and then all four pairs are wrapped in another shield. STP is very popular in Europe, where historic structures prohibit the use of ground rods, making installation more difficult.

Figure 8–5
STP Cable

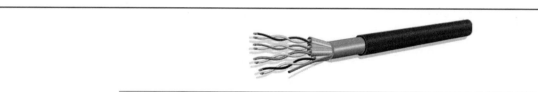

Two varieties of STP are available:

- **100-ohm STP:** Simple STP, or 100-ohm STP, is used primarily for Ethernet networks. Like UTP, it has a 100-ohm impedance. The shield is not part of the data circuit, so it is grounded at only one end, typically at the telecommunications room or hub. In this way, the shield acts as a protective sleeve that intercepts EMI and RFI and then sinks it to ground before it can affect the data signals in the cable. 100-ohm STP improves the EMI and RFI resistance of twisted-pair wire without a significant gain in size or weight.
- **150-ohm STP:** The most common form of STP, introduced by IBM and associated with the IEEE 802.5 Token Ring networking architecture, is known as 150-ohm STP because of its 150-ohm impedance. The shielding on 150-ohm STP is not part of the signal path, but is grounded at both ends. 150-ohm STP can carry data using very fast signaling with little distortion, but all the shielding causes signal loss that increases the need for greater spacing—that is, more insulation between the wire pairs and the shield. The increased amount of insulation and the large amount of shielding increase the size, weight, and cost of the cable. 150-ohm STP, with an outside diameter of approximately 0.98 mm, quickly fills wiring ducts.

Screened Twisted-Pair (ScTP)

Screened twisted-pair (ScTP) has only a single shield, usually a layer of foil that protects all the pairs of the cable, as illustrated in Figure 8–6. Each pair is not wrapped in a shield like STP. Because it does not have extra shielding around the pairs, ScTP is less expensive, has lighter weight, is smaller in diameter, and is easier to ground than STP.

Figure 8–6
ScTP Cable

High Pair-Count Twisted-Pair

Also known as multipair cables, high pair-count telecommunications cable (more than four pairs) comes in many configurations, up to 4200 pairs of wires. Typical configurations include 5, 6, 8, 12, 25, 50, 75, 100, 150, 200, 300, 400, 600, 900, 1200, 1500, 1800, 2100, 2400, 2700, 3000, 3600, and 4200 pairs. High pair-count twisted-pair cabling is used for backbone service in multi-dwelling unit environments.

In most cases, running cables with more than 900 pairs may cause difficulties, especially between buildings. Most countries have wiring codes that require the use of surge-protecting fuses at the wires' point of entrance. This is necessary because external sources of electricity, such as lightning strikes or downed power lines, can present hazardous voltages to the building, its occupants, and its equipment. The installation of 900 or more protection devices would occupy a great deal of space and take time, which increases labor costs. As an alternative, a single fiber-optic cable can carry just as much traffic, is not affected by lightning or other induced voltages, and does not need to be grounded.

Color-Coding Scheme for Twisted-Pair

When working with twisted-pair cables, it is important to be able to identify individual wires within the cable. The wires can be identified by the colored insulation. The color-coding scheme is the standard that all manufacturers follow to ensure that the correct wires are terminated.

Four-pair cable is the most common, but it is not the only configuration. In multiple-pair cables, the color code adheres to the same general color scheme. In fact, the four-pair wiring scheme is a subset of this larger color-coding system. Pairs 1–4 of a four-pair cable use the same color system that is used in a 25-pair cable.

For each pair of wires, one wire is the tip and the other is the ring, as shown in Figure 8–7. The origins of the tip and ring were discussed in Chapter 4.

Figure 8–7
Tip (Green) and Ring (Red) on Telephone Wire

The primary color in a four-pair cable such as Category 5e is the tip, which is usually white with a tracer or stripe that is the same color as the pair's solid color wire, which is the ring. On some cables, the tip wires are opaque and the rings are translucent. Designating the white wire as the tip is a matter of convention. It is rooted in the fact that this wire is the first wire of each pair to be punched down to a punch or termination block, particularly when working with larger pair-count cables, such as cables with 25 or more pairs.

> **NOTE**   Manufacturers have taken liberties with the marking system. Some manufacturers do not put a colored stripe on the tip wire. Others do not use a solid color, but rather a translucent shade that tints the color of the wire that shows through it. In still other cases, the manufacturer distinguishes wires by using periodic splotches of the companion color.

When indicating colors for a pair, the tip colors come first because that is the order in which the cables are punched down on a termination block. Ring colors are just the opposite; that is, if pair 22 has tip colors violet-orange, orange-violet would be the ring colors for that pair.

Cables that have more than 25 pairs group the wires in 25-pair units, each of which is wrapped with colored tape to form binder groups. The binder groups follow the same color code as the wires of each pair; that is, the first binder is blue/white, the second binder is orange/white, and so on.

Coaxial Cable

Coaxial cable

Coaxial cable was once the cable of choice for all networking applications, and it is still commonly found in older networks. Today, its role is more likely to be in cable TV or video (surveillance and security) applications. 75-ohm coaxial Series 6 (RG-6) cable is the accepted cable type for cabling the outlets for video/TV signals. Coaxial cabling is used in conjunction with coaxial patch cords, equipment patch cords, and distribution device cords.

Coaxial cable gets its name from the way it is constructed. When the cable is viewed from either end, the copper conductor is at the core and is encircled by a layer of insulation, then a layer of shielding, and finally an outer jacket layer, as shown in Figure 8–8. All these layers are built around the central axis (the copper wire); hence the cable is called coaxial, or simply coax.

Figure 8–8
RG-6 Quad Shield
Construction

Coax cable was historically designated as RG (radio frequency) cable. It is now referred to as Series-X cable. The "X" relates to the construction of the cable based on several factors, including the center conductor diameter, outer cable braid percentage of coverage, dielectric composition, impedance, and whether the center conductor is solid or stranded. Common examples are Series 6 coax (RG-6) cable and Series 11 coax (RG-11) cable.

Coax cable generally provides a much higher bandwidth and more efficient protection against EMI than twisted-pair cable. A dual-shielded coax cable, for example, uses aluminum foil for the high-frequency shield and aluminum braid for the low-frequency shielding. Coax cables can also have additional shielding. A tri-shield cable has two foils and a braid, whereas a quad-shield cable employs two foils and two braids. A quad-shield coaxial cable construction is recommended because it provides the best shielding characteristics.

Residential structured cabling systems (RSCS), discussed later in this chapter, should employ Series 6 coaxial cabling for transmission of digital video signals from the distribution device to the outlets. Series 6 coaxial cable has the following features:

- More bandwidth than twisted-pair cable, which is needed for digital video signal transmission
- Coated/aluminum foil shield over the dielectric, which is needed to shield against high frequencies
- Braided shield over the coated/aluminum foil shield, which is needed to shield against low frequencies
- Solid-center conductor
- Characteristic impedance of 75 ohms

Coaxial cabling requires that all the outlets and equipment serviced by coaxial cabling use coaxial patch cords as well as equipment cords. Cords that are provided with consumer audio/video equipment or devices are only good for use as equipment cords. Additionally, there is a distinction between patch cords and equipment cords.

- **Patch cords:** Used to connect outlet cabling to the distribution system. They must be the same cable type used in cabling the outlets, but they can be either factory- or field-terminated with F connectors.
- **Equipment cords:** Used to connect equipment to the outlets. They may be Series 59 or Series 6 coaxial cabling, but should be factory-terminated with an F connector.

ANSI/EIA/TIA Series Cables

Devices that connect to computers require cables other than those already discussed. The most common computer cable is ANSI/TIA/EIA-232. Recently, this specification has been improved so that products deliver data at faster speeds with less interference. For the gauge ratings of the following cables, see the manufacturer's documentation.

ANSI/TIA/EIA-232   ANSI/TIA/EIA-232 supports a point-to-point connection over a single copper cable, so only two devices can be connected. Historically, almost every computer has one ANSI/TIA/EIA-232 serial port, although it is becoming a common trend with laptop manufacturers to not include these serial ports on new computers. Modems, monitors, mouse devices, and serial printers are designed to connect to the ANSI/TIA/EIA-232 port on computers. This cable is also used to connect modems to telephones. There are two types of connectors: a 25-pin connector (DB-25) and a nine-pin connector (DB-9).

There are several limitations to using ANSI/TIA/EIA-232. The most significant limitation is that the signaling rates are limited to only 20 Kbps. The rate is limited because the potential for crosstalk between signal lines in the cable is very high.

ANSI/TIA/EIA-422 and ANSI/TIA/EIA-423   Although ANSI/TIA/EIA-232 is still the most common standard for serial communication, ANSI/TIA/EIA-422 and ANSI/TIA/EIA-423 are expected to replace it in the future. Both support higher data rates and have greater immunity to electrical interference. In addition, ANSI/TIA/EIA-422 supports multipoint connections, whereas ANSI/TIA/EIA-423 supports only point-to-point connections. The new standards are backward-compatible so that ANSI/TIA/EIA-232 devices can connect to an ANSI/TIA/EIA-422 port.

ANSI/TIA/EIA-485 ANSI/TIA/EIA-485 supports multipoint communications and several types of connectors, including DB-9 and DB-37. ANSI/TIA/EIA-485 is similar to ANSI/TIA/EIA-422, but can support 32 nodes per line because it uses lower-impedance drivers and receivers. It satisfies the need for faster rates, up to 50 Mbps; and longer distance capabilities, up to 1200 meters. By using balanced differential mode, which uses a matched pair of wires to carry complementary signals, noise is limited because the noise is the same for each wire.

FireWire (IEEE 1394)

FireWire is Apple's version of IEEE 1394 High-Performance Serial Bus. Other companies have similar products called Lynx and iLink. This is a high-speed serial connection for devices that connect to computers. FireWire provides a single plug-and-socket connection on which up to 63 devices can be attached with data transfer speeds up to 400 Mbps. The serial bus functions as if devices were in slots within the computer, sharing a common memory space. A 64-bit device address allows a great deal of flexibility in configuring devices in chains and trees from a single socket.

FireWire is extremely fast, flexible, reliable, and hot-pluggable. Its speed makes it ideal for devices that need to transfer high levels of data in real time, such as video devices. Like USB, IEEE 1394 supports both plug-and-play and hot-plugging, and also provides power to peripheral devices. The main difference between IEEE 1394 and USB is that IEEE 1394 supports faster data-transfer rates and is more expensive. For this reason, it is primarily used for devices that require large throughputs, such as video cameras.

FireWire and other IEEE 1394 implementations provide the following:

- A simple common plug-in serial connector on the back of the computer and on many different types of peripheral devices
- A single 1394 port that can be used to connect up to 63 external devices
- A thin serial cable rather than the thicker parallel cable you now use to your printer, for example
- A very high-speed rate of data transfer that will accommodate multimedia applications (100 and 200 megabits per second today with much higher rates later)
- The capability for any device (for example, a video camera) to be plugged in while the computer is running
- The capability to chain devices together in a number of different ways without terminators or complicated setup requirements

The standard requires that a device be within 4.5 meters of the bus socket. Up to 16 devices can be connected in a single chain, each with the 4.5-meter maximum (before signal attenuation begins to occur), so theoretically you could have a device as far as 72 meters away from the computer.

There are two levels of interface in IEEE 1394: one for the backplane bus within the computer and another for the point-to-point interface between device and computer on the serial cable. A simple bridge connects the two environments. The backplane bus supports 12.5, 25, or 50 megabits per second data transfer. The cable interface supports 100, 200, or 400 megabits per second. Each of these interfaces can handle any of the possible data rates and change from one to another as needed.

IEEE 1394 implementations are expected to replace today's serial and parallel interfaces, including Centronics parallel, ANSI/TIA/EIA-232, and SCSI. The first products to be introduced with FireWire include digital cameras, digital video disks (DVDs), digital video tapes, digital camcorders, and music systems.

USB

The Universal Serial Bus (USB) provides the same hot-plugging and plug-and-play capabilities as the IEEE 1394 standard at a less expensive price, but its data speed is much slower—only 12 Mbps. This speed still accommodates a wide range of devices, including MPEG video devices, data gloves, and digitizers. Used for computer peripheral devices such as joysticks, mouse devices, keyboards, scanners, and printers, USB has become the standard—replacing serial and parallel ports. A single USB port can be used to connect up to 127 peripheral devices. Like FireWire, the new device can be added to your computer without having to turn the computer off. The USB standard was developed by Compaq, IBM, DEC, Intel, Microsoft, NEC, and Northern Telecom.

Optical Fiber

Fiber-optic cable is cabling medium that allows the transmission of light impulses instead of electrical signals. The concept of light transmission in optical fiber is centered on two key elements of the cable: the core and the cladding. The *core* is the innermost part of the fiber through which light pulses are guided. The core is surrounded by the *cladding,* which keeps the light in the center of the fiber. Light traveling along the core that strikes the cladding at a glancing angle is reflected and confined in the core by internal reflection. Figure 8–9 illustrates the construction of optical fiber cable.

Fiber-optic cable

Figure 8–9
Optical Fiber Cable

The core and cladding of fiber-optic cabling are very small, and the diameter varies. The core and cladding diameter or size is measured in micrometers (µm) or simply microns. A micron is one one-millionth of a meter (0.000001 meter or 10^{-6} meter). The most commonly used fiber cables today are the 50/125 µm and 62.5/125 µm multimode fiber and the 8-9/125 µm single-mode fiber. The number to the left is the core diameter, and the number to the right is the cladding diameter. For example, in the 50/125 µm fiber, the core size is 50 µm and the cladding is 125 µm.

Fiber-optic cabling may be described as single-mode, multimode, or composite. Each has specific characteristics and applications.

Single-Mode Fiber   Single-mode fiber is used in outside plant fiber-optic systems. Outside plant is the fiber-cabling system outside the home; for example, telco and CATV systems (fiber-to-the-home). Single-mode fiber has the following characteristics:

- Depending on the manufacturer, it can have 8 µm to 9 µm diameter
- Typically uses a laser light source
- Supports wavelengths of 1310 nanometers (nm, which is equal to 10^{-9} meter)
- Can transmit signals for distances up to 3000 meters (9840 feet) for RSCS; longer distances possible with service providers

Multimode Fiber   Multimode fiber is used inside the home for applications such as LANs and security. Multimode fiber has the following characteristics:

- Has a 50 µm or 62.5 µm core and a 125 µm cladding diameter
- Has a distance limitation of 2000 meters (6560 feet) for structured cabling systems

- Typically uses a light emitting diode (LED) as a light source
- Supports wavelengths of 850 nm or 1,300 nm
- Is the most commonly used fiber-optic cabling type for backbone and horizontal runs within buildings and campus environments

Composite Fiber   Composite (hybrid) fiber is a combination of single-mode and multimode cable.

Other Characteristics of Fiber-Optic Cable   Many other characteristics are associated with fiber-optic cabling, the knowledge of which is useful in the selection of appropriate cabling products for residential system installations.

- **Simplex:** Provides one fiber in the cable. It is commonly used for unidirectional (single-direction) transmission of signals (for example, video cameras).
- **Duplex:** Comprised of two fibers in the cable. It is commonly used for bidirectional (two-way simultaneous) transmission of data signals.
- **Multiple fibers:** Cabling that has multiple fiber strands. Typical constructions include 6, 12, and 24 strands. It is commonly used for backbone cabling.
- **Tight-buffered:** Fiber cable that protects the fiber strands by supporting each strand of glass in a tight-buffered coating, which increases the strand's diameter to 900 μm. It can be single-mode or multimode fiber, but is primarily used inside buildings.
- **Loose-tube:** Unlike the tight-buffered type used primarily inside the building, loose-tube fiber cabling is used primarily outdoors. When specified for outdoors, it contains a water block to prevent water from penetrating into a fiber cable, freezing, and expanding. This cable can also be multimode or single-mode fiber.

Speaker Wire

Speaker wire, as shown in Figure 8–10, is copper wiring that is typically 14- and 16-gauge, and is used for entertainment systems such as stereos and speakers. Speaker wire has either one or two twisted-pairs per cable. The wire size is a relevant issue because the impedance of speakers is usually low (typically 4 or 8 ohms). This means that the resistance of the wiring that feeds the speakers determines how much power actually reaches the speakers. Cables with AWG size 14 and 16 represent the best compromise between line loss, cost, and the ease of installation. Additionally, most whole-house audio system connectivity equipment is designed for these wire sizes.

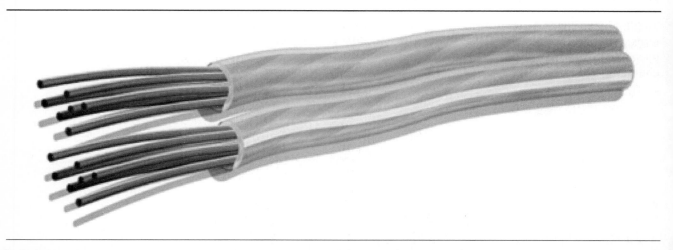

Figure 8–10
Speaker Wire

The use of non-standards-compliant wire size immensely increases the difficulty of installation. Usually, installers can turn to a speaker wire power loss budget table (see Table 8–2) for guidelines on selecting the right size speaker wire.

Speaker Ohms	dB Loss	Power Loss	16 AWG Run (feet)	14 AWG Run (feet)
4	0.5	11	60	100
4	1	21	130	210
4	2	37	290	460
4	3	50	500	790
8	0.5	11	120	190
8	1	21	260	410
8	2	37	580	930
8	3	50	990	1580

Table 8–2
Speaker Wire Power Loss
Budget Table

Thermostat Wire

Many of the cables that run from the HVAC (heating, ventilating, and air conditioning) system to the thermostat on the wall are unusual in that they have odd numbers of wires rather than the standard pairs. Thermostat cables have two, five, or seven wires. Furthermore, their color-coding scheme is unique in the cabling industry.

A typical 18-gauge thermostat wire is 300 volts and comes in three varieties:

- **2 Conductor:** Used for "millivolt" heat-only systems or damper motor control
- **5 Conductor:** Standard HVAC systems
- **7 Conductor:** Typical heat-pump systems

Other applications for thermostat wiring include touchplate systems, burglar alarms, intercom systems, doorbell systems, remote control units, signal systems, and other low-voltage installations.

Security Wire

A security system also uses cables that are specifically designed for the subsystem. A variety of gauges are used, each with a specific purpose. In addition, security systems may use LAN cabling, such as CAT 5 or coaxial.

The types of security system wire used are the following:

- **22 AWG two-conductor copper wire:** Connects inside sounding devices to control panel terminals and to connect external contacts to transmitters.
- **22 AWG four-conductor copper wire:** Connects consoles to control panel terminals. Also, it is used to connect the RJ-31x jack to the phone company's interface box.
- **18 AWG two-conductor copper wire:** Connects external power supply to control panel terminals and outside sounding devices to control panel terminals.
- **14 AWG single-conductor copper wire:** Used to ground the security system.
- **CAT 5 or coaxial cable:** Used to connect the security camera.

Composite Cabling (Bundled Cable Systems)

The media types discussed in this chapter also come as bundled cable systems. A bundled or composite cable is an assembly of two or more cables continuously bound together to form a single unit, as shown in Figure 8–11. Typically, four or more cables are tied together with cable-ties or speed-wrap and pulled throughout the building as one bundle. Common bundled systems include:

- Two coaxial and one CAT 3 or 5 cables
- One CAT 5/5e and two speaker (14- or 16-gauge) cables
- Two CAT 5/5e and two RG-6 (coaxial) cables
- Two CAT 5/5e, two RG-6, and two fiber-optic cables
- Two CAT 5/5e, two RG-6, and two pairs of 14-gauge speaker cables

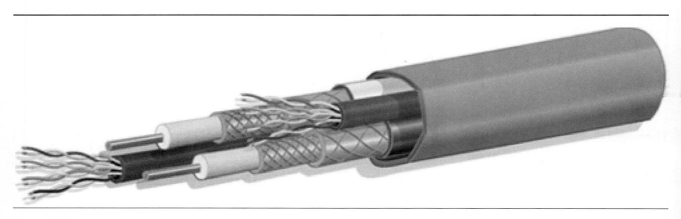

Figure 8–11
Composite Cable Composed of Two Series 6 (Quad Shield) and Two CAT 5e Cables

It is advisable to use a structured wiring package whenever possible. Using structured wiring by the same manufacturer ensures that it is compatible. In addition, if the cabling malfunctions, the manufacturer will be more likely to replace it because you used compatible equipment. In some cases, the HTI might even be required to use the manufacturer's cabling in order for the devices' warranties to be honored.

Residential Structured Cabling Systems (RSCS)

RSCS

A residential structured cabling system (*RSCS*) is the collective configuration of cabling and the associated hardware installed in a residence to provide a comprehensive telecommunications infrastructure for the home network. An RSCS should not be device-dependent or application-dependent.

The range of uses for such an infrastructure may include the following:

- Basic voice service
- Data networking
- Security services
- Entertainment

A typical RSCS begins at the point where the local exchange carrier (LEC), also known as the access provider (AP), ends. This termination point, usually evidenced by a network interface device (NID), is called the demarcation point (DP), or simply the demarc. In a high-speed Internet or telephone service, for example, the AP or LEC furnishes one or more service lines based on the customer's request and connects these to the NID at the demarc. However, all installations, including the cabling system and maintenance beyond the demarc, become customer-owned and the customer's responsibility. In effect, this is the point at which the job of the HTI begins.

The equipment hookup and configuration in a home network integration project starts with the configuration of the residential structured cabling system to be used. There are many possible configurations of an RSCS, determined mainly by variations in the following:

- Cabling and connectivity products installed
- Customer requirements versus actual needs
- Configuration of any existing wiring infrastructure, including upgrades and retrofits
- Present and future needs of the homeowner
- Nature and characteristics of the devices to be supported by the RSCS infrastructure installed
- Manufacturer warranties

The structured wiring used for a home network integration project will vary, depending on the types of wires needed for successful subsystem networking. The types of wiring used in a home network project are discussed in detail later in this chapter.

CABLING EQUIPMENT

A structured cabling system such as the residential structured cabling system (RSCS) infrastructure encompasses several elements:

- Low-voltage media (coax, twisted-pair copper, and fiber)
- Outlet/connector types (coax connectors, twisted-pair connectors, and fiber connectors)
- Termination blocks (mainly 110 block, 66-style block)
- Distribution devices

The totality of these infrastructure components and the various patch cords connected together at a site, following existing specifications and standards, constitutes structured cabling.

In order to complete a successful home technology integration project, the HTI should have a good understanding of the fundamentals of RSCS products. The sections that follow give more information on each of these products.

Structured Cabling Panel

A structured cabling panel is contained within the distribution device (*DD*), or distribution panel, the central element of the home technology integration infrastructure.

The roles that the cabling panel plays in a residential network are the following:

- Terminates and connects wires between voice and data subsystems
- Acts as the center of a star-wired residential network, playing the crucial role of signal distribution throughout the system, allowing devices to be controlled and to communicate with each other
- Connects the access providers to the residence by connecting the home network to the NID, which designates the demarcation point
- Facilitates moves, modifications, add-ons, and changes of the cabling infrastructure within the home

DD

Figure 8–12 shows a structured cabling panel containing a telephone line distribution module (on the left), two Category 5e voice and data modules (center), and a six-way video splitter (on the right).

Figure 8–12
A Structured Cabling Panel

A telephone line distribution module is shown in Figure 8–13. Note that the security line is separate from the other lines.

Figure 8–13
Telephone Line Distribution
Module

An expansion board is shown in Figure 8–14. This is installed to accommodate additional lines for the residence.

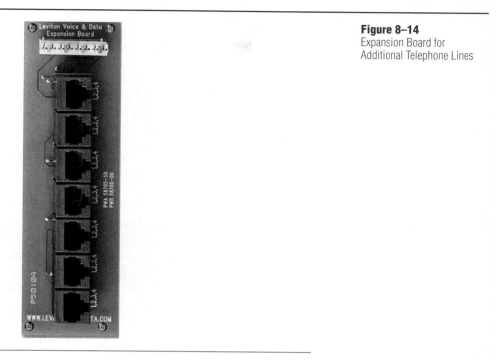

Figure 8–14
Expansion Board for
Additional Telephone Lines

Figure 8–15 shows the distribution panel that contains the structured cabling panel. Note the bundled cabling connected and the power outlets plugged in at the bottom.

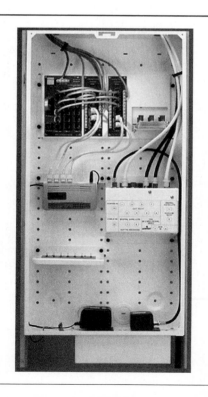

Figure 8–15
Distribution Panel Containing
the Structural Cabling Panel

Patch Panels

Patch panels, as shown in Figure 8–16, are used to interconnect data networking or voice systems to the physical cable network. Patch panels are also used to interconnect backbone cable systems to network distribution cable systems. The rear of the patch panel has network cables that are punched down; for example, on a 110 block. The front of the patch panel has a factory-terminated interface. Patch panels can be used for UTP, STP, or fiber-optic connections. The most common patch panels are for UTP. These patch panels use RJ-45 jacks, which constitute data ports. Patch cords with RJ-45 plugs connect to these ports. One patch panel can have any number of ports depending on the need. A typical configuration is 24 ports.

Figure 8–16
Patch Panel

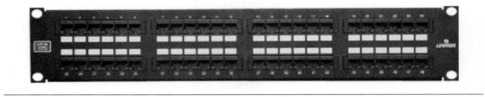

A uniform wiring plan must be used throughout a patch panel system. All jacks and patch panels should be wired using the same wiring plan. If the T568A wiring plan is used for the information outlets or jacks, T568A patch panels should be used. The same is true for the T568B wiring plan. Patch panels are mainly used in multidwelling units (MDU).

Twisted-Pair Patch Cords

Patch cords

Twisted-pair *patch cords*, as shown in Figure 8–17, are used to connect outlets to the distribution system. The patch cords must be of the same category-verified cable as the one used in cabling the outlet. For example, a CAT 5e patch cable must be used if the cable to the outlet is CAT 5e.

Figure 8–17
CAT 5 Patch Cord

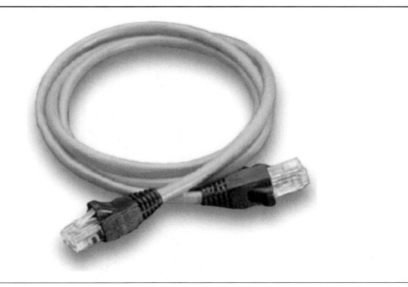

Twisted-pair patch cords have the following characteristics:

- They have stranded conductors for flexibility, unlike regular twisted-pair cabling, which has solid conductors for strength.

- They are allowed to have up to 50 percent more attenuation than regular cabling with solid conductors.
- They typically have eight-position, eight-contact (8P8C) connectors, also known as RJ-45 jacks, on the ends.

Insulation Displacement Connectors (IDCs)

Twisted-pair cable requires a special type of connector known as an insulation displacement connector (*IDC*). An IDC is a connector type that permits the termination of an insulated conductor without stripping insulation from the conductor. The wire insulation is displaced as the insulated conductor is inserted between two or more sharp edges of the IDC contact. This allows contact to be made between the copper conductor and the connector terminal. Two common examples include the following:

IDC

- **66-blocks:** These are insulation displacement termination blocks that are not rated for high-performance data installations and are commonly used in voice (telephone) applications. 66-blocks, as shown in Figure 8–18, have four columns of 50 pins so that they can accommodate either 25-pair termination or 50-pair terminations, depending on their configuration. In the more common 50-pair termination, 25 pairs can be terminated on the left side of the block and an additional 25 pairs can be terminated on the right side of the block. These blocks are mounted on backboards with 89-style brackets. These parts must be ordered together.

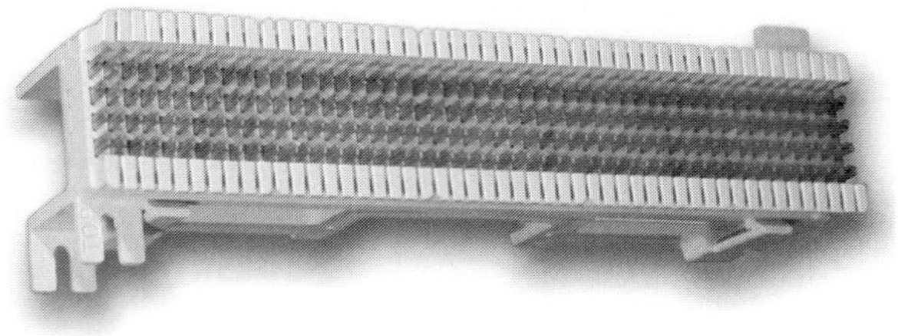

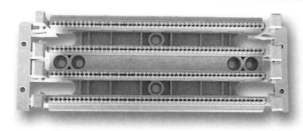

Figure 8–18
66-Block with Four Columns of 50 Pins

- **110-blocks:** These are high-density termination blocks suitable for either voice or data applications. The insulation displacement connection provides a low-resistance, gas-tight connection. 110-blocks come in 100-pair and 300-pair configurations. The 100-pair configuration, shown in Figure 8–19, has four rows of 25-pair terminations. Blocks are designed to be stacked in different combinations to accommodate different size requirements. The 110 system includes wire-management troughs that also act as spacers between the blocks.

Figure 8–19
110-Block with 100-Pair Configuration

Jack

Other types of high-density, high-performance insulation displacement termination blocks include the BIX and Krone blocks. Both are suitable for voice and high-performance data applications. Each requires its own type of punch-down tool.

Ports, Jacks, and Plugs

A *jack* is the female end and a *plug* is the male end of a jack and plug system. The plug is inserted into a jack in order to establish electrical contact. A combination of jacks on networking equipment such as patch panels or switches constitutes (data) ports.

8P8C or RJ-45   8P8C jacks are modular jacks that have eight conductor positions occupied by eight wires. They are also known as RJ-45 jacks (RJ stands for Registered Jack). Figure 8–20 shows a Category 5 (8P8C) jack. 8P8C modular jacks are used to interface twisted-pair cables with equipment lines or patch cords.

Figure 8–20
Category 5 (8P8C) Front and
Back Views of a Jack

Cables are terminated on the rear of the jack, whereas the front of the jack provides access for the modular 8P8C or an RJ-45 plug, shown in Figure 8–21. The term *positions* refers to the disposition for conductors, while the term *contacts* refers to the actual number of wires installed in the jack or plug.

Figure 8–21
RJ-45 Plug

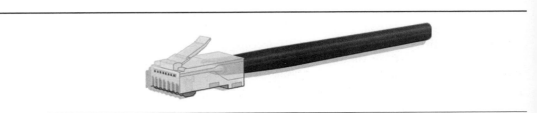

RJ-45 jacks are eight conductor jacks that are designed to accept either RJ-45 plugs or RJ-11 plugs. Jacks are normally wired to the T568A or T568B standards shown in Figure 8–22 and Figure 8–23, respectively. RJ-45 plugs have eight pins that will accommodate up to four pairs of wires. As with RJ-11 plugs and jacks, pair 1 is always terminated on the center pins, in this case, pins 4 and 5. Pair 4 (white/brown) is always terminated on pins 7 and 8. Pairs 2 and 3 may differ, depending on the wiring plan. Using T568A, pair 2 terminates on pins 3 and 6, whereas pair 3 terminates on pins 1 and 2. T568B reverses pairs 2 and 3 so that pair 2 (white/orange) terminates on pins 1 and 2. Pair 3 (white/green) terminates on pins 3 and 6.

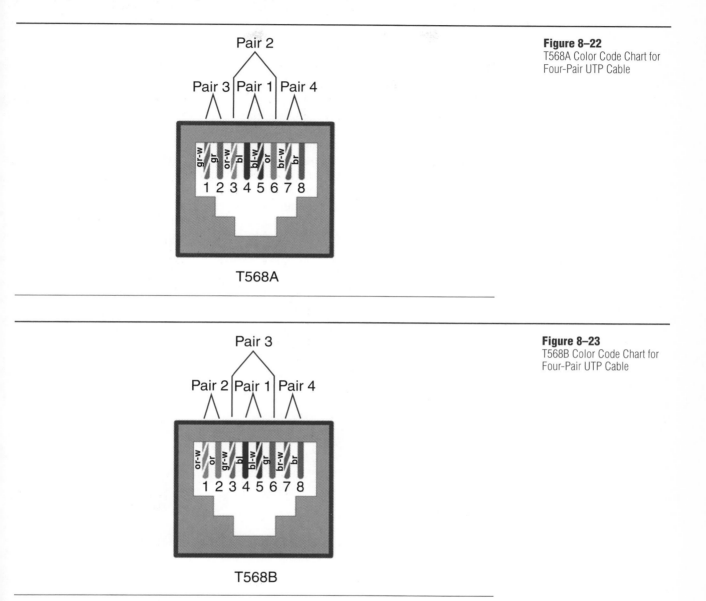

Figure 8–22
T568A Color Code Chart for Four-Pair UTP Cable

Figure 8–23
T568B Color Code Chart for Four-Pair UTP Cable

As with RJ-11 plugs, RJ-45 plugs are available for both stranded and solid wire. Only stranded plugs should be used on stranded wire, and only solid plugs should be used on solid wire.

6P6C or RJ-11   Unlike the 8P8C modular jacks, the 6P6C (RJ-11) jacks have only six conductor positions occupied by a total of six wires. The RJ-11 connector, which can be either a jack or plug, is used for terminating Category 3 cable, previously the most common type of telephone wire. This common connector has six pins. Pair 1 (white/blue) is terminated on pins 3 and 4. Pair 2 (white/orange) is terminated on pins 2 and 5. Pair 3 (white/green) is terminated on pins 1 and 6.

Copper wire can be either stranded or solid. An RJ-11 plug or jack should be chosen for the type of wire that the cable is made of, either stranded or solid. If a jack or plug intended for another type of wire is used, the connection may be poor or nonexistent. The jack or plug may even be damaged.

> **NOTE**   None of the six position plugs meet the requirements of ANSI/TIA/EIA-570-A.

RJ-2IX   Many key and private branch exchange (PBX) telephone systems commonly found in multidwelling units (MDUs) use 25-pair copper wires to interface the entrance facility and cross-connect fields. These are special jacks, and very often telephone system manufacturers specify the demarcation point to be an RJ-2IX.

SC and ST   Although fiber-optic cable connectors perform the same functions as copper connectors, fiber-optic cable connectors transmit light impulses through optical fiber (as opposed to electrical signals through a metallic copper wire conductor). As such, fiber-optic connectors must be able to ensure a maximum signal transfer between connectors. This is the most important consideration in selecting fiber-optic cable connector products. The most commonly used connectors include:

- Straight-tip (ST) connectors
- Subscriber (SC) connectors

Figure 8–24 shows a straight-tip connector.

Figure 8–24
ST Connector

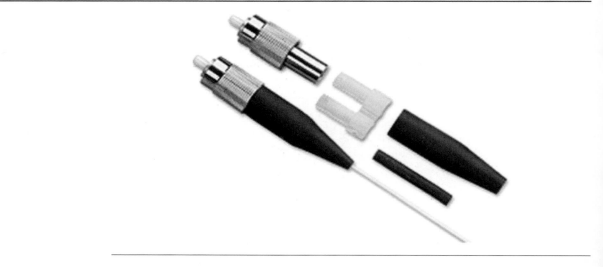

Figure 8–25 shows a subscriber connector.

Figure 8–25
SC Connector

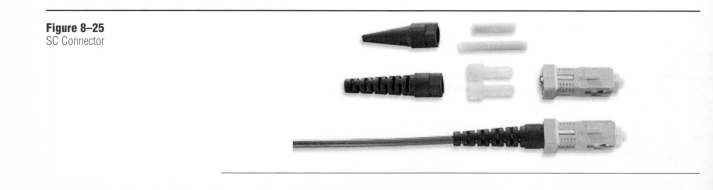

Other fiber-optic connectors may also be used, as long as they conform to the ANSI/TIA/EIA-570-A standard. A good example is the small form factor (SFF) connector, which is particularly suitable because it requires less space in the outlet and the distribution device (DD).

S-Video   S-Video connectors allow a higher picture quality in video displays than most video standards. The quality of the S-Video connector plays a very important role in final signal quality. The QuickPort S-Video Module allows you to run any S-Video signal through any high-quality CAT 5, 5e, or 6 cable for high speed and high quality. This allows you to run S-Video signals for a considerable distance without dual runs of RG-6 cable.

Figure 8–26 shows the front and rear views of an S-Video QuickPort Module.

Figure 8–26
S-Video QuickPort Module

RCA   Radio Corporation of America (RCA) connectors are used for line-level audio and/or composite (component) video routing. An RCA connector is sometimes known as a phono plug and jack. Figure 8–27 illustrates an RCA connector.

Figure 8–27
RCA Connector

BNC, F, and N   The most common types of connectors used with coaxial cables are the British Navel Connector (BNC) connector (shown in Figure 8–28) and the Type F connector (shown in Figure 8–29). The F series connector is used for modulated radio frequency applications, and the BNC connector is usually used for networking and video applications. The F connector is used to terminate Series 6 coaxial cable (RG-6). A third type of coaxial connector that is sometimes used is the N coaxial connector. It consists of a threaded coupling, very similar in design to an F connector, except that it is larger.

Figure 8–28
BNC Connector

Figure 8–29
F Connector

Speaker Terminals   Older speaker terminals may have *binding posts*, which are screw terminals for wires. The wire has to be stripped and wrapped around a screw, and a nut is tightened. Binding posts are no longer used in communications, but they are occasionally used in sound and in retrofit projects. Often, speakers use spring clip connectors to grab, hold, and make good electrical contact with the speaker wire.

Amplifiers

As mentioned earlier in the chapter, some amount of signal is lost in the form of energy from the transmission media when signals are transmitted on a network. This is defined as *attenuation,* and it can cause a signal to become unintelligible and not properly interpreted by the receiving device. This is especially prevalent with long cable runs. To mitigate this problem, a device called an *amplifier* may be deployed between a network appliance and the signal distributor (or source) on a residential network. Figure 8–30 shows a Leviton amplifier.

Figure 8–30
Amplifier

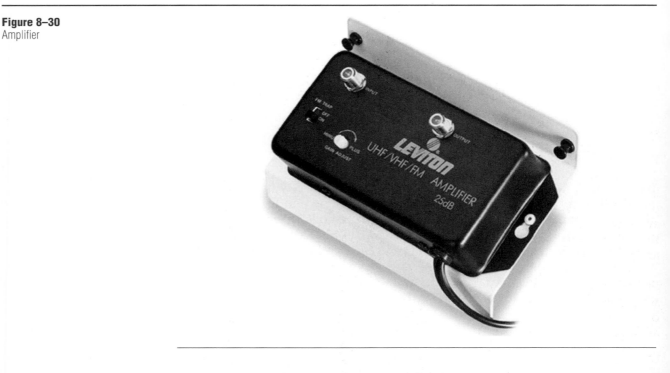

Amplifiers are active devices on the network, and they boost the levels of signals in the accepted frequency range or magnitude specified by the technology in use. They are commonly used to boost A/V signals in whole-house audio or multichannel home theater.

Filters

A filter, as shown in Figure 8–31, is a device that selectively sorts signals and passes through a desired range of signals while suppressing the others. Filters may be employed to suppress noise or to separate signals into bandwidth (how fast data flows on a given transmission path) channels in a home technology integration installation.

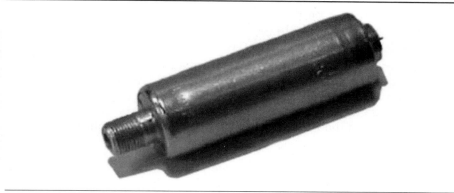

Figure 8–31
Filter

Some filters have amplifiers to boost the level of signals in a specified frequency range. Known as *active filters*, they require an external source of power. Other filters, known as *passive filters*, do not amplify and consume no power in doing their job.

EQUIPMENT LAYOUT

The equipment layout is largely determined by the cabling system configuration and topology. The equipment layout plan is an important element of the overall residential system's infrastructure installation. A diagram of the equipment layout for the whole system is useful for orienting installers, especially if the HTI is using subcontractors such as cable installers, who depend on the specifications provided to complete the installation.

Telecommunications Room (TR)

The telecommunications room (TR) is also known as a wiring closet. Some very large single-family homes, especially in new construction areas, can accommodate a TR. Even in retrofits, an area can be designated as the TR. These small areas or closets house telecommunications cabling system equipment such as the DD, routers, gateways, switches, relay racks, and mechanical terminations and/or cross-connect for the horizontal and backbone cabling systems.

The TR must be centrally located in the residence. A central location provides cabling runs to the outlets and network-enabled devices and appliances so that they do not exceed the standards-compliant maximum length. The TR or wiring closet must be located in a well-ventilated and climate-controlled area that is accessible in case of future repairs, maintenance, or upgrades.

NOTE An MDU-TR is the telecommunications room in a multidwelling unit. Multidwelling units are discussed in Chapter 15.

Distribution Panel and Equipment/Relay Racks

There are five important considerations in the location of the DD and equipment racks in a residence:

1. **Accessibility:** The DD shall be installed inside the home in a location that is accessible for cabling maintenance. Additionally, such a location should be centralized to minimize the length of outlet cables. The DD and associated equipment/devices can be mounted on a backboard, or recessed between stud spaces or on equipment racks.

2. **Proximity to breaker panel:** The media center location must be a minimum of 1.5 meters (five feet) from the breaker panel.

3. **Must be within 90 meters (295 feet) of the farthest jack.**

4. **Length of outlet cable:** A standards-compliant installation requires that space should be allocated adjacent to or within the DD for the installation of a surge-protection device for each conductive cable entering or leaving the building. Also, access to the building electrical ground shall be provided within 1.5 meters (five feet) of the DD, and in accordance with any applicable codes.

5. **Must be grounded.**

A good point at which to decide on the location of relay racks is during the planning phase of an RSCS. One of the most important factors in deciding its location is the demarcation point (where service provider services come to the home) and the potential need to plan for additional peripherals that may require alternating current (AC) power.

The HTI must pay careful attention to where a DD is eventually installed within the home when planning equipment layout.

Cable Termination Points

The main termination points involved with residential cabling configuration are outlets and termination (punchdown) blocks.

According to the TIA/EIA standard for residential wiring in the United States, a minimum of one outlet location shall be installed within each of the following rooms in a residence:

- Each bedroom
- Kitchen
- Family/great room
- Den/study

In addition to these specifications, the standard adds that a sufficient number of telecommunications outlet locations should be provided in order to prevent the need for extension cords. An outlet location should be provided in each room as well as additional outlet locations within unbroken wall spaces of 3.7 meters (12 feet) or more.

Additional outlet locations should be provided so that no point along the floor line in any wall space is more than 7.6 meters (25 feet), measured horizontally, from an outlet location in that space. However, the outlet mounting heights have to be in accordance with applicable codes of the area in which the project is located and in accordance with the needs of the homeowner. For example, handicap installations may require outlets to be raised to make them easy to reach.

The ANSI/TIA/EIA-570-A standard details the outlet locations, but not the number and types of media that are brought to an outlet. This is due to the following possibilities:

- **Grade 1 installation:** One twisted-pair cable run, and one coaxial Series 6 cable run for one television and one phone.

- **Grade 2 installation:** Two CAT 5e twisted-pair cables, one two-fiber cable, and two coaxial Series 6 cables to feed the telephone, TV, computer network, and networked printer.

Based on the preceding list, the determination of the cable configuration (number of media) to each outlet is determined by the design of the home network.

In general, it is recommended that each outlet location have a minimum of Grade 1 (Level 1) wiring, that is, one coax and one CAT 5 for VDV. This would be appropriate for a bedroom. Other locations such as the living room may require outlet locations with a Grade 2 cabling configuration: two CAT 5/5e cables and two coaxial Series 6 cables.

Termination blocks are typically located centrally, and are inside or close to the DD to stay within recommended maximum lengths for patch cords and other cabling distances.

> **NOTE**   For each channel, a total of 10 meters (approximately 33 feet) is allowed for equipment cords and patch cords or jumpers.

Certain fixed devices, such as intercoms, security system keypads, sensors, and smoke detectors may be hard-wired (terminated) to the fixed device controller. All outlets should be directly connected to the distribution device in a star or home run wired topology.

Cabling Layout or Topology

The cabling topology of an RSCS or integrated network refers to the features, shape, configuration, or physical appearance of the whole system. Network topologies can have physical or logical configurations. There is a distinction between the physical and logical topologies of a network (distribution system):

- **Physical topology:** The physical appearance of the network.
- **Logical topology:** How the network functions. The logical topology of a network is determined by how the information (signal) is transmitted from one device to another on the network.

There are times when a network has a certain physical topology, but logically transmits information in a completely different topology. The two fundamental RSCS topologies are *bus* and *star*.

Bus or Daisy-Chain Topology

A bus topology reflects a linear configuration of equipment and devices on the network. (See Chapter 3 for a discussion of bus topologies.) In an Ethernet (IEEE 802.3) setting using CSMA/CD, a bus topology means that when one device tries to communicate with another, all the devices on the network "see" the information, although only the device for which the information is destined will recognize and process it.

In a bus topology arrangement, the ends of the transmission channel are not connected to any network devices. Problems can occur on the network when signals reach these ends, bounce back into the cable, and confuse other oncoming signals. To solve this problem, each end of a cable must be connected to a terminator.

In residential networking, the most common applications of a bus topology include:

- Intercom systems
- Fire alarm systems
- MDU video and broadcast satellite (DBS) distribution systems

Star or Home Run Topology

A star topology is a wiring scheme in which telecommunication cables are distributed from a central point. In residential networking, the star topology has as its physical and logical center a distribution device (DD). (See Chapter 3 for a discussion of star topologies.) Unlike the bus topology, all the devices are connected to this central hub like the spokes of a bicycle. Instead of sharing the same bus (data channel), each device in the star cabling topology has its own direct dedicated line to the DD, which is known as a *home run*. A star cabling topology presents several advantages over a bus cabling topology:

- Much easier to install and maintain (especially for the do-it-yourself homeowner)
- Easier to locate and isolate faults in the system; problems can be repaired without shutting down the whole system
- Any device that is disabled or isolated from the DD is the only device affected
- Provides a central location for managing the network

The networking topology is an important consideration in the design and installation of residential networks. The star cabling topology is recommended for the residential network. Knowing which cabling topology to use for wiring for any given application is a must for the HTI.

Connectivity Documentation

Connectivity documentation is crucial for a successful residential wiring installation. In structured wiring, the documentation provides valuable information for the homeowner. Your documentation should include the following:

- Building floor plans or blueprints that are obtained from the homeowner or the building contractor, as shown in Figure 8–32.

Figure 8–32
Example of Floor Plan

- Wire charts and schematics showing wire-routing details, not only for the computer network but also for all the subsystems that need to be integrated into the residential network.
- Signal flow diagrams that designate device locations with respect to the location of the distribution device (distribution panel) and the demarcation point, as shown in Figure 8–33.

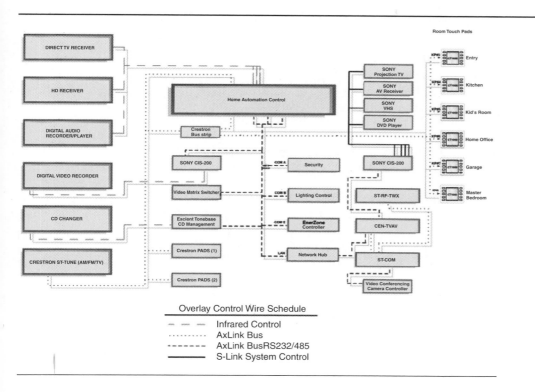

Figure 8–33
Example of Signal Flow Diagram

In addition, the following should be provided with connectivity documentation:

- Documents showing the locations of data ports, jacks, and other outlets.
- Blueprints and schematics that use installation symbols and icons to identify cable runs, different types of raceways, and information about outlets and jacks. They provide a uniform method to graphically identify requirements on a blueprint.
- As-built drawings showing cable routes, termination points, and cable types as they were actually installed for existing homes. In most cases, some cables are not installed as originally planned because obstructions or problems are encountered. As-built drawings are very important for retrofits. They help locate terminated and blind-terminated cable termination points.

 TESTING AND CERTIFICATION

Testing and certification are an integral part of a home integration cabling infrastructure installation. Without testing and certifying each circuit from end to end (distribution device to outlet), there is no way to verify that the equipment and devices to be supported by the cabling system will operate properly. The testing must be tailored to the specific installation in order to be comprehensive. The goal of testing is to flush out any bottlenecks and potential problems with the installation.

Basic Testing and Certification

The simplest way to perform basic testing is to use a multimeter or volt-ohm meter to test for both shorts and continuity. To check for shorts, simply use the ohmmeter to check for an open condition between each of the wires in the four pairs of Category 5 UTP cable. Repeat this test for each cable run.

You should conduct two types of tests on Category 5 cables:

- Continuity/wire map or channel test
- TIA certification test

Perform both tests on the length of cable from the distribution panel to a junction box. Each cable length should not exceed 90 meters (295 feet).

The continuity/wire map or channel test verifies that end-to-end connectivity exists, that both ends of a cable are terminated in the proper pin configuration, and that no opens or shorts are present on the cable.

The TIA certification test measures the following:

- **Length:** The link length should not exceed 295 feet (90 meters).
- **Attenuation:** How much signal degradation occurs from one end of the link to the other.
- **Near-end crosstalk (NEXT):** How much signal from one pair of wires is picked up by all other pairs.

Both the continuity/wire map and the Category 5 certification test should yield a pass or fail result. If you get a fail rating, use a cable tester to pinpoint the location of the problem.

> **NOTE**   A non-TIA/EIA-570-A system that passes Category 5 testing is not TIA/EIA-570-A-compliant.

Testing Twisted-Pair

Twisted-pair cable, such as Category 5 cable, should be inspected throughout its installation. Doing so is much easier than trying to trace down a problem later within a bundle of cables.

Good testing equipment will do the following:

- Help keep your installation Category 5-compliant
- Find any errors or disruptions
- Point out problems in your installation techniques so they can be corrected
- Add value to your installation, to the builder, and to the homeowner
- Make it possible to provide the manufacturer extended warranty on installation

When field-testing Category 5 cable installed for data transmission, use a Category 5 field test set. Select a test set from one of those listed or from another reputable manufacturer, and follow the manufacturer's instructions.

Only cable runs with Category 5 jacks available on both ends of the run (wall plate Category 5 jack and Category 5 module at the SMC end) can be tested for Category 5 compliance.

> **NOTE** Your test instruments must meet or exceed the applicable requirements in TSB-67. Commercially available instruments will specify whether they meet Level 1 or Level 2 accuracy. You will want to use those that meet Level 2 accuracy. To meet TSB-67 standards, the Category 5 testers must be able to test for length for all pairs of an installed link, attenuation, and near-end crosstalk (NEXT). Testers may also measure for delay on an installed link; that is, the time it takes for a signal to travel from one point to another.

The next section deals with testing and troubleshooting for wiring in the entertainment subsystem. You will need to test audio cable, speaker cable, and coaxial cable.

Testing Audio Cable

When testing for unterminated audio cable, check for an opening between the wires in the speaker or volume control audio cable.

For terminated audio cable, check to see that the resistance reading of the cable from the audio module in the distribution panel is nearly the same as the resistance of the terminals on the volume control or speaker to which the audio cable run is terminated. A measurement close to the resistance of the volume control or speaker indicates that the cable is continuous and without shorts.

Testing Speaker Cable

When testing speaker cable, do the following:

1. Look for damage in the cable run.
2. Do a standard continuity test. Check all connections for loose strands and make sure the terminal screws are secure:
   - Check all audio module connections.
   - Check all volume control connections.
   - Check all speaker connections.
3. Turn the system on with the volume turned all the way down and slowly raise it to test performance.

Testing Coaxial Cable

Coaxial cable can be tested for continuity between the distribution panel and the termination point. It can also be tested for breaks in the cable itself and for cable resistance. Breaks can be checked using a multimeter or coax cable tester. The multimeter also checks for resistance.

- Be sure connections at coax splitters are sound.
- Make sure that all connections are hand tight.

Connecting a small portable television to each video F-connector jack will indicate cable continuity and whether the signal level is adequate. A snowy picture is an indication that the signal level is too low. You may need to add a video amplifier to correct the problem. If your picture shows several wavy lines and "ghost images," check to make sure that all terminations without equipment attached have the 75-ohm termination caps in place. Hand-held coaxial cable testers are available from several manufacturers to simplify the above testing. Some cable testers for telephone applications offer twisted-pair and coax testing in a single unit, enabling you to locate faults in coaxial cables.

Testing RG-6 Cable

The link length of RG-6 cable should not exceed 200 feet. On RG-6 cable, a cable tester searches for opens and shorts. RG-6 cable attenuates between 3 dB and 7 dB per 100 feet, depending on frequency.

For unterminated RG-6 quad shield cable, use a multimeter or volt-ohm meter to check the resistance between the shield and center conductor. If the meter shows a finite resistance reading (below 100 kohms), the cable has a short (and that cable run fails the test).

For RG-6 coax that is terminated in a splitter or in a run with a 75-ohm termination cap, the resistance measurement should be near 75 ohms to pass the test. A coax cable tester will indicate faults and breaks in coax cable. Prior to the installation of the endpoint device (for example, television, VCR, and so on), the cable can be tested by plugging it into the back of a portable device such as a small television or radio.

Test Rules

The following are the rules that must be followed during the test to make it valid:

- Cabling and components cannot be moved during testing.
- Both pass/fail indications and the actual measured values (frequencies) should be recorded.
- Reconfiguration might require retesting.
- Qualified adapter cords should be used to attach the test instruments to the link under inspection.
- If you perform a channel test, the end-user patch cords should be tested in place.
- The field tester must meet TSB-67 accuracy requirements for Level 1 or 2.

Most companies that manufacture distribution panels and elements will require the HTI to provide testing documentation for guarantee purposes. It is crucial that you use the forms that come with the product or at least keep the test documents that the testers can print out.

INFORMING THE HOMEOWNER

After installing, testing, certifying, and troubleshooting any problems with the systems installed, it is helpful to train the homeowner so that he or she understands the residential wiring infrastructure and its components. Such training can center on giving the homeowner basic knowledge and understanding of aspects of the installation such as the following:

- **Location of the auxiliary disconnect outlet (ADO):** Allows the homeowner to switch between different services or disconnect service to different rooms of the home for maintenance and troubleshooting.
- **Basic system configuration:** Provides information so that the customer can move equipment around. It also shows the wiring for specific devices and appliances.
- **Which circuit serves which room:** Enables the homeowner to check the circuit breaker to turn appliances on and off; for example, in reaction to electrical faults or lightning surges.
- **Control of some sensitive areas:** Fully equips homeowners so that they can control all circuits in the home.

These are just some of the equipment and system-configuration issues that may be included in the client training and orientation process. Always give your clients the chance to ask as many questions as possible and give them clarification in your answers.

SUMMARY

This chapter discussed the following specifics of low-voltage wiring:

- A residential structured cabling system (RSCS) is the collective configuration of cabling and the associated hardware installed in a residence to provide a comprehensive telecommunications infrastructure for the home network.
- The three common methods of signal transmission are electrical, optical, and wireless.
- Factors that cause internal or external noise that results in EMI, RFI, resistance, inductive reactance, capacitive reactance, and impedance.
- Attenuation, which refers to any reduction in the strength of a signal, can affect a network because it limits the length of network cabling.

- Plenum cable, which is made of fire-resistant materials, must be used when running cable through ventilation plenums.
- The ANSI/EIA/TIA-570-A standard defines two grades of residential cabling.
- A distribution device (DD) acts as the center of a star-wired residential network and connects the access providers to the home by connecting the home network to the NID.
- Twisted-pair cabling consists of pairs of insulated copper wires that are twisted together and then housed in a protective sheath.
- Coaxial cable wraps the center conductor in insulation and covers it with one or more layers of braid and foil, followed by a protective sheath.
- The simplest way to perform basic testing is to use a multimeter or volt-ohm meter and test for both shorts and continuity.

GLOSSARY

Category 5 or 5e   A category of performance for unshielded twisted-pair (UTP) copper cabling. Used in support of voice and data applications requiring a carrier frequency of up to 100 MHz for a distance of up to 100 meters.

Coaxial cable   A transmission medium that consists of a central copper wire conductor. Coaxial cable is usually used for CATV, CCTV, and broadband video.

DD   Distribution device.

Fiber-optic cable   A transmission medium with a conductor made of glass or plastic. Fiber-optic cable carries information as light and can carry much more information over greater distances than copper cabling.

IDC   Insulation displacement connector. A special type of connector used with twisted-pair cable.

Jacks   The low-voltage receptacles, mounted within a wall plate, into which telephone, audio, and video equipment are plugged.

Patch cord   A cable that connects circuits together.

RSCS   Residential structured cabling system. The collective configuration of cabling and associated hardware installed in a residence to provide a comprehensive telecommunications infrastructure for the home network.

Twisted-pair cable   A cable made up of one or more separately insulated twisted-wire pairs.

CHECK YOUR UNDERSTANDING

1. What is the most commonly installed media?
   a. Category 3
   b. Category 5
   c. Coaxial
   d. Optical fiber

2. What is the name of the cable that has an outer coating made of a fire-resistant material?
   a. ScTP
   b. Plenum
   c. Screened shield
   d. Foil

3. What forms a circuit that can transmit data?
   a. A pair of wires
   b. A single wire with a foil shield
   c. Both a and b
   d. Neither

4. How many wires are in Category 5 UTP?
   a. Two wires
   b. Two pairs of wires
   c. Four wires
   b. Four pairs of wires

5. Copper wire with a screening layer to protect the wires from outside interference is called:
   a. unshielded twisted-pair (UTP).
   b. shielded twisted-pair (STP).
   c. screened twisted-pair (ScTP).
   d. high-count twisted-pair.

6. What kind of cable is used for backbone service in multi-dwelling unit environments?
   a. Unshielded twisted-pair (UTP)
   b. Shielded twisted-pair (STP)
   c. Screened twisted-pair (ScTP)
   d. High-count twisted-pair

7. What causes signal degradation?
   a. EMI and RFI
   b. EMI only
   c. RFI only
   d. None of the above

8. What is the advantage of using single-mode fiber over multi-mode fiber?
   a. Data travels further before requiring regeneration.
   b. It is less prone to interference.
   c. It does not need to be grounded.
   d. It radiates fewer signals.

9. Typically, what is the primary color in a four-pair cable?
   a. Orange
   b. White
   c. Red
   d. Blue

10. Which of the following types of wiring is not prone to problems with interference?
   a. Power line
   b. Phone line
   c. Fiber-optic
   d. Coaxial

11. The access provider's termination point is also known as the:
   a. LEC.
   b. demarc.
   c. FCC.
   d. distribution panel.

12. What is the term that refers to any reduction in the strength of a signal?
   a. Crosstalk
   b. Attenuation
   c. Noise
   d. Dispersion

13. What is the unwanted transfer of signal from one or more circuits to other circuits?
   a. Crosstalk
   b. Attenuation
   c. Noise
   d. Dispersion

14. To interconnect data networking or voice systems to the physical cable network, use a(n):
   a. distribution panel.
   b. patch panel.
   c. auxiliary panel.
   d. distributing patch.

15. A device that selectively sorts and passes through a desired range of signals while suppressing the others is called a(n):
   a. splitter.
   b. distribution device.
   c. amplifier.
   d. filter.

16. The link length of RG-6 cable should not exceed:
   a. 300 feet.
   b. 500 feet.
   c. 200 feet.
   d. 100 feet.

17. Why is it important for the homeowner to know the location of the auxiliary disconnect outlet (ADO)?

18. A residential structured cabling system (RSCS) is the collective configuration of cabling and the associated hardware installed in a residence to provide a comprehensive telecommunications infrastructure for the home network. What is the range of uses for such an infrastructure?

19. Why should a residential structured cabling system (RSCS) employ Series 6 coaxial cabling for transmission of digital video signals from the distribution device to the outlets?

20. What are the three distinct methods of wireless transmission?

9

High-Voltage and Electrical Protection Systems

OBJECTIVES

Upon completion of this chapter, the HTI will be able to accomplish the following tasks:

- Adhere to NEC, bonding, and grounding standards
- Understand the fundamentals of electricity
- Identify electrical equipment
- Identify electrical wiring present in a home
- Connect the home automation system to the electrical wiring

INTRODUCTION

In this chapter, you learn about electrical protection systems including design considerations and the products used, such as electrical outlets, light switches, and light fixtures. Upon completion of this chapter, you should have a clear understanding of the National Electrical Code (NEC) and the guidelines for installing and working with high-voltage systems. You also learn the safety precautions that need to be used when working with high-voltage wires.

ELECTRICAL STANDARDS

Several organizations have established standards for fire resistance that apply to network cables. There are also standards that apply to grounding and bonding. Finally, there are standards on how firewalls should be treated. Together, these standards lay the groundwork for the HTI to build upon as the installation proceeds.

Cable-Burning Standards

The National Electrical Code (NEC), published by the National Fire Protection Association (NFPA), sets the minimum safety requirements for electrical wiring and grounding. Revised every three years, the NEC is the most widely used and accepted criteria for all electrical installations. It is adopted as law by most states, cities, and municipalities. NFPA is a worldwide leader in providing fire, electrical, and life safety to the public. Since 1896, the mission of this international nonprofit member organization is to reduce the worldwide burden of fire and other hazards on quality of life by developing and advocating scientifically based consensus codes and standards, research, training, and education.

As you learned in previous chapters, the NEC contains the standards for electrical wiring and is supported by most local licensing and inspection officials. One of the standards defined by the NEC is how a cable burns. The standards specify the amount of time that it takes a cable to catch fire after exposure to flames and how long it will burn after it starts. Additionally, because the outer coverings of cables and wires are typically plastic and can create noxious smoke, the standards state where the cables can be used.

NEC-type codes are listed in catalogs for cables and cabling supplies. Table 9–1 lists the types of cables and their uses.

Table 9–1
Cable Types

Cable Type	Use
CM	Communications
CL2, CL3	Class 2, Class 3 remote control, signaling and power-limited cables
FPL	Power-limited, fire protective signaling cables
MP	Multipurpose
PLTC	Power-limited tray cable

Some companies choose to test their cables as remote control or power-limited circuit cables. Due to the difference in electrical power that the cable is exposed to, they get CL2 or CL3 (Class 2 or Class 3) ratings instead of the more general type of ratings. The flame and smoke criteria, however, are generally the same for all tests.

Plenums and Firewalls

Standards developed by the National Fire Protection Association (NFPA) and adopted by the American National Standards Institute (ANSI) describe the amount of smoke and the type of gasses a burning cable can generate. This is important because cables are often used in ventilation systems. Such ventilation spaces, called plenums (which were introduced in Chapter 8), must not allow toxic gasses to spread to other areas of the building in the event of fire.

Firewalls

Penetrating

Coring

Backfilled

Firewalls are constructed of fireproof material to help prevent the spread of fire. Wires often need to go through these firewalls. Puncturing or *penetrating* a firewall should not be undertaken without proper approvals because it could affect the building's fire-worthiness rating. Puncturing a floor, called *coring*, has fire-stopping requirements and building structural considerations as well. When punctures cannot be avoided, the holes must be *backfilled* with fire-stopping materials to restore the firewall's integrity.

Grounding and Bonding Standards

To have some commonality in grounding and bonding systems, a national standard has been formulated.

"Commercial Building Grounding and Bonding Requirements for Telecommunications," also known as TIA/EIA-607, supports a multivendor, multiproduct environment; as well as the grounding practices for various systems that may be installed on customer premises. TIA/EIA-607 specifies the exact interface points between a building's grounding system and the telecommunications equipment grounding. This standard also specifies the building's grounding configurations needed to support this equipment. Proximity of network and high-voltage wiring, lightning grounding wire, and grounding rods are important considerations after the network is properly bonded and grounded.

 ## ELECTRICAL POWER FUNDAMENTALS

A residential network and its associated equipment need electrical power to operate. Although the HTI does not do the residential power wiring, it is crucial to understand the fundamentals of electrical power in order to properly design residential network installations and to ensure a smooth crossover of deliverable processes. Two of these fundamentals include voltage and current.

In this course, high voltage refers to the nominal 120/240 volts that may be carried in residential power lines (120 volts in North American homes and 240 volts elsewhere). High-voltage wiring should be understood vis-à-vis low-voltage wiring (for example, 48 volts on phone lines), which is discussed in Chapter 8.

High-Voltage Terms

The HTI should know the following terms related to electrical load requirements:

- **Circuit:** The wiring controlled by one fuse or circuit breaker.
- *Ampere:* A measure of the number of electrically charged particles that flow past a given point on a circuit per second. An ampere is a basic unit of electric current (its abbreviation is *A* or *amp*).   Ampere
- *Circuit breaker:* A protective device used for each circuit. It automatically cuts off power from the main breaker in the event of a short or overload. A circuit breaker is designed to tolerate only a regulated amount of current.   Circuit breaker
- *Main breaker:* Turns the power entering the home through the breaker box on or off. The main breaker may be located in a breaker box or it may be in a separate box at another location.   Main breaker
- **Breaker box:** Also known as the breaker panel, it houses the circuit breakers or fuses, and distributes power to various parts of the home.
- *Neutral bus bar:* A metal bar to which all neutral wires are connected in a breaker box.   Neutral bus bar
- **Roughing-in:** The placement of outlets, switches, and lights prior to actual electrical hookup.
- *Volt (V):* A measure of the current pressure at receptacles and lights. The average household voltage in the U.S. is 120 V AC.   Volt (V)
- *Watt (W):* The rate at which an electrical device (a light bulb or another home appliance) consumes energy. Watts = volts * amps. An appliance rated at 1000 watts is more expensive to operate than a similar appliance with a 600-watt rating.   Watt (W)

Power, Voltage, and Current

Power (P) is measured in units called watts (W). *Power* is a term that applies to the energy consumed to operate electrical devices and appliances. Power is composed of two main components:

- Voltage (V) as measured in volts
- Current (I) as measured in amperes

Power requires that there be both current and voltage. No amount of voltage without a concurrent current will provide power. You can think of voltage in terms of water pressure, and current as the amount of water delivered per unit time.

Power equals the amount of voltage multiplied by the amount of current, as shown in the following formula:

$$P = V * I$$

Electrical power fundamentals also include alternating current (AC) and direct current (DC), which are discussed next.

AC and DC Current

Alternating current (AC)

Direct current (DC)

Current occurs in two forms, *alternating current* (*AC*) and *direct current* (*DC*). AC is the electricity used in a home. It is defined as electric current that reverses its direction of flow (polarity) according to a frequency that is measured in hertz, or sine waves per second. *Sine waves* show the cycles that the current travels. In the U.S., the standard commercial power frequency is 60 hertz, which translates to 60 sine wave cycles per second. Each sine wave cycle starts with the voltage at the zero level. The voltage then rises until it attains a positive maximum value; then it decreases until it passes the zero level and becomes negative. The voltage then continues in the negative direction until it attains a maximum negative value, at which point it changes direction and becomes less negative until it again reaches the zero level, completing the cycle. Figure 9–1 illustrates AC.

Figure 9–1
Alternating Current (AC)

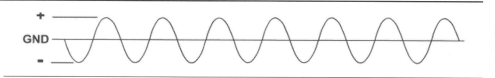

Direct current (DC) refers to a steady value of current that does not change direction of flow. Unlike AC, DC rises to its maximum value when switched on and remains there until the circuit is interrupted. Batteries are examples of DC sources.

Electricity is brought to the home by power lines, which carry electricity in the form of alternating current. The voltage of an AC power source can be easily changed by means of a power transformer, which enables the voltage to be increased (stepped up) for transmission and distribution, or decreased (stepped down) to values desired by different appliances. High-voltage transmission is more efficient than low-voltage transmission over long distances because the loss caused by conductor resistance decreases as the voltage increases.

The sections that follow explore more detailed aspects of power connectivity/electrical protection systems as they relate to residential system installation.

Residential Load Requirements

Load requirements determine the size of circuit breakers or fuses in a residential electrical distribution panel. Load requirements also dictate the gauge

(AWG) of the wire feeding the outlet. For instance, a large motor may be rated as a 20-amp motor, which requires a 20-amp circuit breaker in the electrical distribution panel. Wire gauge charts are available to determine the proper wire size for a 20-amp circuit.

Load requirements generally have a built-in safety factor. A single 15-amp circuit breaker and 14-gauge wire will generally feed a number of outlets in a home. The selection of the breaker size is based on two factors: the known high-amperage devices plugged into the circuit and the assumption that it is unlikely that all devices will be in use at the same time.

It is possible to exceed the load of the circuit for a short period of time without tripping the circuit breaker. Too many lights and appliances operating on a single circuit can cause the lights to appear dim. This is a clear signal that some devices must be unplugged or a newer, higher capacity circuit is required.

Most lighting and utility outlets in a home will be 15- to 20-amp circuits. The amperage of the circuit is defined by the rating of the breaker. If the amperage on an appliance is not listed, the wattage is usually listed. To determine amperage, divide the wattage by the nominal voltage, as shown in the following formula:

$$A = W/V$$

 **HIGH VOLTAGE AND SAFETY**

Learning basic safety principles that will apply in work situations is extremely important. The HTI works with wiring designed for low-voltage systems. The voltage applied to a data cable would be hardly noticeable to most people. The voltage used for network devices, however, can range from 120 to 240 volts. If a circuit failure allowed this voltage to become accessible, it could give the HTI a dangerous shock, or could even be fatal. The HTI should not become complacent about high-voltage wiring just because most of the work is with low voltage. If high voltage is contacted suddenly, it may be difficult to control muscles or pull away. An HTI should always be prepared for such events.

Safety Practices
The HTI must learn how to detect unsafe voltages and how to deal with them safely until a trained electrician arrives. The following are guidelines that should be followed when working near electricity:

- Electricity can cause serious injury and even death. Most of the cables installed by the HTI carry only low signaling voltages, but these cables may travel near wires that carry much higher voltages. Similarly, the network devices that will be installed may not use a high voltage to process signals, but they are connected to power mains all the same, and failures can happen.
- Hazardous voltages are likely to exist in buildings in which the HTI works because these work areas are often in unfinished spaces and access areas.
- Keep an eye out for abnormal circumstances, such as circuit breaker boxes with the covers removed, frayed or loose cords, and damaged equipment.
- Pay particular attention to other devices and workers in the work area. In the workplace, cables may have to be run in areas with heavy motors, compressors, or other equipment that can draw considerable current and in turn can cause the voltage to decrease.
- Follow the "one hand rule." Never touch a cable that might be "live" with both hands because a complete circuit could be formed. Always keep one hand in your pocket. Safe work habits around electricity should be developed now and carried through to the workplace.

Cutting and Drilling

Quite often, the HTI or residential system installer needs to make holes in walls and floors so that wires can be routed. Safety must always be observed when cutting or drilling. Drilling into a live wire or a gas pipe can have fatal consequences. Although drilling or cutting into existing data and telephone circuits or water pipes is not dangerous, it can be costly to repair the damage. Before cutting or drilling through a wall or floor, make a small hole in the wall or floor, or consult the building blueprints.

This section of the chapter has focused on protecting the HTI and the network from electrical damage. Most electricity is generated by high-voltage sources such as power lines or lightning. But these are not the only sources that can damage a network. Electrostatic discharge, or ESD, can in fact be the most damaging form of electricity. It is discussed later in this chapter.

Lightning Danger

High voltage is not restricted to power lines; lightning is another source of potentially fatal high voltage. Lightning can damage network equipment if it enters the residential network cabling. The following steps help avoid personal injury and damage to network equipment:

- All outside wiring must be equipped with properly grounded and registered signal circuit protectors at the entrance point (the point at which they enter the building). These protectors must be installed in compliance with local telephone company requirements and applicable codes. Telephone wire pairs should not be used without authorization. If authorization is obtained, do not remove or modify telephone circuit protectors or grounding wires.
- Never run wiring between structures without proper protection. In fact, protection from lightning effects is probably one of the biggest advantages to using fiber-optic cable between buildings.
- Avoid wiring in or near damp locations.
- Never install or connect copper wiring during electrical storms. Improperly protected copper wiring can carry a fatal lightning surge for many miles.

 WORKING WITH ELECTRICITY

When working with electricity, you need to plan for power interruptions and failures. This section of the chapter details the types of power interruptions and failures that can happen. You should protect the home subsystems from these problems by installing backups and surge protectors. Furthermore, it is important to properly ground and bond all cables.

Power Interruptions and Failures

Blackout

Brownout

Sag

Noise

Spike

Surge

Blackouts, brownouts, sags, noise, spikes, and *surges* are all examples of transient power interruptions or events that can cause system malfunctions or even complete system failure.

Examples of power events that can cause problems to electronic devices and appliances in the home include the following:

- **Blackout:** The complete loss of power for any amount of time. It is usually the result of a strong weather event, such as high winds, lightning, or earthquakes.
- **Brownout:** A drop in power. Brownouts occur when voltage on the power line falls below 80 percent of the normal voltage. Overloaded circuits can cause brownouts, and they can also be caused intentionally by a utility company seeking to reduce the power drawn by users during peak

demand periods. Like surges, brownouts account for a large proportion of the power problems that affect networks and the computing devices that are attached to them.

- **Sag:** A brownout that lasts less than a second.
- **Noise:** Caused by interference from radio broadcasts, generators, and lightning. The ultimate effect is unclean power, which can cause errors in a computer system or network.
- **Spike:** A sudden increase in voltage that is much higher than the normal level. Spikes are usually caused by lightning strikes, but can also occur when the utility system comes back online after a blackout.
- **Surge:** A brief increase in voltage, usually caused by high demands on the power grid in a local area. Surges can last for considerable amounts of time. When a computer receives too much power, it may cause the power supply to run hot or even stop working.

The HTI can prevent many problems from occurring by having a thorough understanding of the types of power events that can cause disruption of the computer network and the integrated residential systems.

Uninterruptible Power Supply (UPS)

Virtually all networked homes have computers and/or microprocessor-driven electronic equipment that can be vulnerable to the power events explained previously. An uninterruptible power supply (*UPS*) is a device that allows a computer to keep running for at least a short time when the primary power source is lost. It also provides protection from power surges/spikes and brownouts. A UPS contains a battery that "kicks in" when it senses a loss of power from the primary source. If a computer is being used when the UPS signals the power interruption, there will be time to save any data and exit before the secondary power source (the battery) runs out. When all power runs out, the data could be lost if not properly saved. When a power surge occurs, a UPS intercepts the surge, preventing damage to the computer and network-enabled appliances.

UPS

Software is available that automatically backs up (saves) any data that is being worked on when the UPS becomes activated. Backing up data is the responsibility of the user.

Residential power consumption is measured in watts. Power consumption becomes important if an uninterruptible power supply (UPS) is deployed. A UPS is rated in wattage and time, and it must be selected to handle the equipment attached. Essentially, a UPS is a battery backup for the system.

A UPS may be rated at 1000 watts for one hour. This means that the UPS will operate an appliance drawing 1000 watts for one hour. They come in a variety of sizes. A small UPS will run a computer for about 10 minutes, giving the user enough time for an orderly shutdown. Critical applications may require a UPS to provide power for eight hours or longer and therefore require a larger unit.

Surge Protection

Surge suppressors are commonly used in network installations to protect against surges in electrical power. Typical surge suppressors have circuitry designed to protect the connected equipment or devices from spikes and surges. They can mount on a wall power socket or can be built into protected outlet strips. The most common method used in commercial applications is a metal oxide varistor (MOV). This device protects the equipment by diverting excess voltage to a ground. However, recent research at the National Institute of Standards and Technology indicates that the commonly used divert-to-ground scheme can still result in damage to data and equipment. Whole-house surge protection is a more comprehensive approach to protecting the computer network and other residential subsystem devices and appliances from electrical surges.

Today's popularity of home offices, the Internet, and home theaters means that more home electronics are vulnerable to voltage surges than ever before. A whole-house protection plan must protect against voltage surges on *both* the power line and the telephone or communication lines. Bear in mind that the homeowner may not be aware that surges can enter his/her expensive electronic equipment (modems, fax machines, answering machines, TVs, and VCRs) through telephone lines and cable TV or satellite dish connections. Most surge-protective devices designed for AC power lines do not prevent telephone or communication line surges from damaging such valuable equipment. Figure 9–2 shows an example of a whole-house surge-protection network.

Figure 9–2
Example of Whole-House
Surge-Protection Network

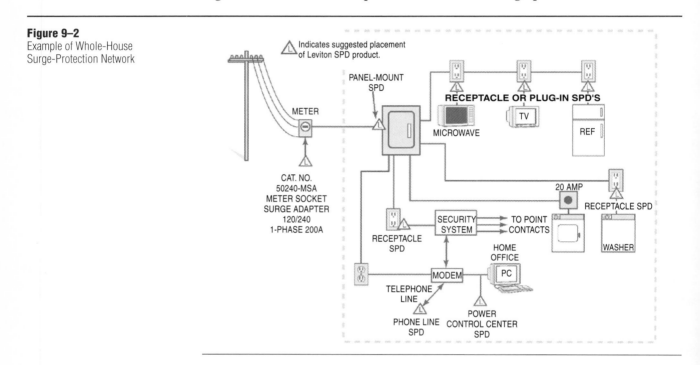

Power Adapters

A *power adapter*, whose importance cannot be underestimated, is typically supplied by the equipment manufacturer to ensure that the nominal 120 volts is stepped down to the level that the equipment is rated to use. Not using this mini-transformer could damage the equipment. Power adapters should never be swapped between two different devices, even if they are rated for the same output power.

Grounding

Grounding is the process of connecting an electrical signal to the earth (ground) by using a grounding conductor. The NEC defines a grounding conductor as "a conductor used to connect equipment or the grounded circuit of a wiring system to a grounding electrode or electrodes." The HTI must establish a separate telecommunications ground to provide a path for safe flow of unwanted voltages. If a piece of equipment fails and high voltage makes its way to the outside chassis of the equipment, a potential exists for dangerous or fatal shocks. A ground wire should connect the chassis to the earth. If a failure occurs on a piece of equipment that is properly grounded, the hazardous unwanted voltages pass safely to the ground.

The HTI can attach telecommunication grounding in two places. You can select one of the following:

- The nearest accessible location on the building ground electrode system
- An accessible electrical service ground

The preference for the telecommunication grounding is an electrical service ground that is a direct attachment to the closest point in the building's electrical service grounding electrode system. This ground effectively equalizes the telecommunications cabling and the electrical power cabling. In all new construction, the electrician is required to provide an accessible means to the electrical service ground, which may be a nonflexible metallic raceway (using an approved bonding conductor), an exposed grounding electrode conductor, or any approved external connection on the electrical power service panel. However, if no electrical service ground exists, you can use one of the following methods:

- A driven ground rod
- Another grounding electrode system installed for this purpose

Examples of how to install a grounding electrode or electrode system can be found in the *NEC Handbook*, Section 250.52. Some manufacturers may include additional instructions for grounding their equipment, and local codes must be considered. To ensure that the grounding actually works, it should be verified and tested by a licensed electrician.

Bonding

Bonding is the process of interconnecting grounded equipment. The NEC defines bonding as "the permanent joining of metallic parts to form an electrically conductive path that ensures electrical continuity and the capacity to conduct safely any current likely to be imposed." Bonding can therefore be thought of as extending the safety net provided by grounding. The combination of bonding and grounding culminates in a system that equalizes the difference of electrical potential among all the components of the residential telecommunications network to be as close to zero volts as possible. Figure 9–3 demonstrates the attachment of the bonding conductor and the ground wire.

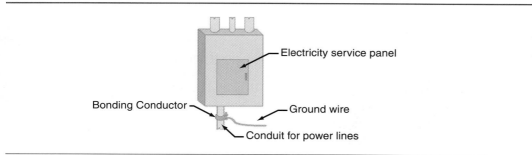

Figure 9–3
Attachment of the Bonding Conductor and the Ground Wire

The telecommunications bonding and grounding that the HTI provides is the additional bonding and grounding installed specifically for the residential telecommunications system. The bonding conductor for telecommunications grounding must be a minimum 6 AWG (that is, 4.1 mm or 0.6 inches). Additional bonding and grounding are provided by the building ground electrode system and the electrical service ground established by the electrician.

Keeping low-voltage wiring (for example, data and telephone) a safe distance from power line wiring is an important practice in all residential telecommunications installations. Do not allow data or voice wires to get too close to wires with electricity or anything else that is electrical. The data- or voice-transmission properties of copper cabling can be seriously affected by the close proximity of electrical circuits or electrical devices such as motors, fans, and lights. The noise from electrical circuits or devices, as you learned earlier, is known as EMI (electromagnetic interference). EMI can interfere with signals to cause distortion. Another reason for the separation of wires is to minimize the possibility of accidental contact with hazardous voltages. You must take the following precautions:

- Never install cabling near bare power wires, lightning rods, antennas, transformers, hot water pipes, or heating ducts.
- Never place data or telephone wire in any conduit, box, channel, duct, or other enclosure containing power or lighting circuits unless the duct or channel has special dividers to prevent these wires from coming together and to provide electromagnetic shielding.
- Never bundle or tie data cables with power cables or extension cords.
- Always provide adequate separation between low-voltage wiring (especially data/voice) and electrical wiring, according to the NEC standards. When in doubt, use the "rule of sixes." This rule requires 1.8 meters (six feet) of separation between low-voltage wiring and open high-voltage wiring, lightning grounding wire, or grounding rods. It requires six inches of separation from all other high-voltage wiring unless in conduit.

Electrostatic Discharge

A common example of an electric charge is called static electricity. Static electricity is the shock experienced after walking across a carpet on a cold, dry day and then touching a metal object. When static electricity discharges in the electronic environment of cables and networking, it is called electrostatic discharge (ESD). ESD can damage or destroy sensitive electronic components, especially computers and other networking equipment. Part of a good network wiring plan includes taking steps to prevent the occurrence of ESD and providing a harmless path for it to travel when it does occur.

Although static electricity is not harmful to living beings, such shocks can be disastrous for computers. If the spark of an ESD discharge can be felt, it is already many times greater than that required to destroy electronic components. Static electricity can destroy semiconductors and data in a random fashion as it passes through a computer. Good grounding and bonding can prevent damage from static electricity because they pass static bursts to the ground before the bursts can damage anything. Steps must be taken to deal with ESD in order to protect sensitive electronic equipment. It is far better to prevent static electricity than to cope with it. Personal grounding devices such as wrist straps are not required while cabling and terminating circuits. However, they should be worn any time when installing, inspecting, upgrading, or changing modules in a device. In addition, set computer equipment on ESD-protective work surfaces and place an ESD protective mat beneath the chair of a computer user (they absorb ESD). Temperature and humidity also play a role in ESD. Cooler and drier environments are especially prone to ESD, so consider using a humidifier.

To prevent ESD damage, sensitive devices should be handled using a ground strap. Once grounded, these charges quickly equalize and lose their strength.

Test Outlets

Sometimes, HTIs install devices and plug them in to find that they are not working. Many will think that the device is faulty, when in reality the power outlet is faulty. That is why it is so important to test outlets before relying on them to supply power.

Most modern networking and other equipment installed with residential systems has power indicator light emitting diodes (LEDs) that flash to indicate that the piece of device or equipment is receiving power. Typically green, these LEDs indicate that the equipment is receiving power and working normally. Sometimes there is a red LED to show that the device is "on" or is charging. Some devices may have a flickering LED to show that the equipment is not properly plugged in. There are all kinds of power functions that manufacturers build into their equipment to show its functional state (normal function, off, on, hibernating, standby, and so on).

ELECTRICAL EQUIPMENT

The location of equipment in the home should be closely coordinated with power outlet locations. One important reason to carefully coordinate the location of power outlets is to accommodate future installations, thereby minimizing the use of extension cords. In addition to outlets, this section covers what the HTI needs to know about light fixtures, light switches and dimmers, and power modules.

Electrical Outlets

An *electrical outlet* is a point on the residential electrical power wiring system at which current is taken to supply the devices installed. Typically, an outlet has a contact device called a receptacle, which is installed for the connection of an attachment plug. You cannot combine low-voltage wiring outlets and power outlets into one assembly.

Future locations of wired phone and other devices should be considered when determining the location of power outlets in the home. The kitchen is a high-traffic area of the home in which you want to avoid having power lines crossing the kitchen floor. You need to indicate certain locations for power outlet installation. This is true for both new construction and remodeling projects.

Each electrical wiring termination point or outlet has three wire conductors running to it: neutral, hot, and ground wires. In the United States, each outlet in a residential setting should provide the nominal 120 volts, with a 15-to-20-amp output. In other parts of the world, the voltage may be different, usually 220 to 240 volts. Figure 9–4 shows a typical electrical outlet in the U.S. labeled to show neutral, live, and ground.

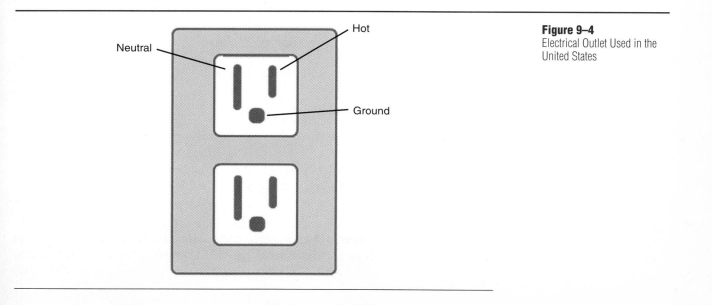

Figure 9–4
Electrical Outlet Used in the United States

Power receptacles or outlets need to be grounded by means of a grounding wire because this is the only grounding for the residential devices plugged into them.

In many retrofits, it is useful to test the voltage in the power outlets before plugging in networking devices. A device such as a computer can develop problems when plugged into a wall outlet in which the power is significantly lower than the nominal 120 volts (this can happen in older homes). The preferred method for testing is to use a circuit tester. Alternately, a multimeter with the AC voltage range selected can be used. Plugging the black probe (negative terminal) into the ground slot and the red probe (positive terminal) into the "hot" slot provides the actual voltage in the outlet. Make sure that all safety guidelines are followed.

X-10 Outlets

X-10, as discussed in detail in Chapter 7, is a common architecture for communication. X-10 is largely software-based and uses a set of commands that are transmitted by modulating the 60 Hz AC electricity used in North America. X-10 outlets or receptacles are the same size and shape as power outlets, and function just like X-10 appliance modules. When installing an X-10 receptacle, the main consideration is whether the X-10 will control one or both outlets. There are different models on the market. Some models, for example, have an X-10 outlet on top and a regular outlet at the bottom. The regular outlet is always "live." Duplex X-10 receptacles offer control over both outlets. For example, if a homeowner wants to control how long the kids stay up watching television or playing music, an X-10 outlet can be installed in the kids' room. The television plugs into the X-10 outlet, and the controller can effectively turn off the outlet and disable the television set at the programmed bedtime.

Light Switches and Dimmers

There are a wide variety of light switches and dimmers for residential electrical applications. Most of these residential devices are rated at 15 amps. Electronic dimming modules are special-purpose switches that operate on the principle that reducing the voltage dims the lighting. The most common type of dimmer module for residential use is the electronic dimmer rated at a 600-watt maximum, 120 volts. These dimmer switches find the most applications in low-voltage outdoor decorative lighting or track lighting.

The HTI and the electrician should coordinate the location of the devices. The optimal location of switches and dimmers is dependent upon the code requirements for location of these devices, future system integration issues, and the requirements of the homeowner. The HTI should coordinate with the electrician at the planning stages so that these needs can be more easily integrated into the electrical wiring plan.

Certain security issues will need to be taken into account regarding the location of these devices. In particular, devices for controlling certain lighting areas (security lights, front door, flood lights, and so on) may be located in areas that have controlled access. Otherwise, it may be desirable to use keyed or locking switches to prevent unauthorized individuals from having access.

Switch Ratings   Underwriters Laboratories (UL) lists switches that are used for general-purpose lighting control as *general use snap switches*. There are two listings:

- **AC general use snap switches:** The most frequently used switch in residential power wiring is the snap switch, sometimes called a toggle switch. It has many applications, which are all restricted to AC only. AC general use snap switches are marked "AC Only."

- **AC/DC general use snap switches:** Unlike the other rating, AC/DC general use snap switches have both AC and DC related applications.

Types of Switches   The types of switches available for residential installations include the following:

- **Interchangeable switch types:** Small wiring devices allowing up to three devices to be mounted on one mounting bar in a single-gang box.
- **Door switches:** Mounted into the doorframe on the hinged side of a door. The junction box is usually furnished with this switch.
- **Key switches:** Also referred to as lock switches, these are used when there is a need to restrict who turns on the particular load controlled by the switch.
- **Wet location switches:** Protected by a weatherproof cover, these are mounted outdoors or in other potentially wet locations.
- **Low-voltage switches:** Incorporate a step-down transformer and are used in low-voltage residential applications. For example, they are used in an application in which there is a need to input 120 volts and output 24 volts.

Light Fixtures

Light fixtures in residential electrical installations use light bulbs. A wide variety of light fixtures are available on the market. The choice of which to install comes down to the needs of the homeowner and the requirements for integrating and controlling the residential lighting. Although the electrician for the most part is involved with the installation of the fixtures, the HTI will find it useful to have sufficient knowledge of these products in order to influence the choice of fixtures for certain locations of the house based on the requirements for integrating the lighting system control with the overall home network.

Light fixtures are rated for the wattage of the light bulbs they use. To avoid overload problems, the rated wattage must be used.

Light fixtures for a home are generally grouped into two main categories: fluorescent and incandescent. Both can be surface mounted, recessed, or suspended from the ceiling. Light fixtures are designed to carry various types of light bulbs. The location of light fixtures is based on the design plan for the home network and the requirements of the homeowner. This should be coordinated with the electrician to ensure that the proper fixture is installed in the proper location. Certain rules guide residential lighting installations. They include the following:

- For security, it is useful to light all sides of the house, especially the dark corners.
- Outside floodlights should be aimed at a down angle to avoid unnecessary light reaching neighbors.
- Light fixtures should be sized in proportion to the size of the home: smaller ones for small homes and larger ones for large homes or multistory buildings.
- Use waterproof fixtures or provide cover for fixtures located in potentially wet areas.
- Illuminate pathways with spreads of light in order to accent the home landscaping.

Important code requirements for installation and location can be found on the labels attached to the lighting fixtures. Some examples of this information include the following:

- Ceiling mount only
- Access from behind wall required
- For chain or hook suspension only
- Suitable for use in suspension ceilings
- Suitable for installation in poured concrete

- Suitable for wet locations
- Maximum lamp wattage

All light fixtures must be listed for the location in which they are to be installed. Specific fixtures must be considered for wet, damp, and dry locations. Any outdoor location that is exposed to the weather should be considered a wet location.

Power Modules

Most Structured Media Centers (SMCs) are designed for low-voltage wiring only; they are not designed for the installation of standard AC cable. When power is required for subsystem modules, power modules are installed. Most power modules have the additional benefit of providing surge protection. Figure 9–5 shows an SMC power module.

Figure 9–5
SMC Power Module

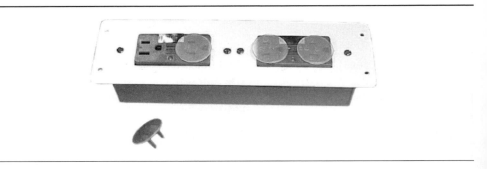

Depending on the SMC, the location of the AC power module may vary. Generally, it is located at the bottom of the panel to allow easy access to a power line. Low-voltage cabling typically comes in from the top of the SMC. Figure 9–6 shows where the power module is located within the SMC.

Figure 9–6
Location of the Power Module
within the SMC

The power outlet supplying the SMC should be a separate circuit on at least a 15-amp breaker, as illustrated in Figure 9–7.

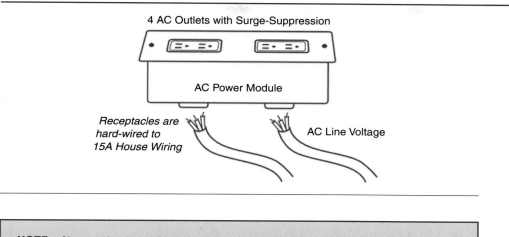

4 AC Outlets with Surge-Suppression

AC Power Module

Receptacles are hard-wired to 15A House Wiring

AC Line Voltage

Figure 9–7
Connecting the Power Module

> **NOTE**   Never wire live AC cable. The circuit must be disconnected at the power source and confirmed with a tester.
>
> **WARNING**   *Communications wiring and components should not be installed during lightning storms or in wet locations. Never touch uninsulated wires or terminals unless the wiring has been disconnected at the network interface.*

ELECTRICAL WIRING

Although the connectivity of devices for data and information exchange is assured by low-voltage cabling, other cabling types designed for power transmission must be installed to supply the power needed by the network equipment. Some of the devices, such as simple phone sets, use line power (48 volts) supplied directly by the telephone company; whereas other devices, such as cameras, require low voltage sometimes supplied over CAT 5 cables. Other equipment and devices, however, such as advanced telephone sets and computers, may need power from the residential power line to function.

Types of Residential Power Wiring

Although residential power wiring is installed by an electrician, it is important for the HTI to understand the process of installing electrical wiring. The HTI needs to coordinate with the electrician regarding the proper wiring and outlet locations that connect power to the networking equipment and devices. Because residential power wiring will likely be encountered by the HTI when working with retrofits or remodels, the ability to identify and handle it in the proper way is critical.

Three examples of electrical media types are NM cable, MC cable, and AC cable (discussed in the following sections).

Nonmetallic (NM) Cable

NM cable is a nonmetallic-sheathed cable type. It has an equipment-grounding conductor that is wrapped with fiberglass. The conductor size depends on whether the cable is copper or aluminum. The conductor size of a copper cable is 14 AWG to 2 AWG. The conductor size of an aluminum cable ranges from 12 AWG to 2 AWG.

> **NOTE**   Remember from earlier chapters that wire gauge increases as the AWG number decreases. Hence, a 2 AWG wire is larger in diameter than a 4 AWG wire.

NM cable has two conductor combinations: The two-wire cable has one white wire, one black wire, and one bare equipment grounding conductor; the three-wire cable has an additional red wire.

The equipment-grounding conductor may be insulated with a green insulating material, but the usual practice is to leave it bare.

> **NOTE**   Most electricians refer to sheathed cable as *Romex*, named after the Rome Wire and Cable Company.

Metal Clad (MC) Cable

Type MC cable is a metal clad cable. It is defined by the NEC as one or more insulated circuit conductors with or without optical fiber members enclosed in an armor of interlocking metal tape, or a smooth or corrugated metallic sheath. The conductors in MC cable are made of copper, aluminum, or copper-clad aluminum. However, the conductors may be solid or stranded.

The minimum conductor size for MC cable differs, depending on the type of conductor, and must be as follows:

- Copper: 18 AWG
- Aluminum: 12 AWG
- Copper-clad aluminum: 12 AWG

This means that MC cable will appear physically smaller in diameter when it is made of copper than when it is made of aluminum or copper-clad aluminum.

The sheath on MC cable is never used as a current-carrying conductor and provides an adequate path for equipment grounding.

Some of the permitted uses for MC cable and places where MC cable can be installed include service entrance cable (aerial or direct buried); power lighting, control, and signal circuits; dry locations such as embedded in plaster finish; and wet locations. When used in wet conditions, certain conditions must be met, as follows:

- There must be a moisture-proof jacket or lead covering under the metal covering.
- The insulated conductors of the cable must be listed for wet locations.
- The metallic covering or the cable must be moisture-proof.

Armored Cable (AC)

Armored cable type AC, simply called AC cable, was one of the first standards to be developed. The NEC defines AC cable as insulated conductors in a flexible metallic enclosure. Some electricians still refer to it as BX, a reference to the Bronx in New York City, where AC cable was once manufactured. AC cable is a type of armored cable. The armor can be steel or aluminum. The words "aluminum armor" or "steel armor" appear on the armored cable's marking tape as well as on the tag attached to each coil.

The conductors of armored cable can be either copper or aluminum. The conductor size depends on whether it is copper or aluminum:

- Copper: Conductor size is 14 AWG to 1 AWG
- Aluminum: Conductor size is 12 AWG to 1 AWG

Similar to nonmetallic-sheathed cable, AC cable has two conductor combinations:

- Two-wire: One white, one black
- Three-wire: One white, one red, one black

AC cable is also available with four conductors (black, white, red, and blue); or even five conductors, with the fifth conductor being a green insulated equipment-grounding conductor. The additional conductor may be used when an isolated equipment ground is required.

Electrical Wiring Chart

An electrical wire chart and building blueprints can help in planning the routing of low-voltage wiring. The electrical wiring chart is also useful for determining the location of outlets for the structured cabling system. In practice, both electrical and low-voltage cable termination (at least for some locations within the house) must be a coordinated effort between the electrician and the HTI. Such coordination is necessary to ensure that power is available where it is needed in the home. This will help to avoid the use of extension cords.

Typically, the electrician develops and provides the electrical wiring connectivity documentation. The building contractor can also provide documentation for electrical wiring.

SUMMARY

The HTI and the electrician should work together on a design/installation plan. The HTI must understand the fundamentals of basic electrical power terminology in order to communicate the client's needs to the electrician in a clear and precise manner.

This chapter discussed the following topics:

- Load requirements determine the size of circuit breakers or fuses in a residential electrical distribution panel and the gauge (AWG) of the wire feeding the outlet.
- Wire gauge charts are available to determine the proper wire size for different circuits.
- Power is voltage times current: $P = V * I$.
- AC is a form of electricity with a current and voltage that continually changes in *amplitude* and periodically changes in *direction*. In the U.S., it has a frequency of 60 hertz.

- DC has a steady value that does not change voltage direction or current flow.
- Surge protectors are vital, not only for computers but also for the complete home system.
- Whole-house surge protection is a more comprehensive approach to protecting the computer network and other residential subsystem devices and appliances from electrical surges.
- A whole-house protection plan must protect against voltage surges on both the power line and the telephone or communication lines.
- It is vital to understand the safety issues as well as the required standards and codes necessary for a licensed electrician to practice.

GLOSSARY

Alternating current (AC)   Electric current that reverses its direction of flow (polarity) periodically according to a frequency measured in hertz, or cycles per second.

Ampere   Unit of measurement for the number of electrically charged particles that flow past a given point on a circuit per second.

Backfill   Process of placing fire-stopping material into a punctured firewall to restore its integrity.

Blackout   A complete loss of power that is usually the result of a strong weather event such as high wind, lightning, or an earthquake.

Brownout   A drop in power for an extended period of time that is intentionally caused by the power company during peak demand periods.

Circuit breaker   A protective device for each circuit that automatically cuts off power from the main breaker in the event of a short or overload.

Coring   Puncturing a floor.

Direct current (DC)   A current that has a steady voltage. Batteries are a source of DC.

Firewall   A wall that is constructed of fireproof materials to prevent the spread of fire.

Main breaker   Turns the power entering the home through the breaker box on or off. The main breaker may be located in a breaker box or it may be in a separate box at another location.

Neutral bus bar   A bar to which the neutral wire is connected in a breaker box.

Noise   Interference caused by radio broadcasts, generators, and lightning.

Penetrating   Puncturing a firewall.

Sag   A drop in power that lasts less than a second.

Spike   A sudden increase in voltage, usually caused by a lightning strike or when the utility system comes back online after a blackout.

Surge   A brief increase in voltage, usually caused by high demands on the power grid in a local area.

UPS   Uninterruptible power supply. A device that allows a computer to keep running for at least a short time after the primary power source is lost.

Volt   A measure of current pressure at receptacles and lights. The average household voltage in the U.S. is 120 V.

Watt   A unit of measurement for power.

CHECK YOUR UNDERSTANDING

1. Which of the following is a standard defined by the NEC?
   a. How a cable should be terminated
   b. How a cable burns
   c. How a cable is pulled
   d. None of the above

2. What is the ventilation system's return space above ceilings or below floors called?
   a. Air duct
   b. Plenum
   c. Attic
   d. Crawlspace

3. What is puncturing a firewall during an installation called?
   a. Penetrating
   b. Coring
   c. Puncturing
   d. Cutting

4. What is puncturing a floor during an installation called?
   a. Penetrating
   b. Coring
   c. Puncturing
   d. Cutting

5. What is another name for the TIA/EIA standard "Commercial Building Grounding and Bonding Requirements for Telecommunications?"
   a. TIA/EIA-568-A
   b. TIA/EIA-569
   c. TIA/EIA-570
   d. TIA/EIA-607

6. Which equation is correct, where P = power, I = current, and V = voltage?
   a. $P = V * I$
   b. $V = P * I$
   c. $P = V/I$
   d. $P = I/V$

7. What is the unit of measure for power?
   a. Amp
   b. Ohm

   c. Watt
   d. Volt

8. Which term is used to describe a sudden brief increase in voltage that is usually caused by lightning strikes?
   a. Surge
   b. Spike
   c. Brownout
   d. Sag

9. Which term is used to describe a brownout that lasts less than a second?
   a. Surge
   b. Spike
   c. Noise
   d. Sag

10. According to TIA/EIA's standard on grounding and bonding, which of the following grounding methods is preferred?
   a. Driven ground rod
   b. Electrical service ground
   c. Bonding conductor
   d. None

11. The process of interconnecting grounded equipment is called:
   a. Bonding
   b. Terminating
   c. Grounding
   d. Filtering

12. Which type of electrical wiring is enclosed in an armor of interlocking metal tape or a smooth or corrugated metallic sheath?
   a. NM
   b. AC
   c. MC
   d. Coaxial

13. Which type of electrical wiring has an equipment-grounding conductor that is wrapped with fiberglass?
   a. NM
   b. AC
   c. MC
   d. Coaxial

14. What should be considered when determining the location of power outlets in the home?
a. Type of outlets
b. Future locations
c. Number of light fixtures
d. Number of phones

15. Which type of outlet allows a homeowner to automatically turn off the outlet and disable a device such as a television set at a specified time?
a. Electrical
b. Receptacle
c. X-10
d. Dimmer

16. What does ESD stand for?
a. Electrostatic discharge
b. Electrostatic degradation
c. Electrical discharge
d. None of the above

17. What is the "one hand" rule?

18. Power is composed of two main components. What are they?

19. What is an electrical wire chart used for?

20. What are the two main categories of light fixtures?

Residential Systems Integration

OBJECTIVES

Upon completion of this chapter, the HTI will be able to complete the following tasks:

- Identify elements of the infrastructure

- Identify the components of the Structured Media Center

- Describe the role of the control processor

- Recognize the need for user-friendly interfaces

- Understand how web portals work

- Understand how residential gateways work

- Be able to document the entire system to meet the customer's needs

INTRODUCTION

In this chapter, you learn about the processes and techniques required to integrate the various systems of an automated building into a unified, automated whole. You identify structured wiring and those situations in which it is and is not the best wiring solution. You learn about control processors, the central "brain" of the automated building. You identify the main issues in the various types of user interface available to the building automation integrator. You learn about the two major types of electronic entry into the residential system: home portals and residential gateways. Finally, you learn the importance of completely documenting the entire system, both hardware and software, and explaining this documentation to the homeowner.

WHOLE-HOME INTEGRATION INFRASTRUCTURE

The infrastructure, or supporting hardware and software, makes home automation possible. Home automation ranges from simple devices to turn lights on and off to the more complicated programming that allows the homeowner to control multiple functions over the Internet. Infrastructure falls into six categories:

- Structured wiring
- Structured Media Center (SMC)
- Control processor
- User interface
- Home portal
- Residential gateway

Each plays a specific role in the total infrastructure and must be designed and installed properly if the total automation system is to function properly. This chapter will explore each of these categories in the following sections.

Structured Wiring

Structured wiring

Structured wiring brings cabling for audio, video, telecommunications, computer networking, security, and home automation together into unified distribution points and cable runs. This means gathering coax, speaker, Category 5, Category 3, and even fiber-optic cables together in bundled runs into distribution panels.

Structured Media Center

Structured Media Center (SMC)

The *Structured Media Center* (*SMC*) is the command center for all the subsystems. Essentially, it is the metal box that holds all the terminations and devices needed to run each subsystem and the integrated home network as a whole. The SMC is also where outside services such as telephone, Internet, and even power are harnessed.

Control Processor

Control processor

For many homeowners, the combined control of all household systems is a major goal of home automation. At the center of any combined system sits a *control processor,* which may be a proprietary dedicated controller or a personal computer running Microsoft Windows, Apple's Macintosh OS, or Linux. This is the central computer that controls the entire building. Choosing the proper control processor means looking at a variety of issues, including functions, expandability, reliability, and ease of use.

User Interfaces

Homeowners can be provided with a range of interfaces, from simple light-switch-style controllers to hand-held touch-screen panels with Windows-style graphical user interfaces. The HTI must consider cost, control processor power, and ease of use in deciding which to offer.

Home Portal

The Internet has brought the expectation that any system can be controlled from any location—on or off the property. A *portal* provides a single interface access to all controlled systems. In consultation with the customer, the HTI must decide whether remote access is desirable and how different users will be able to see and manipulate the control of individual systems from the single portal interface.

Residential Gateway

When the systems inside the building must interface with systems outside the building—as telephone, cable television, Internet access, security, and electrical services usually do—they must pass through a *gateway* that provides security, access control, and usage-monitoring capabilities. Residential gateways can be vital components in the security and usability of a total automation system, and must be carefully designed with the specific needs of the homeowner in mind.

Gateway

Each of these categories is explored in more detail in the following sections.

 ## STRUCTURED WIRING

The infrastructure for integrating home automation begins with a plan for wiring each subsystem. Controlling systems in any building requires sending and receiving information and commands between control processors, sensors, and activation modules. Although some information flows through radio or infrared signals, most installations use cables or wires to connect the various components of the system. To avoid the confusing tangle of cables that can result from multiple installed systems, the HTI uses structured wiring; that is, cable sets designed to carry signals for all automation, communications, and entertainment systems to every room in the house in a neat and orderly fashion.

Because structured wiring frequently makes use of single aggregated cableways, it has been used primarily for new construction. Although some types of structured cabling, such as those binding multiple cables in a single jacket, are much more easily installed before interior wall surfaces go up in a new building, the principles of structured wiring can be applied to both new construction and remodeling projects.

Selecting Structured Wiring

Within allowable code limits, there are other factors to take into consideration when choosing cable for a structured wiring installation.

Low-voltage structured wiring cables *cannot* be bundled with AC electrical service wiring. Not only does this bundling violate building codes, it almost guarantees poor performance of the networking, communications, and entertainment systems because of interference caused by AC electrical wiring. This type of bundling also allows for the possibility of dangerous currents passing into the low-voltage wiring.

It is important to consult local building codes for clarification of issues such as allowable voids or empty spaces in structural framing members and required cable jackets for runs through attics, plenums, and other fire-sensitive areas.

Using CAT 3 Cable

Category 3 (CAT 3) cable and connectors are certified for up to 16 MHz of bandwidth, making them suitable for voice and low-speed (10 megabits per second) computer networking. Although most single- and two-line telephones use four-conductor RJ-11 plugs, an HTI using CAT 3 cable in a structured wiring installation should consider placing at least six-conductor jacks in all wall plates to allow for future phone expansion. It is important to note that CAT 3 specifications allow up to three inches of untwisted wire at each connection. The HTI should take care to minimize the untwisted length to avoid interference to and from the signals carried on the cable.

Using CAT 5 or 5e Cable

Category 5 (CAT 5) cable and connectors are certified for signals such as Ethernet with up to 100 MHz of bandwidth, making them the preferred choice for computer networking installations. CAT 5 cables are terminated in pairs according to a specification called T568A. The specifications for CAT 5 allow up to one-half inch of untwisted wire at each connection. The untwisted length should be kept to a minimum to avoid interference to and from the signals carried on the cable. CAT 5 cable is the preferred medium for voice and data transmission. In addition to its reliability, CAT 5 cable ensures that the cable plant will be able to handle any future telephone and most anticipated data applications.

Some home-automation sensor and command systems use two- or four-conductor cables for signal transmission. In all of these systems, Category 3 or Category 5 cable can serve as transmission wire. If economy is a prime consideration, four- or six-conductor telephone-style cable may be used, although the HTI should be certain that all conductors are at least size 24 AWG (American Wire Gauge).

Using Coaxial Cable

Video may one day be carried throughout the residence by unshielded twisted-pair (UTP) cables such as CAT 5. Today, however, coaxial video cable is the standard and must be included in any structured wiring installation. RG-6 coax cable should be used for highest bandwidth over the distances encountered in most residences. RG-59, although lower in cost than RG-6, should not be used because it can cause a reduction in signal strength known as attenuation.

Using Fiber-Optic Cable

Fiber-optic cable is also included in a structured wiring installation. Although relatively few systems use fiber-optic cable for distributing media or communications throughout a residence, it is widely assumed that entertainment and computer networking systems will each move to fiber-optic cabling in the future. When installing fiber-optic cabling, the two largest concerns for the HTI are cable termination and bend radius. There are several termination systems available. The HTI should become proficient in termination techniques to avoid creating cable with excessively high attenuation (signal strength loss). Optical fiber cable requires a gentle bending radius that is a minimum of 10 times the diameter to protect the high performance of the cable.

Installing Structured Wiring

Structured wiring is installed in a home run configuration, with all cables led to a central point of concentration. (For more information on home run configuration, see Chapter 8.) In individual rooms, multiple cable types may be led to single-wall panels with multiple media insertion points. Single-wall panels may have multiple cable outlet types in a structured wiring plan.

STRUCTURED MEDIA CENTER (SMC)

The Structured Media Center (SMC), otherwise known as the distribution device (DD) or distribution panel, is a central element of the home technology integration infrastructure. As shown in Figure 10–1 and Figure 10–2, the SMC is typically a metal box in which controlling components of each subsystem are housed for interconnection or cross-connection of the various components of the residential network. An SMC can be characterized by the following:

- Terminates and connects outlet cables, distribution device cords, equipment cords.
- Acts as the center of a star-wired residential network with home run cabling, playing the crucial role of signal distribution throughout the system, thereby enabling devices to communicate with each other or be controlled.
- Connects access providers to the residence by connecting the home network to the network interface device (NID), which designates the demarcation point.
- Facilitates moves, modifications, add-ons, and changes of premises cabling infrastructure within the residence.

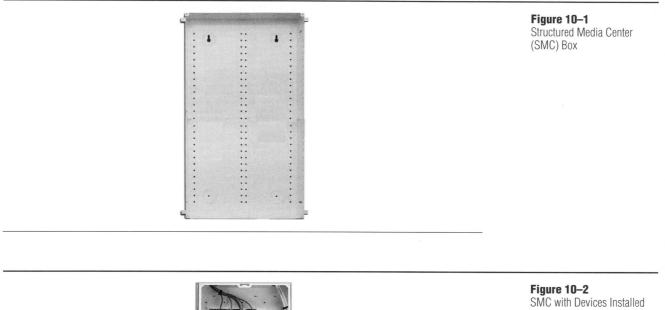

Figure 10–1
Structured Media Center (SMC) Box

Figure 10–2
SMC with Devices Installed and Cables Connected

Passive versus Active Cross-Connect Facilities

The SMC may consist of a passive cross-connect facility, an active cross-connect facility, or both. A passive cross-connect, as shown in Figure 10–3, is a simple signal relay device that distributes signals without the capability to manipulate or modify them; it has no intelligence.

Figure 10–3
Passive Video Splitter

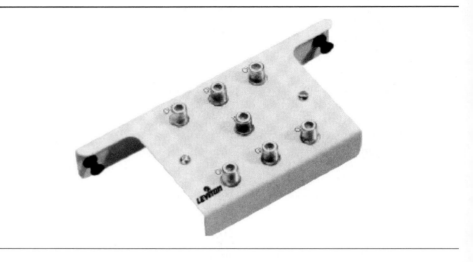

Conversely, an active cross-connect, as shown in Figure 10–4, has intelligence and can both distribute signals and manipulate or modify them by amplification or switching.

Figure 10–4
Active Video Splitter

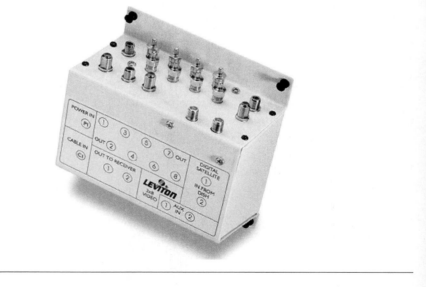

SMC Configuration

The SMC configuration depends largely on the number and type of devices to be supported on the network. It is usually configured as the center of the network in a star topology. All devices in the system have home run cables with direct channels to the SMC. One set of channels can be repaired without the entire system needing to be taken down. Figure 10–5 shows a typical SMC configuration.

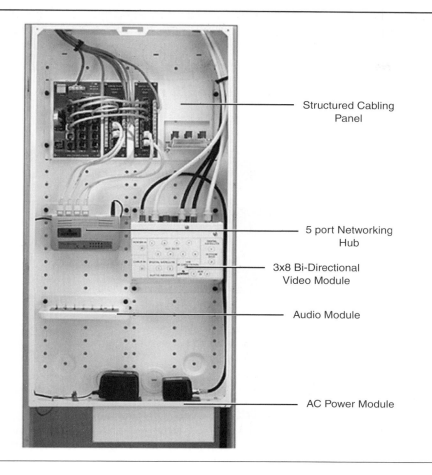

Figure 10–5
Typical SMC Configuration

Structured Cabling Panel

5 port Networking Hub

3x8 Bi-Directional Video Module

Audio Module

AC Power Module

The auxiliary disconnect outlet (ADO) is housed within the SMC. This option provides the homeowner with the capability to unplug the network from the access provider (local exchange carrier) network. Unplugging the access provider allows repairs or tests of various circuits in the residential network to be conducted without interfering with the service coming into the home. It is also useful for maintenance and upgrades.

The typical recommendation for a single-family residence is for one SMC to have sufficient capacity and functionality to accommodate the following minimum system requirements:

- CAT 5e data distribution
- CAT 5e telephone distribution
- Coaxial Series 6 (quad-shield) cable for video distribution
- Sufficient size conduit from the SMC to the nearest accessible space

There is no requirement for cabling to be in conduit unless this is specified by the local codes. In new construction, the builder typically installs conduit that will accommodate additional cable runs in the future.

Network Interface Device (NID)

The *network interface device* (NID), as shown in Figure 10–6, is a device installed by the phone company that connects the inside telephone wiring to the exterior telephone lines. It is a box located on the outside of the house, typically mounted near the electrical meter. There are two sides to the NID: the homeowner side and the phone company side. As you learned in Chapter 4, the point where the two sides meet is known as the demarcation point, or

Network interface device (NID)

Demarc

demarc. If there is a problem on the phone company's side of the demarc, the phone company is responsible for its repair. But if the problem is on the homeowner's side of the demarc, the homeowner is responsible for the repair. The homeowner's side of the NID contains a modular plug that allows the homeowner to disconnect all inside wiring and connect a working phone to test the line and determine who is responsible for the repairs.

Figure 10–6
A Network Interface Device
(NID)

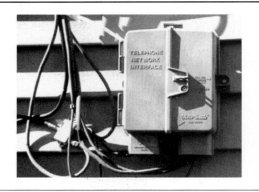

In some older installations, there may not be an NID. Instead, there is a protector block, which is a small plastic-covered box in which the cable coming out of the house connects to the phone company's cable. The protector block serves the same purpose as the NID, but does not provide a modular jack. The phone company can install an NID, which may be free of charge if a second line is installed.

Auxiliary Disconnect Outlet (ADO)

An auxiliary disconnect outlet (ADO) provides the means for a homeowner to disconnect from an access provider. The residential telecommunication standard recommends that in a single residential unit, the ADO should be installed where a means of disconnecting it is not otherwise provided or if the location is not easily accessible. The ADO should be located within the SMC.

The service company cable extends from the demarcation point to the ADO. In the case in which a single residential unit is part of a multiresidential building, the ADO cable may extend from the floor-serving terminal to the ADO in the homeowner space.

The cable types recommended for the ADO include the following:

- **Network connectivity:** One fiber-optic cable and conduit for fiber-optic cable
- **Data/voice:** Two CAT 5e cables
- **Off-air antenna:** One coaxial Series 6 (quad-shield) cable
- **Video (e.g., DBS):** Two coaxial Series 6 (quad-shield) cables (prewire only); some systems may require up to six quad-shielded coax cables. (DBS, or Digital Broadcast Satellite, was discussed in Chapter 7.)
- **Cable TV (CATV):** Two coaxial Series 6 (quad-shield) cable with F connectors

The ADO is installed by the HTI, usually in the SMC. Service disconnects are provided by the access provider at their own demarc.

For multidwelling units (MDUs), the minimum ADO cables are determined by the needs of the residents. More information on MDUs is found in Chapter 15.

CONTROL PROCESSOR

The control processor is the "brain" of any integrated home automation installation. It is sometimes referred to as the central control unit (CCU). The control processor may be a dedicated unit in a wall mount or a stand-alone personal computer running control software applications. The processor may be a Microsoft Windows computer; a Macintosh computer; a Linux computer; or a proprietary, stand-alone controller. In each of these cases, the interface on the control processor will primarily be seen by the person programming the system or making changes to the preset lighting scenes. In most cases, the HTI will want to provide an alternative user interface (see the section "User Interfaces") for the homeowner and family because the homeowner will not be changing the system on a regular basis.

Selecting a Control Processor

The control processor must have sufficient input and output capacity to deal with the various controlled subsystems, and it must have sufficient processing power to keep the programs running. In general, the control processor should be a dedicated computer rather than a software suite running on a family PC. The reasons for keeping automation control on a separate computer center are tied to reliability—it's best not to have a child's computer games cause the home automation system to crash.

There are dozens of control processor hardware and software products available on the market. In choosing a control processor for a particular installation, there are several issues that must be weighed:

- **Control Language:** X-10 is a common standard for controlling home systems. There are others, however, and you must ensure that the control program you choose will fully support the standard used by the sensors and control modules of the total system.
- **Input/Output Ports:** Does your automation system communicate via Ethernet, RS-485 serial, AC power modulation, or by some other means? Dedicated control processors purchased as part of an integrated system are likely to have I/O ports matched to the needs of the controllers and sensors, but PCs used as control processors may not have them.
- **User Interface:** If the homeowner wants Microsoft Windows as the user interface, the control processor will be more easily integrated if it also runs Microsoft Windows. Some dedicated controllers are designed to support wall switches as the primary control mechanism, whereas others require specific control units that run a proprietary interface. In many cases, it is possible to program an intelligent remote control unit to work with a given control processor, but doing so may require extensive software design and programming if the wall or remote control unit and the control processor work to radically different standards.
- **Programming Interface:** If you do have to create a link between the control processor and the various sensors, user control units, and activation modules, you should consider whether you can "point and click," or work in a specific programming language is required. If you do not have programming experience, don't underestimate the time required to learn a programming language.

Housing the Control Processor

In most cases, the control processor can be located in the SMC or alongside it in the wiring and equipment closet that houses networking equipment, entertainment media components, and other home automation devices. The benefits of locating the control processor in the SMC include access control, environmental control, and ready access to the wires and cables terminated at the distribution panels in the closet.

Configuring the Control Processor

Configuring the control processor takes on the user interface of the host computer. The programming interface of a control processor may allow for setting many parameters for each scene and system.

Configuring the control processor is as much a procedure of software as of hardware. An HTI should ensure that software requirements (for example, the processor speed, RAM, and hard drive space needed) can be accommodated by the hardware available. Be prepared to test both hardware and software after installation. Configuration of the control processor should center on making sure that the programs for each automation subsystem function properly, that subsystems interact properly with one another, and that the system responds gracefully to erroneous input or other error conditions.

One of the more crucial areas for configuration involves checking to make sure that control software is communicating properly with the I/O port used for control of the particular subsystem. Serial ports must be matched to the transfer rate, flow control type, and data configuration of the controlled application subsystem.

Once I/O ports are properly configured, care must be taken to ensure that software is communicating only with the ports designated. In a simple system with only one control medium—for example, in a system using X-10 control modules, to which the control processor communicates via an RS-232 serial port—this is a straightforward task. In more complex environments—those that see the control processor communicating with various systems through RS-232, RS-485, Ethernet, and parallel ports—keeping signals and interrupts straight can be the subject of significant debugging time.

An HTI must consider the consequences when the process controller's environment changes in unexpected ways. Control systems must be able to reject command sequences that would cause harm to equipment, and must be able to recover environmental data and resume processing if power is lost. The system's behavior after a power loss must be configured and defined before the system is turned over to the user.

Connecting to Devices

Controlled subsystems can communicate with the control processor in many different ways, but most subsystems communicate in only one way, making it possible to segregate the different subsystems by the types of media used. Integrated home automation systems may use modulated power line AC serial communications (either RS-232 or RS-485), Ethernet, or telephone connections to the sensors and control modules of the total system. Determine which media are needed to connect to each subsystem and ensure that the control processor has the respective ports to accommodate those types of media. Adapters can be used for ports that are incompatible.

Protecting the Control Processor

Uninterruptible power supply (UPS)

The control processor should also be protected by an *uninterruptible power supply* (*UPS*) that guards against long reboot times and lost data in the case of a power outage. As discussed in Chapter 9, a UPS allows for the continued operation of the control processor during brief power outages and protects the control processor from electrical surges, spikes, sags, and noise. Depending on the type of control processor deployed, rebooting after a power failure can take up to several minutes, and program information may be lost. Most UPSs have software that will trigger an orderly shutdown of the processor in the event of an extended power loss, saving data and program states and preparing for a successful reboot when the power is restored.

 **USER INTERFACES**

Integrated user interfaces can encompass computer terminals, wall switches, or cell phones. The user interface is a crucial element in the design and installation of any home automation system. When the user interface relates to multiple systems, it becomes even more crucial. Although integrated systems can have multiple user interfaces, the primary motivation behind the integration of multiple systems is the capability to have a single user interface through which all major home systems may be controlled.

There are many home automation software programs available. HTIs should research the ones available and select one that would best serve the homeowner, taking into consideration the types of subsystems that are part of the home automation system and the level of technical competency of the homeowner. Microsoft Windows XP is one such user interface that is being used to make home automation easy for the homeowner because it is familiar to most existing computer users.

Fixed Solutions

Wall switches and panels are still popular choices for many installations be-cause they offer immediate and direct control with a familiar interface. The major issue for the HTI comes when wall panels and switches are part of a multi-interface system that includes hand-held remote controls, wall-mounted touch screens, cell-phone or telephone system control, and other input methods. In these complex interface systems, the HTI must ensure that input from one device is reflected to other input mechanisms in order to avoid conflicts. Figure 10–7 shows wall-mounted switches, and Figure 10–8 shows a wall-mounted lighting scene programming device.

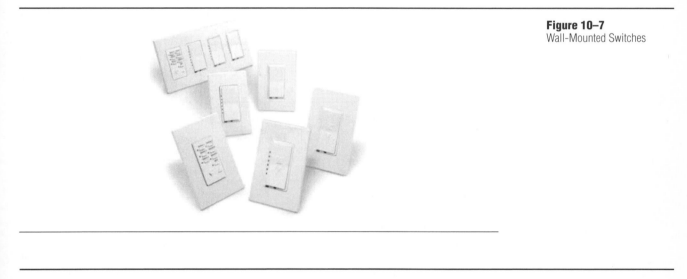

Figure 10–7
Wall-Mounted Switches

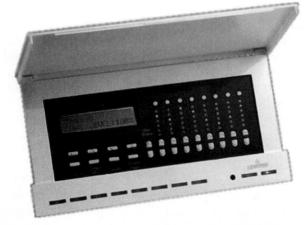

Figure 10–8
Wall-Mounted Lighting Scene
Programming Device

Workstation Solutions

Many installations use a Microsoft Windows interface for control of complex integrated systems. Windows CE is based on the Windows 32-bit operating system. Windows does not specify what the CE stands for, but some reports state that it originally stood for "Consumer Electronics." It is designed for implementation on mobile, hand-held, and other space-constrained devices that are suitable for deployment as automation controllers. If the system's control processor is also running under a Windows operating system such as Windows XP, Microsoft networking can be used as the communications mechanism between the two units. More information on Microsoft networking can be found at www.microsoft.com.

An HTI may still choose a windowing interface even if the operating systems of both the hand-held controller and the control processor are non-Microsoft based. It is important to verify that non-Microsoft products are compatible by checking product requirements. Graphical controllers allow multilevel menus to be built for individual control of many different systems, as well as scene controls that tie lighting, A/V, and other settings together into unified scenes.

Hand-Held Solutions

Hand-held remote control units without graphical screens are common for integrated automation applications because of their similarity to familiar A/V remote control units. These devices frequently provide the advantages of lower cost and programming that is easier than that of graphical hand-held remotes, with much of the functionality of the more-expensive units. But these hand-held devices cannot offer context-sensitive buttons and menus to help users make choices appropriate to the device and situation being controlled or configured. Referring to user manuals and manufacturer's websites will provide any additional guidance needed.

> **NOTE**   Microsoft Windows CE-based hand-held remote control devices can provide preset control screens as well as direct control of automated devices.

Interface Configuration

Configuring user interfaces is a process of combining user expectations, allowable system input, and characteristics of the particular user interface. Complex systems must be controlled through interfaces that are straightforward, unambiguous, and easy for everyone in the household to learn. One of the most common and effective means of doing this is to combine multiple system states into complete scenes by combining lighting levels, HVAC settings, and A/V commands.

It is important to build scenes and control combinations carefully, testing each controlled aspect fully before adding the next. Care should also be taken to make each control point aware of levels set by other control points when multiple control points are available (as they almost certainly will be for lighting). Confusion between control points can result in unanticipated device behavior and unsatisfied homeowners.

HOME PORTAL

With the advent of the World Wide Web, opportunities for integrated home automation user interfaces increased dramatically. Many homeowners who are comfortable with shopping, learning, and being entertained on the web are eager to control their home systems through their personal computer's web browser. The resulting marriage of web and home automation provides the users with "point-and-click" control of their home's systems, and may offer them the ability to modify the preset controls without having to become expert in the control language of the individual system.

Integrated home automation systems can provide a web browser interface through the use of a residential portal, which takes the command structures of the automation systems and translates them into control codes that can be understood by a browser. The portal may be a piece of software running on a home PC or an application running on a stand-alone server appliance. To set up the home portal, you need to install a web browser and portal software on your PC. Figure 10–9 shows an example of a simple home portal with icons for entertainment, computer, phones, security, and so on. Clicking on these icons opens a window that enables the options to be set.

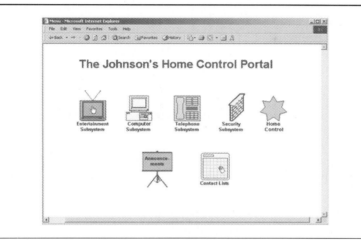

Figure 10–9
Example of a Home Portal

Web Browser

The first requirement for a home portal is that a web server be installed. A *web server* is an application that understands a request for information in the standard http://www.xxx.yyy form and returns the requested file or files in response. Server software uses HTTP to return HTML documents and any associated files and scripts when requested by a client, such as a web browser. A web server can run on a dedicated computer or it can be an application running on another computer.

Portal Software

The next requirement for the home portal is an application that takes the home automation control and sensor data streams and translates them into information that can be displayed on and acted upon by a web browser. It is possible for an HTI to write this translation software using software development tools such as Python. Python is an interpreted, interactive, object-oriented programming language often compared to Tcl, Perl, Scheme, or Java.

Developing personal software is not for everyone. It is far easier and more cost-effective to purchase an application that bridges the gap between automation and the web. There are programs for both PC and Macintosh; one such program is HomeSeer, found at www.homeseer.com. HomeSeer works with the PC for X-10 technologies and offers a free 30-day trial. These types of programs can be set up with the basics, and modules can be added that provide more functionality.

Selecting a Portal

There are several key factors to be considered when choosing any product to provide a portal for residential customers. Answering the following questions can help ensure that the project is successful from both the HTI and home-owner points of view.

- How integrated is the portal?
- Does it already understand the controls being used in the implementation, or does it provide a mechanism for readily adding the controls to its interface?
- Does the portal have a web server integrated into the basic application, or does it require a separate web server to be installed and configured?
- Does the portal provide the same interface for both programming and using automated systems?
- Is the user protected from making changes to scene presets, or can the homeowner make changes to any scene at any time?
- If there are two separate application screens for programming and use, are both readily accessible by any user?
- How secure is the portal?
- Does it require users to log in to control home systems?
- Does it require users to log in to program home systems?
- Is the portal optimized for one particular browser? If so, is it the browser the customer currently uses?
- Is the portal built on industry standards such as XML, SOAP, or Java? If not, are you entirely comfortable with the software publisher's chances for long-term survival?
- How user friendly and intuitive is the portal? What amount of time will be required by the homeowner to learn the system?

Each project may require a different set of correct answers to these questions. If you make sure that each question is answered, however, you stand a better chance of creating a successful portal for your customers. Home portal products may offer varying functions, but all operate within the familiar interface of a web browser. Table 10–1 lists the websites for some of the software products available for home automation.

Table 10–1

Software Providers for Home Automation

Home Automation Site	Description
Automated Future Home	Automated Future Home sells Windows software for X-10 home automation that enables your home to be aware of you by communicating with lamps, appliances, motion detectors, and all X-10 devices. It has speech, e-mail, and security capabilities. (Windows 95/98/2000/XP)
Flipit: Unix support for the X-10 Firecracker (the CM17a)	Flipit is a simple command-line program that supports the X-10 Firecracker (model CM17A) on a Unix machine.
Heyu X-10 Linux Home page	Gadget lover's page, featuring a program (heyu) to operate the X-10 CM11a interface from a Linux system.
Home Automated Living	Voice recognition software for Windows 95/98 that supports HVAC, telephony, infrared, Internet, X-10, and security. Voice-controlled digital home systems integration software, including HAL1000, HAL2000, HAL3000, and Virtual Voice Assistant.
Homeseer Software Plug-Ins	Provides home automation software plug-ins, including caller identification units, security systems, and weather stations.
HomeSeer Technologies	Provides remote access to your lights, appliances, security system, and e-mail from any web browser.

HomeTool Pro	Enables installers to configure and program individual SmartOne products or an entire SmartOne integrated control system.
Hometouch Inc.	Designs software for the automation and controls industry.
IEC Intelligent Technologies	Provides LonWorks ® network management solutions and embedded software and hardware products. (Producers of ICELAN G and the Peak Components API.)
InHome Solutions	Provides automation and surveillance solutions for homes and businesses.
The Java X-10 Project	Java APIs and Forum for managing an X-10 hardware network.
Macintosh Home Automation	Macintosh software for two-way X-10 home automation and security.
Master Control Panel	PC software for building graphical onscreen control panels for home theater and home automation equipment.
MiDaTek International	Provides SCADA MMI HMI DDE factory automation software for the industry—to control and monitor production lines and machines.
Scott Crevier's Home Automation	Offers an X-10 web interface and X-10 Windows 95 QuikMenu techniques.
Wipro CommEngine Communication Software	Software infrastructure solution for OEMs, addressing the end-to-end communication needs of Home Gateway devices.
XPLAB	Research laboratory for industrial automation, real-time software, data acquisition, and I/O board.

Configuring the Portal

Configuring the home portal requires dealing with several aspects of the application:

- **Connection between portal and controlled systems:** As previously mentioned, many portal applications designed for home automation deployment come with the translators necessary to control common automation modules. If the installation is built around a common command protocol such as X-10, there may be many portals that come preconfigured for control. If the installation uses a less-common protocol, make sure that the correct command set is included in the software. This is typically provided on a CD included by the manufacturer. If it is not, a mechanism for adding command sets should be included in the product, and the HTI needs to understand the programming language required for adding command sets. For help, visit the manufacturer's website; it provides up-to-date information, technical support, and current downloads.
- **User interface:** The basic user interface for a home portal is set—it is the appearance of the web browser on the user's computer. Within the browser window, however, decisions may have to be made. Some portals come with an established set of buttons, slides, and other controls; whereas others may enable you to choose from sets of controls. In either case, you have to decide whether the portal will be a web-based representation of a hand-held remote control or a window into the programming interface for the integrated home system.
- **Connection between portal and other computer applications:** Many portals come with the control application and web server integrated into a single package. These portals have a clear advantage for keeping software integration and programming efforts to a minimum. If the portal application requires a separate web server, you need to understand the basic server setup. The most commonly used server on the web is either Microsoft's IIS or Apache, the open source web server.

- **Security:** Security becomes an issue with portals in two cases. First, if the portal allows access to individual device control and scene reprogramming, these advanced development sections should be protected by password to keep curious guests and mischievous children from wreaking havoc on the home environment. Second, if the portal allows access to home control from the Internet, all control functions must be password-protected. More sophisticated systems can discern the difference between connections made from the local network (inside the home) and those coming from the Internet, but blanket password protection is the minimum requirement for security.

Connecting the Portal

If the home portal is on a stand-alone computer or server appliance, it connects to the control processor, the residential gateway, and other computers through a standard computer networking protocol. In today's homes, that may mean Ethernet, in either 10-megabit or 100-megabit speed. Given the current market, the HTI should not install 10-megabit Ethernet in a new network—there is no significant cost advantage in the slower network speed, and users will be limited as new technologies come to market.

If the portal is to be accessible from the Internet, a residential gateway is required. The home portal connects to the gateway via Ethernet, and the HTI has to correctly configure the gateway to allow browser access to the necessary software ports on the portal.

RESIDENTIAL GATEWAY

Residential gateway

In any modern residence, there are numerous points of contact between internal communications systems and the outside world. In an integrated automated residence, it means that there are several points at which the control of the home is open to those outside the home. The points of access include the telephone lines; a broadband Internet connection such as cable modem, DSL, ISDN, or point-to-point wireless; television cable; and satellite television. A busy household may have multiple instances of any of these communications media, and the need to control and secure access is therefore crucial. The answer is a *residential gateway*, which is a device that connects multiple computers in a home to a single Internet connection and connects all internal home networks to their external counterparts. Without a residential gateway in place, there is no security or control between the residential network and the outside world. Figure 10–10 shows a residential gateway.

Figure 10–10
Residential Gateway

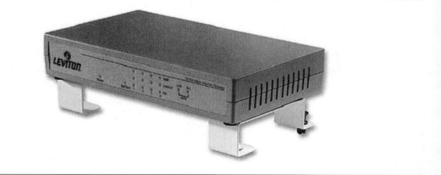

Another appliance may house the home gateway, an application responsible for control and security of the homeowner's connection to the Internet.

The gateway takes a single broadband connection to the Internet and distributes it to the residential network. It should support the operation of a firewall, the application that prevents unauthorized users from gaining access to any of the residential computers. The gateway can also include software that limits the activity of family members to acceptable websites and content.

You can think of a residential gateway as the software answer to the structured wiring closet. Depending on the nature of the systems, a residential gateway can provide control and access for telephone, Internet, and security connections. If the homeowner's Internet access comes via cable modem, cable access may also be included.

The main justifications of the residential gateway are control and security. Properly configured, the gateway blocks access to the home network from those who do not have a legitimate reason to have that access, and they may provide a measure of control to parents over the content that children can bring into the home.

Selecting a Gateway

The first question the HTI must ask when choosing a residential gateway is, "How many systems will be integrated into the gateway?" If the gateway serves only as an Internet access device, there will be one set of decisions. If telephone is added to the gateway, the decisions change.

Most residential gateways are used for secure access to the Internet. The goal is to provide protection to the computers inside the network while allowing controlled access to resources outside the network. Most computer networking gateways provide two functions: security and distribution. The security aspect is handled through an application called a firewall, which is designed to keep unauthorized users or applications from gaining access to the computers inside the network. The distribution function generally takes the form of a 10/100 Base-T switch or hub that distributes the single broadband Internet connection to several computers within the home. In a growing number of products, a wireless 802.11b Ethernet access point is also included in the gateway to make wireless laptop computer networking a possibility.

Configuring the Gateway

Many gateways are configured simply by plugging in the various cables and turning them on. The appliances come preconfigured with the options most users will need. In other cases, the aspects of computer networking need to be understood in order to best configure the gateway.

If gateway configuration is required, you should understand how TCP/IP (Transfer Control Protocol/Internet Protocol) addresses are assigned, and whether you need to assign each computer and network device its own address or allow addresses to be assigned automatically by the gateway. In either case, security dictates that you assign each device an address that is nonroutable in an address space; that is, it cannot easily be fooled into believing that an outside computer is actually attached inside the network. An address in the 192.0.0.0 or the 10.0.0.0 address spaces should be safe.

You also need to know which protocols are used most frequently by the members of the household. HTTP is the protocol used to view web pages, whereas SNMP is the protocol for sending e-mail. Both should be allowed. Other protocols, such as CHAT or instant messaging, may be used by members of the household. If a protocol is not used, it should be blocked at the gateway for security reasons. (For more information, select the Technologies option at http://www.cisco.com.)

Whether they are stand-alone appliances or software running on a server, gateways and other security applications are best configured alone. Unless economics force combinations, don't put the gateway on a server that will be running the home portal. By keeping the applications separate, you increase the likelihood that the computer's processor and I/O subsystems will be able to keep up with the demands of the applications, and reduce the likelihood that a bug or crash in one application will keep all important functions from operating.

TESTING AND DOCUMENTATION

Testing home automation installations is similar to the testing required by any technical project. Each individual component must be tested, followed by subsystem testing, with total system testing at the end of the process. If problems are detected, determine their nature, fix them, and document the repairs required.

It is also important to document each individual subsystem in the controlled residence, and it is even more important to document how the various subsystems connect to each other. In addition, the software installed should be well documented, with clear comments and meaningful routine and variable names. For complex systems, make sure that copies of source code are retained in order to provide updates or upgrades to the system.

Testing

When testing the whole system, each individual component must be tested first, followed by subsystem testing, with total system testing at the end of the process. When complex scenes are programmed in an integrated home, subsystems should first be tested alone, then in pairs, and finally as complete scenes. A thorough testing regimen will include testing of the scene if one or more systems are disabled—individual subsystem failures should not mean the failure of the total system.

Test the components, subsystems, and total system for the results of incorrect and correct input. Remember to test for unusual circumstances so that there can be confidence that the system will continue to function properly, even when the users don't use it correctly.

Ask the following questions when testing for system behavior after a power failure:

1.   Do all the affected systems reboot properly?

2.   Does the process controller wait until all subsystems have successfully rebooted before beginning to query sensors and issue commands?

3.   If the subsystems reboot in an unanticipated sequence (that is, one system returns to full operation before another), does the process controller respond gracefully? Ensuring good power-failure behavior can minimize panic-stricken phone calls at 3:00 a.m.

Documenting Problems and Changes

It is important to keep a careful record of each problem or change made. Record the reason for the change or repair, the person making the change or repair, the time and date of the change or repair, and the complete details of the change or repair. Although tedious to make, these records provide valuable information if subsequent changes are necessary.

Documenting System Design

Documenting the design and implementation of a fully integrated home automation system is a crucial part of the total automation project. This documentation includes the following:

- A wiring diagram that shows where all outlets, modules, switches, and cable runs are located
- A schematic diagram that indicates the logical connections between all components
- A block diagram that shows how commands move through the system, including information about the systems that must activate first in multisystem scene settings
- Labeled photographs and equipment location diagrams that show how connections are made
- The comments made by the programmer for the software that will allow for updates and upgrades to custom software

In every case, a complete set of wire charts, block diagrams, and schematics should be included in the homeowner's packet delivered at the end of implementation and training. A well-designed, professionally packaged set of documentation gives the homeowner an accurate sense of the HTI's competence and professionalism, and will help when the time comes to make changes or additions to the system.

Wire Charts and Schematic Diagrams

Complete diagrams that show how each cable is run through the building, where each terminates, and how each is labeled should be prepared as part of the documentation. In a structured wiring installation, these diagrams should support the labeling of each cable that comes back into the central wiring closet and indicate where each cable terminates at a wall panel or junction box throughout the residence. A list of symbols, colors, and abbreviations used on the wiring diagrams should always be included as part of the documentation package.

Schematics that show the logic of all connections made should also be prepared. These schematics are crucial when debugging or adding to an integrated automation system because they can indicate the consequences of attaching new devices to particular buses, tapping into cables, or reordering the order of device attachment. Figure 10–11 shows a schematic diagram.

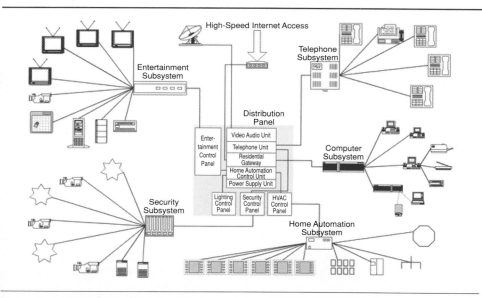

Figure 10–11
Schematic Diagram

In an integrated system, it can be important to know how signals flow from one device to another. Flow charts (such as the one shown in Figure 10–12) provide information about how commands are evaluated in sequence, which systems begin to change state first, and how commands are processed. This information can have critical importance when debugging or changing a system.

Figure 10–12
Signal Flow Chart

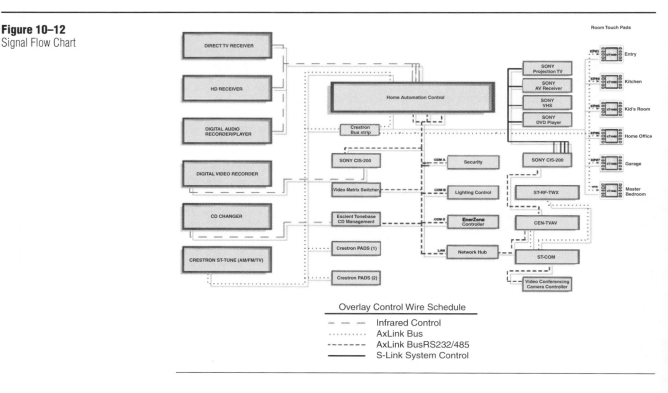

Equipment Layout Diagrams

Documenting the layout of cables and arrangement of equipment is a crucial step in the complete documentation of an installed system. The HTI can label photos of equipment racks with equipment names, cable runs, and other pertinent information. Where cables can be separated by jacket color or connector type, these distinctions should be listed and made clear in the documentation. Photos labeled with information on which equipment is connected and diagrams reinforcing the information in the photos can make wiring changes more certain and less time-consuming. Figure 10–13 illustrates an equipment layout diagram.

It's also important to list power connections on the wiring guide and label connectors at the AC connection end, so that technicians and homeowners have a clear understanding of the consequences of unplugging a particular cord or turning off a specific power strip.

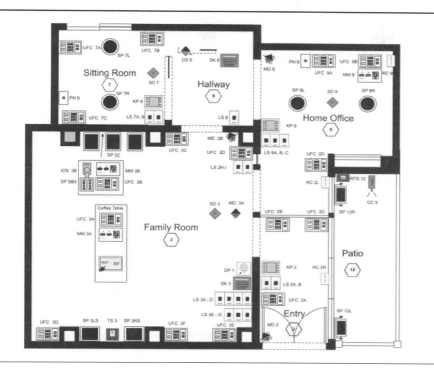

Figure 10–13
Equipment Layout Diagram

SUMMARY

Integrated home automation systems bring control of all major automated systems—communications, lighting, entertainment, and security—into a single user interface.

This chapter discussed the following topics:

- Structured wiring brings cabling for all media and control functions together in single distribution points.
- Integrated home automation systems are controlled through a control processor, also known as the central control unit (CCU).
- The control processor is a computer that may be a stand-alone appliance or a personal computer running Microsoft Windows, the Macintosh OS, or Linux.

- The control processor can be housed in the Structured Media Center, which is the central point of an integrated home network.
- The network interface device (NID) is a device that connects the inside telephone wiring to the exterior telephone lines.
- An uninterruptible power supply (UPS) can protect the control processor from power outages and interruptions.
- User interfaces are user-friendly devices that make controlling the home automation easy for homeowners.
- A home portal provides a single web browser interface to all controlled systems in the home.
- The residential gateway provides security and control between the residential network and the outside world.

GLOSSARY

Control processor   At the center of any combined home-integration system sits a control processor, which may be a proprietary dedicated controller or a personal computer running Microsoft Windows, Apple's Macintosh OS, or Linux. The control processor is the central computer that controls the entire building. Also known as the central control unit (CCU).

Demarc   The point in the NID that separates the responsibility of repairing the telephone lines between the phone company and the customer.

Gateway   A device that controls access and provides security to the residential network from the outside, primarily the Internet.

Home portal   A web-accessible interface that enables the homeowner to control all the subsystems for a PC, PDA, or other web-accessible device.

Network interface device (NID)   A device that connects the inside telephone wiring to the exterior telephone lines.

Residential gateway   Device that connects multiple computers in a home to a single Internet connection and connects all internal home networks to their external counterparts.

Structured Media Center (SMC)   The location in which all the cables and central control devices for subsystems are housed.

Structured wiring   Bundles of wires for all the subsystems that are pulled together and then terminated in the Structured Media Center.

Uninterruptible power supply (UPS)   Device that protects the control processor from power outages and interruptions.

CHECK YOUR UNDERSTANDING

1. What is the wiring plan that runs cables for all home systems to a single distribution point called?
   a. Combined wiring
   b. Structured wiring
   c. Ribbon cable
   d. Home automation

2. Low-voltage structured wiring cables cannot be bundled with AC electrical service wiring because of:
   a. interference caused by AC electrical wiring.
   b. interference caused by low-voltage wiring.
   c. low-voltage wiring is too thick to bundle.
   d. None of the above

3. What is cable and connectors certified for up to 16 MHz of bandwidth and suitable for voice and low-speed computer networking called?
   a. RG-6
   b. Category 3
   c. RG-58
   d. Category 5

4. What is cable certified for signals such as Ethernet, with a bandwidth up to 100 MHz, called?
   a. RG-6
   b. Category 3
   c. RG-58
   d. Category 5

5. What are the two major concerns when installing fiber-optic cabling?
   a. Cable termination and bend radius
   b. Cable termination and length
   c. Cable termination and cost
   d. Cost and bend radius

6. Which of the following does *not* characterize a SMC?
   a. Terminates and connects cables
   b. Center of a star wired configuration with home run cabling
   c. Connects service providers
   d. Restricts any infrastructure change or modification

7. What does ADO stand for?
   a. Auxiliary disconnect outlet
   b. Added disconnect outlet
   c. Active disconnect outlet
   d. Auxiliary diagram outlet

8. Which of the following is an acceptable operating system for a control processor?
   a. Windows
   b. Macintosh OS
   c. Linux
   d. All of the above

9. What is the device called that protects a control processor from loss of electrical power?
   a. UPS
   b. USB

c. SLR
d. GMT

10. Which communication link may be used between a controlled system and a process controller?
    a. Ethernet
    b. RS-232
    c. RS-485
    d. All of the above

11. Home portals allow automated systems to be controlled through:
    a. Microsoft Windows.
    b. Linux
    c. voice commands.
    d. a web browser.

12. What do home portals that allow access from outside the home network require at a minimum?
    a. Their own telephone
    b. Password protection
    c. Video cameras
    d. A cable modem

13. What are the two functions of a computer residential gateway?
    a. Security and dependability
    b. Expandability and distribution
    c. Security and expandability
    d. Security and distribution

14. A residential gateway contains a piece of security software called a:
    a. guardhouse.
    b. stone wall.
    c. portal.
    d. firewall.

15. What does a well designed system test plan do?
    a. Tests each component only once
    b. Tests subsystems first and then components
    c. Tests components, then subsystems, then combinations, and then the total system
    d. Tests the total system first and then each component

16. Why is it necessary for the HTI to be proficient with cable termination?

17. What is the difference between CAT 5 and CAT 5e cable?

18. What is the difference between passive and active cross-connect facilities?

19. Give an example of an active cross-connect facility.

20. The HTI should not install 10-megabit Ethernet in a new network. Why not?

Installation Planning and Implementation

OBJECTIVES

Upon completion of this chapter, the HTI will be able to complete the following tasks:

- Understand the homeowner's needs

- Determine outlet locations and cable requirements

- Complete the design process

- Plan for the SMC including power requirements

- Develop a bill of materials and estimate labor cost

- Construct the timeline to complete the project

- Use the seven steps to create a proposal

- Secure approval with a signed contract

INTRODUCTION

In this chapter, you learn about planning the project for the integrated home network. The first step in the design process is to determine the homeowner's present and future needs and lifestyle. The design process includes conducting the site survey and determining where outlets will be placed. It also includes planning the Structured Media Center (SMC). In this process, the timelines are developed and the estimates are made in order to present a proposal to the homeowner for approval.

The seven phases to developing a proposal help to determine the scope of the project and become the project plan. The planning process is completed and work can begin once a contract is signed. Contracts are binding legal documents that the HTI should write carefully and that the homeowner should understand before signing.

273

DETERMINING THE SCOPE OF THE PROJECT

When designing an integrated home network, there are some general design considerations that determine the scope of the project. These include:

- The homeowner's current and future needs
- The budget
- The timeline

All of these factors affect the size and pace of the project.

Assessing Homeowner Needs

The first step in the design process is determining what the homeowner needs. The homeowner's current needs are largely dictated by the intended use of the home network. You need to know what the homeowner wants the network to do for them and what existing subsystems need to be incorporated. You will need to meet with the homeowner to gather all this information. Some homeowners will be very specific about what they want. Others will not know what home networks can offer them. For the latter, you should be prepared to give a small presentation covering their possibilities.

Tailor your interview and presentation according to the type of project (retrofit vs. new construction). A homeowner whose home needs to be retrofitted may be particularly concerned about the cost of installing new wiring and the physical damage that will be done to the home (e.g., tearing holes in the wall). A homeowner who is having a new house built will not have these concerns. Instead, you can explain that the cost of wiring the entire house now is significantly less than any subsequent wiring. Figure 11–1 shows an example of a Leviton connection planner.

		Other location-based components								
		8-inch speakers	8-inch speaker		6.5-inch speakers	Outdoor speaker	Weather cover	Analog vol. control	Digital vol. control
	Leviton part numbers	40890-WP pair	40890-WS single		40891-W pair	40827 single	5977-GY	40841-D* impedance matching	48211-V* impedance matching
		Total Req'd.	6		1	1	2	1	4
Room type & applications	Number of locations planned								
1. Living Room									
Telephone									
Data (multi-location networking)									
Cable TV									
Satellite TV									
Home theater									
Multi-location stereo									

Figure 11–1
Leviton Connection Planner

Many homeowners do not know what is possible with a home automation system. Use the following list to help them determine their needs and desires:

- Which services do they want? Video, e-mail, Internet, security, pool control, etc.
- Do they want to share devices such as printers, files, scanners, CD-ROMs, and DVDs?
- Do they want the option of moving equipment around (and thus a wireless solution)?
- Do they have a home business or do they telecommute?
- How many people live in the house? Do they foresee additional family members in the future?
- Is security a concern?
- Do they need to control or protect children's access to the Internet?
- Is there a need or desire to monitor a child's room or the home entrance (i.e., a need for video cameras)?

A first-time or technologically shy user requires a user-friendly interface and may need customer support material. Assess the technical competency of the homeowner beforehand so that you can tailor the interface and instructions for them.

- Has the homeowner worked with computers?
  - Work related
  - Playing games
- Does the homeowner own a computer?
  - PC
  - Macintosh
  - Desktop
  - Laptop
- Are they currently connected to and using the Internet?
  - Fast connection (cable or DSL)
  - Dial-up modem
- Do they use e-mail?
  - Daily
  - Occasionally
- Is there a connected home office?
  - Type of Internet connection
  - Additional phone line for fax machine
  - Additional outlets for equipment
  - Proper lighting

Determining Budget Constraints

Most projects will have a limited budget. It is important for you to know the amount of money the homeowner is willing to pay for a home network. It is your job to design a network that fits within that budget. You may have to explain to the customer that they have to choose one feature over another in order to stay within the budget. A home computer network typically costs in the range of $1,000.00 – $10,000.00, depending on its scope. Also, it is important to note that the cost will depend on the geographic location of the customer and how much usable equipment exists among other items.

For example, a computer network including three or four PCs, a shared printer, and an Internet connection would cost on average $5,000.00 ($3,000.00 labor and $2,000.00 for hardware and wiring) in Silicon Valley, California. This would be for five to eight hours of installation, and two hours of follow-up and fine-tuning about two days later. In some cases, you may need to make another follow-up service call, lasting for approximately two hours, about two weeks after installation. This estimate does not include the installation or hardware for any of the other subsystems. The cost will vary by location, the type of equipment installed, and the type of advanced technologies included.

Searching the Internet for sites on home automation will provide information on new technology and equipment. Equipment catalogs are generally free and can be ordered or downloaded. These sites provide a starting point when pricing the equipment needed to complete a project. Labor charges can also be researched on the Internet but will be dictated by the going rate for your specific location.

Web links:
http://www.leviton.com
http://www.hometech.com

Determining Timeline Constraints

Setting up a timeline of the work to be accomplished will help to ensure that deadlines are met. It is important for you to adhere as closely as possible to the scheduled deadlines. Be sure to allocate enough time so that you do not become crazed to meet the deadlines. Remember that it is better to ask for more time at the beginning of the job than to fail to deliver the work due to impossible circumstances; for example, if you find any parts required on backorder or if the homeowner makes any additions. Your work will suffer in quality if you try to cram it into an unreasonable amount of time. Be sure to take into account the fact that you may be serving more than one homeowner at a time. Other factors that can affect the timeline include:

- **Weather.** Typically, work can continue indoors when the weather is bad. Keep current about changes in the weather so that any installation of outdoor components can be handled when the weather is dry.
- **Changes to the initial design.** Any changes requested by the homeowner will require additional time. This provides for any parts that may need to be ordered and updating the documentation including the schematic diagram, flow charts, and equipment layout.
- **Parts availability.** The project can be completed more efficiently if all parts, including the cabling, are organized before beginning the installation.
- **Multiple projects.** Make sure each project is allocated sufficient time to complete including the training phase. Never rush a homeowner who has questions.

DESIGNING THE INTEGRATED HOME NETWORK

The design of an integrated home network is completed after it is determined which types of subsystems are needed. Subsystems, such as a computer, telephone, security, HVAC, and so forth, have been detailed in previous chapters. The design process starts with a survey of the site to determine what is required. The next step is to document existing equipment and outlets. This will help to determine where new equipment and outlets should be placed.

Figure 11–2 shows an example of a telephone needs analysis.

Conducting a Site Survey

The HTI gathers general and specific information during the project site survey. This information includes the type of construction (new, remodel, retrofit, single/multi-story, or single/multi-family dwelling), the current environment (existing equipment, systems, and their location), and the location of the home (rural, suburban, urban, the presence of tall trees, surrounding hills, etc.). You will need to inspect the home to document the physical layout of the home and the existing equipment. The size of the home will affect the cost of wiring, as will the placement of devices. The existing equipment needs to be tested and inspected for its condition and ability to be upgraded.

The information gathered constitutes the building and construction constraints to be taken into account in the network design and engineering process.

Leviton Integrated Networks Home Analysis

How many people will be living in the house?_____
What are their names?_____
No./age of children: 1_____ 2_____ 3_____ 4_____ 5_____
How long do you plan on living in this house?_____

Telephone Needs Analysis

1. How many telephone lines do you currently have?_____

2. For what purpose do you use the various telephone lines?

Line 1	Personal_____	Business_____	Internet Access_____
Line 2	Personal_____	Business_____	Internet Access_____
Line 3	Personal_____	Business_____	Internet Access_____
Line 4	Personal_____	Business_____	Internet Access_____

3. How many phones do you anticipate needing in the new house?_____
   Tell me the primary use for each line:

Line 1	Personal_____	Business_____	Internet Access_____
Line 2	Personal_____	Business_____	Internet Access_____
Line 3	Personal_____	Business_____	Internet Access_____
Line 4	Personal_____	Business_____	Internet Access_____

4. Will your telephone needs change in the future? Yes_____ No_____
   If yes, what are the changes?

5. In which rooms will you want each telephone line?

	Bed 1	Bed 2	Bed 3	Bed 4	Living Room	Family Room	Kitchen	Den	Bath 1	Bath 2	Garage
Line 1											
Line 2											
Line 3											
Line 4											

6. What type of telephone lines will you be using? One_____ Two_____ Multiple line_____

7. Do you currently use or have a need for a key system or PBX? Yes_____ No_____
   If yes, what size do you need?_____ How many trunks will be required?_____

8. Will you have a security system hooked into the phone system? Yes_____ No_____

Figure 11–2
Telephone Needs Analysis

Generally, the most critical features for the design phase include the following:
- Type of project
  - New home construction
  - Remodel or retrofit
- Home materials
  - Wood
  - Stone
  - Stucco

- Number of stories
- Existence of basement and/or attic
- Number of rooms
- Types of rooms
- Total square feet to be networked
- Demarcation point locations (telephone/broadband/electricity/entertainment)
- Proposed location of distribution panel
- Garage or car port
- Is an architect or remodeling contractor needed
- Number of existing outlets and their locations

Determining Demarcation Points

Demarc

The location of the demarcation point or *demarc* is typically determined by the service providers, such as the telephone company and the cable TV company, upon installation. Coordinating with these providers is necessary if a specific location is required due to design requirements. As explained in Chapter 4 and Chapter 10, the demarc is the point where the service provider network meets the internal home network. The demarc divides the responsibility of the service. Problems on the service provider's side of the demarc are the responsibility of the service provider. Problems on the other side of the demarc are the homeowner's responsibility.

The demarc also provides a simple means to disconnect from the service provider. This is important for testing and troubleshooting.

Network interface device (NID)

The telephone company's demarc is often called a *network interface device* (*NID*). This device provides a quick means to connect and disconnect from the telephone circuit for testing and troubleshooting. It also provides lightning protection by means of gas tube elements. Gas tubes are not fuses; in the event of high voltage, like a lightning strike, they do not blow open, but rather short to ground. The advantage of a gas tube element is that it will take repeated lightning strikes before it needs to be replaced.

Most telephone company demarcs are referred to as intelligent devices. The telephone company can remotely disconnect from the inside wiring and test their lines without having to dispatch a service technician to the site.

Ground block

The cable television company's demarc, shown in Figure 11–3, is usually referred to as a *ground block*. This generally looks like a splitter, but it has a ground lug on the housing.

Figure 11–3
Ground Block

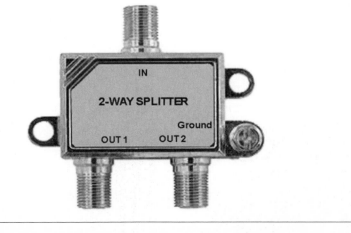

A ground wire is run from the ground block to the building ground. A coax 'F' connector is used at each end of the ground block. The cable television company may also install filters at these locations. Filters prevent unsubscribed channels from entering the home network. Premium channels, such as Home Box Office (HBO), are often filtered either at the pedestal or at the ground block. Cable television companies may also install what they call *high-pass filters* at the ground block. These prevent noise or unwanted RF signals generated within the home from entering the cable television system. Since these filters block all return signals, they must be removed when using two-way services like digital cable or cable modem service.

High-pass filters

Identifying Outlet Locations

Identifying the outlet locations in the home network is another step in the planning process. The requirements of the homeowner are the most important determining factor. Another determining factor is the type of service. For example, an outlet for phone service in the bedroom should be installed near the head of the bed. It is therefore important for the HTI to include the placement of major furniture and appliances in the design. By carefully selecting outlet locations, the length of equipment cords required will be minimized and unsightly outlets can be concealed. Although most outlets are mounted about 15" above the floor, some outlets will require higher mounting. Wall phones are often mounted 60" above the floor. If handicap access is needed, special outlet heights may be required and must then be incorporated into the design. Figure 11–4 and Figure 11–5 illustrate a floor plan with the outlets marked for each room.

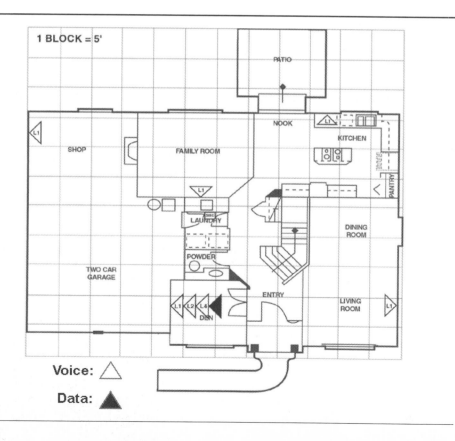

Figure 11–4
Identifying Outlets—
1st Floor Plan

Figure 11–5
Identifying Outlets—
2nd Floor Plan

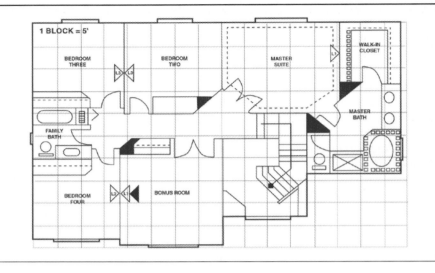

Home theater speakers will require an outlet near the ceiling. Speakers themselves may be mounted in the ceiling. Volume controls should have outlets at the same height as electrical wall switches. Surveillance cameras may also require an outlet that is mounted near the ceiling. Don't overlook outdoor outlets for video or audio applications. The popularity of wireless phones has all but eliminated the need for outdoor phone outlets, but the ability to hear the phone from outside is a consideration.

Determining the Number of Outlets

Once the outlet locations are finalized, a total count of all of the outlet locations is made. The total number of outlets is a good measure of system size. The number of cable runs terminated in the SMC typically determines the size of the SMC. As you learned in Chapter 8, "Low-Voltage Media," one important configuration requirement of a Structured Media Center (SMC) is to have the capacity and functionality to handle the cable terminated.

It takes approximately the same amount of time to pull a cable to an outlet, regardless of the type of service or cable. Determining the number of outlets gives a good approximation of the labor time in the rough-in phase. Planning the rough-in phase early and accurately is critical. Cables must be installed after the electrical contractor has completed the electrical rough-in, but before the drywall contractor has begun hanging sheetrock. Construction schedules are sometimes planned tightly, and the HTI cannot delay the start of the drywall contractor.

Identifying Power Requirements

Identifying power requirements is important in a residential systems installation planning process, because this may affect outlet locations. Identify which services will be network powered and which services will require house power. Video outlets should be located near the power outlet that will feed the television. Wireless phones will require house power, so telephone outlets for these should be located near a power outlet.

Computing equipment that will not be located in the SMC, such as cable modems, DSL modems, hubs, switches, and gateways, will also require power. Computing devices are often run with a power strip with surge protection that requires access to house power as well. Walls are typically constructed with 2 × 4 studs on 16" centers. Communications outlets should not be mounted on the same stud or in the same stud cavity as AC outlets. This means that communications outlets will be a minimum of 16" away from AC outlets.

Final outlet locations may be adjusted after taking power requirements and power outlet locations into consideration. In retrofit projects, be sure to check both sides of a wall for existing AC outlets for the 16" separation rule. What appears to be an excellent location for a video or data outlet in the family room may be the same stud opening that has an AC outlet in a bedroom on the other side of the wall.

Determining Applications Served by Outlets

Up to this point, the HTI has accurately located the outlets, determined the number of outlets and the magnitude of the project, and approximated the size of the SMC. The detail work can now begin. Each outlet should have some type of application assigned to it. Some outlets may be defined as "future use." An EIA/TIA Grade 1 (or Level 1) installation will have one CAT 3 or CAT 5 and a single coax run to it, as shown in Figure 11–6.

UTP Cat 3 Cable

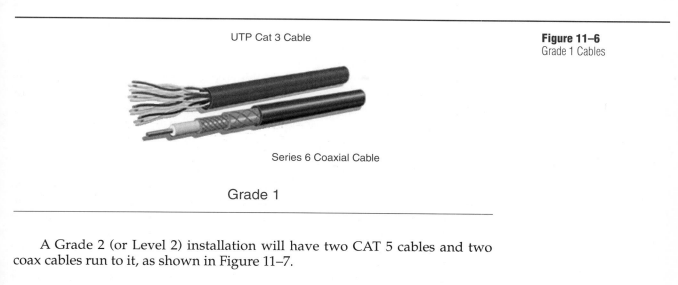

Series 6 Coaxial Cable

Grade 1

Figure 11–6
Grade 1 Cables

A Grade 2 (or Level 2) installation will have two CAT 5 cables and two coax cables run to it, as shown in Figure 11–7.

2 UTP Cat 3 Cable

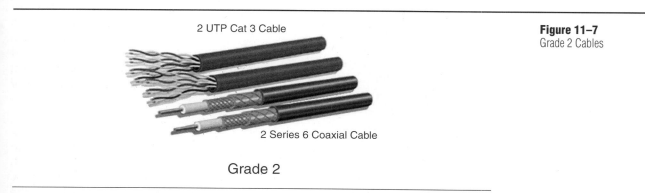

2 Series 6 Coaxial Cable

Grade 2

Figure 11–7
Grade 2 Cables

Most installations will be a combination of grade 1 and grade 2. For example, the children's bedrooms may typically have a grade 1 outlet, while the master bedroom may have both a grade 2 and a grade 1 outlet. The kitchen may have a grade 1 outlet or even a grade 2 (depending on the homeowner's requirements), while the home office may have two grade 2 outlets.

Entertainment centers may have two or more grade 2 outlets, plus outlets for specialty cables like speakers. Sound systems may require a CAT 5 cable for volume control and speaker wire for the actual speakers. Some internal video systems will require a coax and a CAT 5 cable. The coax would deliver the video signal and the CAT 5 would carry low-voltage power to the camera. Other internal video systems will use baluns and operate entirely on CAT 5.

> **NOTE**   The term *balun* is derived by combining balanced and unbalanced. A balun is a device that joins a balanced line (one having two conductors, with equal currents in opposite directions, such as a twisted-pair cable) to an unbalanced line (one having just one conductor and a ground, such as a coaxial cable). A balun is a type of transformer that is used to convert an unbalanced signal to a balanced one or vice versa. One typical use for a balun is in a television antenna. In a balun, one pair of terminals is balanced, (the currents are equal in magnitude and opposite in phase); the other pair of terminals is unbalanced (one side is connected to electrical ground and the other carries the signal). Leviton does not use baluns. Instead, they produce connectors that serve the same function as a balun.

Determining Cable Type and Quantity

Once the outlet applications have been clearly defined, the type and the amount of cable required for each outlet can be determined. Always keep in mind that in addition to types of services, cable distances can affect the cable selection. Speaker wire that is 16-gauge (AWG) may be a recommended standard; however, long runs may require a lower gauge.

Standard video distribution cable is a coax Series 6 cable. Sensitive video application may require a quad-shielded coax.

Once the cable types and quantity per outlet are determined, the actual quantity of each cable type is calculated. Using the blueprints, the HTI should approximate the distances from the SMC to each outlet. There is no basic math equation to use when determining the amount of cable to run, but the following example shows how cable from the SMC is totaled.

Example of determining total cable length

SMC to Shop	50'
SMC to Family Room	15'
SMC to Den	30'
SMC to Kitchen	35'
SMC to Living Room	50'
SMC to Master Suite	50'
SMC to Bedroom 2	35'
SMC to Bedroom 3	35'
SMC to Bedroom 4	55'
SMC to Bonus Room	55'

Total: 410' + 20% = 500'

The example shows 20% added, but you can also add approximately 12 to 16 inches to the final measurement in order to have enough cable to properly terminate at the SMC and the outlet. The excess is then pushed back into the wall. This excess cable will be necessary should the outlet need any future repair. It is better to have the additional cable available than to have to re-run cable because an outlet is damaged.

Matching Cable Types with Applications

The type of cable used for an installation depends on the type of application required. While new construction poses few obstructions, a retrofit installation can be more challenging. Space is often limited. The newer composite cables that combine two CAT 5 cables and two coaxial cables in a single jacket can be used when limited space allows only a single cable pull.

Sometimes it may be necessary to go outside the house to reach a particular room. In this instance, outside plant rated cables should be used, even for shorter runs. Outside plant cables have special jacketing that is not affected by the weather or UV rays from the sun.

High-end video and audio applications often use fiber optics. Fiber optics can also be useful when planning for future upgrades.

When applications cannot be clearly identified, it can be helpful to run a flexible PVC conduit to outlet locations. This provides a path to an outlet when future applications are defined.

Providing Redundancy

Redundancy is also known as fault tolerance. In this context, it is a backup cable that is installed to ensure that service can continue even if one cable fails. It is highly recommended that at least two CAT 5 cables be run from the SMC to the telephone company's NID. Two four-pair cables can support up to eight incoming telephone lines. This arrangement is typically more than a homeowner would need. However, two cables in separate sheaths provide redundancy should one cable fail. They also allow for future growth.

Most cable systems use a single coax to provide cable TV services to the home. Some cable television companies elect to use separate coax cables for the cable modem to the computer and for the television. Running two coaxial cables from the SMC to the cable television company's ground block provides redundancy and ensures that future services can be supported.

Running fiber cable between the SMC and the location where application service providers enter the building is another installation option. As service providers consider advanced services and advanced concepts like fiber-to-the-home, a pre-installed fiber allows for future integration. Keep in mind that while the service provider may provide a housing for the NID or ground block, they will generally not provide a housing for a future fiber cable. This will require some type of small enclosure provided in the design of the home network.

Redundancy

PLANNING THE STRUCTURED MEDIA CENTER

The Structured Media Center (SMC) location should also be determined in this phase. As mentioned in previous chapters, the SMC, or distribution device (DD), should be centrally located whenever possible. This minimizes the length of cable runs and keeps your installation within prescribed standards. It also minimizes cable loss for applications sensitive to this. The SMC should not be located in areas subject to temperature extremes or high humidity. A well ventilated and/or climate controlled location is also recommended.

The following is a list of do's and don'ts for the SMC.

Do install the Structured Media Center

- Centrally in the house
- In a location that ensures CAT-5 cables are less than 295' (conductor length)
- On a dedicated 15A circuit
- At a serviceable height
- In a well-lit location

Do *not* locate the Structured Media Center

- On an outside wall
- On an uninsulated garage wall

- Within 5' of the main AC service panel
- In a moist location
- In a location without easy access
- In a fire-rated wall

The SMC should be located where it is easily accessible. In a retrofit, a utility room or closet is the best option when a dedicated SMC is not available. Avoid using the garage and laundry room unless an enclosure within the space can be built that controls extreme temperatures and humidity.

Determining Distances from the SMC

Outlet distances from the SMC should be accurately determined for all standards-compliant installations. Most designs will incorporate a grade 1 or grade 2 cable according to EIA/TIA standards and the design specifications.

The strength of the video signal diminishes as the cable gets longer. The need for RF amplifiers can be assessed if the distance is accurately determined. Since many services, like cable modem and digital television, are *two-way services*, always use an amplifier that will amplify on the return path as well as the forward path.

There are other services that are distance sensitive; some may not work at all. Services like S-video, Firewire, and universal serial bus (USB) will not work beyond the 100-meter distance defined by the EIA/TIA.

Determining the Internal Layout of the SMC

As mentioned earlier in the chapter, the SMC deployed must have sufficient capacity and functionality to support the system. The size or capacity of the SMC is determined by the number of cables terminated, the types of cables, the types of applications, and the active devices planned for the SMC.

As discussed earlier, common walls are constructed with 2 × 4 studs on 16-inch centers. Almost all SMCs are designed to fit within the 2 × 4 spacing. This means the typical SMC will be about 14.5" wide. They come in varying lengths—14", 20", 24", and 36" are common lengths. Planning a slightly larger SMC housing than is required will accommodate future growth.

Active devices that require electricity to work and that can be housed in an SMC are hubs, switches, gateways, and even cable modem or DSL modems for Internet access. Video devices include amplifiers, modulators, notch filters, and video switchers. Audio systems may include amplifiers. Generally the security system controller is not included in the SMC.

It is sometimes necessary to provide two or more SMCs for large systems. If this is the case, it is best to separate systems by application or type of devices. For instance, one SMC could be used to house a data system, while another could be used to house an audio/video system. Another approach would be to use one SMC to house all cable terminations, while using another SMC to house active equipment. This would require an interconnection between the devices.

SMCs are designed to have cables enter from the top. Once an installation is completed it is very difficult to add cables. It is a good idea to stub a short section of flexible PVC conduit from the top of the SMC to the open area above the ceiling. This provides access to the SMC if additional cabling is needed at a future time for upgrades or add-ons.

Using an Auxiliary Disconnect Outlet (ADO)

The *ADO* or *auxiliary disconnect outlet* provides a simple means to disconnect the inside cabling from the service provider's network. Although the service provider provides a disconnect at the NID, it is more convenient to have an ADO disconnect at the SMC. Circuits can be disconnected and tested without leaving the work location. An ADO is shown in Figure 11–8.

Two-way services

Auxiliary
disconnect outlet
(ADO)

Figure 11–8
ADO

The NID may not be easily accessible to the homeowner, or it may be too complex. A well labeled ADO is essential for troubleshooting activities. These can even be accomplished with the homeowner over the phone.

Using the RJ-31x

The RJ-31x is the standard interface to an actively monitored alarm system. When the security system is not plugged in, the RJ-31x delivers a dial tone like any other ADO or telephone jack. When a security system is plugged in, the dial tone passes through the security system before delivery to the rest of the house. This gives the security system control over the telephone service. In the event of an alarm, the security system seizes control of the telephone circuit, and dials the monitoring company. If the circuit is in use, the security system drops the call in use and places its call to the monitoring company. The RJ-31x should be installed in the SMC so that it is accessible.

Determining Power Requirements in the SMC

Almost all SMCs come equipped with a single gang duplex outlet. This should be a dedicated non-switched circuit with at least a 15-amp breaker. Many devices require 12 V DC to operate. The duplex AC outlet may be augmented with a power strip if additional outlets are required within the SMC.

Manufacturers of SMCs have developed 12-volt power supplies that can power multiple devices from a single source. A single 12-volt power supply with multiple leads can power many devices within the SMC.

ESTIMATING COSTS AND TIMELINES

Staying within an allotted budget is a major concern in any project. Proper work planning will help ensure that the project budget is proportionally allocated according to the different phases of the project. Budget mismanagement affects not only the cost of the project, but also the overall satisfaction of the homeowner. For example, plan on the rough-in (laying of data cabling) just after the electrician has put the power cables in place. Doing so will avoid situations where data cable is run close to power cables or the electrician is causing bends or kinks in data cables. It is important to understand that electricians don't necessarily have knowledge of the special rules for running or routing low-voltage cabling; and so there may be instances where they could, without understanding the consequences, route power cables through the same conduits as data cables.

Going back to reroute or pull certain data cables will add time and money to the project. Put a contingency plan in place to deal with these situations without putting an unbearable strain on the budget. Good coordination with the building contractor is imperative when creating a timeline for installation and testing.

Developing a Bill of Materials

After all outlet locations are identified and finalized and all applications have been defined, a detailed bill of materials can be developed. A bill of materials, as shown in Figure 11–9, is a document that lists all of the materials (cable, outlets, connectors, etc.) that will be used in the project. It includes the quantity of each and the cost.

Figure 11–9
Bill of Materials

QTY	Part number	Description	Price	Extended price
500		CAT 5e cable	$60.00/1000	$30.00
1	47605-28W	28 Enclosure with cover	98.10	98.10
1	47603-ASO	Advanced small office panel	182.47	182.47
1	47605-DP	AC power module	77.55	77.55
1	47605-EH	10 Base-T hub	82.50	82.50
2	40223-S	Wall phone jacks	4.50	9.00
10	5G108-RW5	CAT 5e jacks	5.00	50.00
10	41644-BW	Four-port decora inserts	3.25	32.50
10	80401-W	Face plates	2.25	22.50
1	40859-BW	Bag of blanks	.16	.16
5	47620-1W	1" CAT 5e patch cord	2.25	11.25
10		Mud rings	1.25	12.50
15		Cable supports	.10	1.50

The distance from each outlet to the SMC can be accurately estimated using blueprints. For each specific cable type, measure the amount required per outlet, or take an average length and multiply by the number of outlets. Some outlets will have different faceplates to address the unique services provided by these outlets.

An accurate bill of materials will reduce shortages on the job site. Shortages increase labor costs and can even result in restocking charges by the distributor.

A typical bill of materials is fairly simple in structure. It should include the following:

- Project information
- Homeowner information
- Type of project
- Components needed
- Support devices
- Type and amount of cable
- Type and number of connectors

In addition, it should include any other information regarding the installation, for example, special conditions that need attention. You should break each subsystem down and list every product needed. Be as specific as possible and include all the suppliers. A very detailed bill of materials will help you keep track of what you have purchased and what you still need to purchase. The bill of materials should remain a part of the project documentation. This is not part of what is included for the homeowner, but remains as part of the project file.

Estimating Labor Costs

Labor costs make up a significant portion of the overall budget for the project. Careful evaluation of labor resources is crucial in large residential technology integration projects involving multi-dwelling units. If installation crews are inadequately staffed or poorly managed, they cannot be productive. Due dates will be missed and costs will escalate. An important step in the planning process is to develop an estimate of the labor that is required.

Labor costs are first determined by dividing the individual tasks into logical units and then assigning the time it will take to complete each individual unit.

The following example is an illustration of how logical task units are put together:

- Labor to install a four-pair jack (RJ-45) = 0.25 hr
- Labor to place 50 m of CAT 5e wire in open ceiling = 1 hr
- Labor to install 3 m of surface mount raceway = 0.25 hr
- Labor to terminate four-pair wire = 0.25 hr
- Labor to test CAT 5e wire = 0.25 hr

Labor for tasks in addition to the prime task is included in the individual unit. For example, the labor to install a jack includes moving a bed and cleaning up wire scraps. Labor to terminate a Category 5e wire will include installing the termination panels.

Special situations will add units. There would not be a labor category to install 110 blocks on a plywood wall field, for example. This would be included in the termination estimate. The additional time required to install floor-mounted racks and patch panels would be provided for, since installing these can be very time consuming. Labor units may not be standard. Contractors establish their own units, based on their own preferences and experience with those tasks. Some installers are faster than others at pulling wire, while others are more efficient at terminating cables.

Some labor units are estimated on a cost per foot or meter. Large riser cables and large distribution cables, for example, are typically estimated on this basis. Remember that labor cost estimates cover only the cost of completing the project. The profit margin will be added to the labor cost figure and that will become the labor bid price submitted to the homeowner.

To do an accurate estimate, the number of components counted, referred to as a "take off," must be accurate and the estimator must read the blueprints precisely. Site survey notes are also included in the final bid you submit to a potential customer.

Constructing a Timeline

The rough-in phase has the most critical timeline because the work must be coordinated with other trades and contractors. The network cabling is installed after electrical and plumbing, but before the drywall contractor hangs the sheetrock. Note that when cables are installed too early, they could be damaged by other trades. The HTI may be held liable for delays in construction and could be required to pay a penalty if other trades are delayed.

When planning the timeline for the rough-in phase, include accessory work such as installing nail plates to protect wires that are run through drilled studs or wall plates. Include time for patching holes if required. It is very costly to make a return trip, and more costly if another trade or contractor has to patch holes made by the rough-in crew.

The trim-out phase includes termination, labeling, and testing of all cables. It also includes system integration or installation of active components.

The finish phase includes installation of faceplates. This phase leaves an attractive and functional cable system, ready for equipment or device installation.

The completion date for the project is typically defined in the bid or Request for Proposal (RFP) and only becomes a contractual obligation when the agreement or contract between the customer and the contractor is signed. Various factors can affect the start date and completion date. These factors include the availability of the materials, the contractor's schedule, and the customer's schedule. Make sure that all of these variables are coordinated before the final contract is signed.

Coordinating with the Building Superintendent

In new construction, both the rough-in phase and the finish phase require careful coordination with the building superintendent. The building superintendent coordinates the schedules of all of the trades working on the project. Working closely with the building superintendent will ensure that the rough-in for the home network installation is planned at the proper time. The building superintendent must be made aware that home network cables are often more fragile than the electrical wires in the home and could easily be damaged by the drywall contractor handling the sheet rock.

Coordinating with the Electrician

Working closely with the electrical contractor will ensure that home network expectations are met. The electrical contractor must run a circuit to the SMC. This is typically done before the SMC is on the job site. The circuit must be dedicated, unswitched, and at least 15 amps. The electrical contractor must know the location of the SMC before his rough-in crew comes to the job site. He must also have a clear understanding of the type of circuit.

The HTI will mount the SMC and finish placing the electrical contractor's cable into the proper portion of the SMC. The electrician must terminate this cable and install the electrical outlet. It is important for the HTI to develop a good working relationship with the electrical contractor so that duties and responsibilities can be easily defined and coordinated.

CREATING A PROPOSAL AND CONTRACT

The proposal is a document that states what you will do for the homeowner. It is created after surveying the home and assessing the needs of the homeowner. This is the document that details the project before a formal contract is drawn up. Once you and the homeowner have agreed to move forward with the project, the contract is signed before work can begin.

Developing the Proposal

Developing a proposal and eventually transforming it into a home integration project design is a systematic process. This process can be summarized in seven phases, including:

- **Phase 1: Requirements Gathering**—Phase 1 is the gathering of the customer's requirements. You need to determine and then document the time, budget, and building constraints.
- **Phase 2: Evaluation of the Requirements**—Phase 2 is the evaluation of the requirements. You will evaluate and then document how each existing subsystem, such as computer, entertainment, telephone, security, and home automation controls, works and what improvements or changes need to be made in order to integrate them into the network.
- **Phase 3: Home Network Elements Determination**—Phase 3 is where you determine what type of network to install, the type of media needed, the type of electronic equipment to be purchased, and whether you will need subcontractors to complete the job.
- **Phase 4: Design Documents Creation**—Phase 4 is the creation of the design documents. In this phase you will create revised floor plans, schematic block diagrams, signal flow diagrams, and installation and testing timelines.
- **Phase 5: Proposal Development and Presentation**—Phase 5 develops and presents the proposal. The proposal contains the scope of the work, the price of the project, the service provided, and the terms and conditions. This phase also covers the subjects of issuing liens, insurance coverage, and conflict resolution.
- **Phase 6: Installation Instructions Creation**—Phase 6 is the installation, which includes ordering the equipment and materials needed, developing a detailed wiring chart, and then actually installing the network.
- **Phase 7: Determining Potential Upgrades**—Phase 7 involves determining future upgrades of the network and customer support.

The extent to which each of these steps is involved depends on the scope of the project. For smaller projects, some of the steps can be skipped or combined. This determination is made at the discretion of the HTI following the initial meeting with the homeowner and the initial evaluation process.

Identifying the Sections of the Proposal

The design and proposal development produce a design document for the home network and systems integration project. Typically the entire project proposal may include several pages of information. Not all of this has to be presented to the homeowner. Normally, you would have to reduce this into a more manageable format including the following sections:

- **Executive summary**—This is provided for the homeowner. It should be no more than two pages and should include only the key points of the proposal. Be sure to focus on benefits that you and your design have to offer. You may include the following items:
    Purpose of the project—One or two paragraphs that state the purpose of this document
    Recommendations—Outline your design strategy. Be sure to relate your recommendations to the homeowner's objectives.
    Implementation considerations—One paragraph that lists implementation considerations, such as integration issues, support, etc.
- **Benefits of the design**—Summarize the overall benefits of the design. Make sure that the benefits relate to the homeowner's objectives.

- **Design requirements**—The design requirements summarize your conclusion as a result of identifying the homeowner's needs. You should include a characterization of the existing network or devices.
- **Design solution**—Describe your solution along with the features and benefits it provides. Organize a list of equipment based on the homeowner's priorities.
- **HTI qualifications**—Include a short resume of your qualifications including other projects completed and specific training or certifications.
- **Summary**—Summarize your design and how it addresses the homeowner's needs. This will also include why your company is the best choice for the homeowner (e.g., reliable, low cost, excellent service, etc.).

This document should be forwarded to the homeowner a few days early to allow them time to read through it. At the presentation, make sure to listen carefully and provide concise answers to the homeowner's questions. Take notes on any issues requiring modifications.

Drawing up a Contract

After the homeowner has approved the proposal, you will need to draw up a contract. The contract is a document that clearly states the expectations and responsibilities of the parties involved in a project. It protects each party's rights.

In any construction or remodeling project, both owner and contractor will follow their own agendas to some degree. All have different understandings of how the project should be carried out. The contract is an attempt to clarify everyone's expectations and make the terms of the project mutually acceptable.

It is beneficial to hire a lawyer for assistance in developing a standard contract and for explaining any details written in legal terms that are difficult to understand. A lawyer should give you a detailed analysis of the contract that points out sections needing clarification or items that are missing. The analysis is designed to identify and resolve potential problems before they develop.

The homeowner should review the contract carefully. After the homeowner reviews the contract and both parties agree to the terms, the homeowner must sign and retain a copy. Do not rely on oral agreements. The signed contract is legally binding. In many states, there is a grace period of three days in which either party may cancel the contract. If one party does cancel, a written notice should be sent by registered mail with a signed receipt requested from the other party.

Confirming the List of Equipment

Once the presentation and signing of the contract is completed, proceed to confirm the equipment list (as per the proposed solution) with your homeowner. The list of equipment is created as part of the design solution of the proposal. This is necessary to confirm that the details have not changed. Always confirm the final equipment list with your homeowner before the equipment is purchased. Some computer stores will not take back parts or computer components. Software that has been opened is a typical example.

SUMMARY

This chapter focused on the following topics:

- Typically, the first step in the design process is determining what the homeowner needs.
- The budget constraints of the homeowner will affect many aspects of the project, from what equipment to buy and which services can be provided, to the amount of labor that can be afforded.

- The site survey is critical in determining possible problems and identifying existing components.
- The demarc is the interface between the service provider's and the homeowner's points of responsibility. It is important to determine where the demarc is located on the property.

- The network interface device (NID) provides a quick means to connect and disconnect from the telephone circuit for testing and troubleshooting.
- The ground block is the demarc for cable television.
- Communications outlets should not be mounted on the same stud or in the same stud cavity as AC outlets.
- High-pass filters are installed at the ground block by cable television companies to prevent noise or unwanted RF signals. They must be removed when using two-way services like digital cable or cable modem services.
- The rough-in phase is when the cables are pulled through the house.
- The trim-out phase involves termination, labeling, and testing of all cables.

- The finish phase involves the installation of faceplates and other finishing touches.
- There are seven phases to developing a proposal. The extent to which each step is involved depends on the scope of the project. The proposal is eventually transformed into a home integration project design.
- The contract is a legally binding document. It is beneficial to hire a lawyer for assistance in developing a standard contract. The homeowner should review the contract carefully. After the homeowner reviews the contract and both parties agree to the terms, the homeowner must sign and retain a copy. Do not rely on oral agreements.

GLOSSARY

Auxiliary disconnect outlet (ADO)   Provides a simple means to disconnect the inside cabling from the service provider's network.

Demarc   The point where the service provider's responsibilities end and the homeowner's responsibilities begin.

Ground block   Term used to describe the cable television demarc.

High-pass filters   Installed by cable television companies to prevent noise or unwanted RF signals.

Network interface device (NID)   The demarc of the telephone company.

Redundancy   Planning for future needs or problems by running extra cables during the rough-in phase.

Two-way services   Services that not only receive signals but transmit information, such as digital cable or cable modems.

CHECK YOUR UNDERSTANDING

1. Typically the first step in the design process is determining:
   a. the number of existing outlets.
   b. where the demarc is located.
   c. what the customer needs.
   d. how much the project will cost.

2. To ensure that deadlines are met, the HTI should:
   a. set up a timeline of work.
   b. cancel other projects.
   c. order extra parts.
   d. refuse to change initial design.

3. The location of the demarc is determined by the:
   a. HTI.
   b. homeowner.
   c. service provider.
   d. city codes.

4. The advantage of using a gas tube element in a demarc is which of the following?
   a. It is more economical than fuses.
   b. It will take repeated lightning strikes before it fails.
   c. It is easier for the customer to repair.
   d. The phone company will replace the element if it fails.

5. What is the term usually used to denote the cable television company's demarc?
   a. Ground block
   b. Splitter

   c. Ground lug
   d. CATV

6. Where should communications outlets be mounted in relation to AC outlets?
   a. On the same stud only
   b. In the same stud cavity only
   c. On the same stud and in the same stud cavity
   d. Not on the same stud and not in the same stud cavity

7. Which type of connector is used at each end of the ground block?
   a. BNC
   b. F
   c. RJ-31x
   d. RJ-45

8. What is the minimum distance communications outlets should be placed from AC outlets?
   a. 0 inches
   b. 6 inches
   c. 12 inches
   d. 16 inches

9. When should high-pass filters at the ground block be removed?
   a. When using one-way services
   b. When using two-way services

c. Never
d. When noise is detected

10. How many four-pair cables are required to support up to eight incoming telephone lines?
a. One
b. Two
c. Four
d. Eight

11. A Grade 2 installation will have _____ CAT 5 cables and _____ coax cables.
a. 1, 1
b. 2, 1
c. 2, 2
d. 2, 3

12. The type of cable used in an installation depends on the:
a. type of application.
b. cost of the cable.
c. length of cable required.
d. obstacles in the way.

13. What are the characteristics of the circuit that the electrical contractor runs to the SMC?
a. Not dedicated and at least 15 amps
b. Dedicated and at least 15 amps
c. Dedicated and at most 10 amps
d. Not dedicated and at least 10 amps

14. Where should the SMC be located?
a. Outside
b. Centrally, in the house
c. In the laundry room
d. In the garage

15. What is the primary purpose of an ADO ?
a. Simple disconnect from the service provider
b. Monitoring services
c. System reports
d. Provides power outlets to the SMC

16. What is redundancy?

17. What is a bill of materials?

18. Why does the rough-in phase have the most critical timeline?

19. Summarize the seven phases of developing a proposal.

20. Why is having a contract important?

12

Rough-In

OBJECTIVES

Upon completion of this chapter, the HTI will be able to complete the following tasks:

- Work with blueprints, meet with the construction superintendent, and coordinate schedules

- Verify local codes for both new construction and retrofits

- Determine placement and install mud rings for proper cable installation in new construction

- Organize cables and stage the installation for new construction and retrofits

- Understand special situations that can take advantage of HomePNA or the use of surface mount raceway

- Make sure that cleanup is done at the end of the rough-in phase for new construction, and as the project progresses for a retrofit

INTRODUCTION

The focus of this chapter is the rough-in process for both new construction and retrofits. New construction allows easier access for running the required communication cables, but requires coordination with the construction superintendent and the electrician. Challenges for a retrofit installation include running cable through an existing wall and establishing a suitable location for the SMC. Preparing the site, drilling all required holes, and pulling cables in bundles allow for a more efficient installation in both new construction and retrofits.

NEW HOME CONSTRUCTION PROJECTS

An installation for a new home requires coordination with the construction company. More specifically, the HTI will report to the construction superintendent who is in charge of all trades on a construction site. The superintendent will report to the construction manager who, in turn, reports to the construction company. Blueprints used to bid on and finalize plans for the home network will be updated and distributed by the construction company. Construction schedules will also be updated based on the timeframes submitted during the bidding process.

The challenges of installing the home network in new construction differ from the challenges of a retrofit in an existing home. In both cases, it is the responsibility of the HTI to know the local codes that dictate requirements for cabling in the home.

Meeting with the Construction Superintendent

Meeting with the construction superintendent is an important step in building rapport and making sure the project goes smoothly. Construction schedules are typically fluid, and it is not uncommon for changes to take place. The construction superintendent will have all of the contact information for the home integration company whose bid was accepted for the job. Any changes in this information—including the main office telephone number; plus the cell phone number, e-mail addresses, and home telephone number for the HTI in charge of the project—should be updated with the construction company and the superintendent.

Mistakes in scheduling are easily corrected, but they invariably cost someone money. Mistakes can be minimized if you communicate well with others. Meeting with the construction superintendent helps define responsibilities and expectations. If the rough-in schedule is not finalized, this meeting is an opportunity for the HTI to let the construction superintendent know how much notice is required to send a crew to the job site.

Many factors can affect a construction schedule. A mechanism for resolving conflicts when things go wrong should be in place. Discuss the following issues:

- **Schedules:** It is important that the HTI install cabling after the electrician has completed rough-in work for the house.
- **Weather:** Construction delays and special circumstances can be caused by inclement weather.
- **Damage by other trades:** This is typically a minor issue, but it can happen in new construction.

Work with the construction superintendent in advance to develop a plan to resolve these issues before they happen.

Securing Blueprints

Blueprints

Blueprints, which show all outlet locations (even if they are entered by hand) are essential for the HTI. Blueprints are provided in order for a company to bid on the project and are used by the field installation crew to accurately wire the project. Any changes to the blueprints will be included on the main copy located in the construction office or trailer with updates sent out to all trades working on the job. Additionally, the superintendent will notify the trades of any changes that require immediate attention. A sample blueprint is shown in Figure 12–1.

Blueprints also show details in the construction of the house. Is the roof construction completed with rafters or trusses? What is the thickness of the walls? Interior walls are normally constructed with 2×4 studs, but exterior walls are typically constructed with 2×6 studs. The walls of an attached garage may be 2×4 or 2×6 construction. The studs in the house may be on 16-inch centers, whereas the studs in the garage may be on 24-inch centers.

Blueprints should provide all of this information. Although the project is sometimes bid with floor plan information only, blueprints identify installation issues before the project is started.

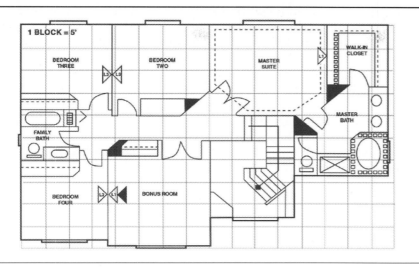

Figure 12–1
Blueprint

Securing the Construction Schedule

Tentative schedules are usually available when the construction company requests bids on new construction. In order to bid a project accurately, it is important to know when construction is expected to start. After construction starts, schedules are updated daily or (at the very least) weekly. New construction employs many different trades, and each has a rough-in and a finish phase. Plumbers, electricians, and HVAC crews all have work to do as soon as the walls are constructed. Most trades working on a job site notify the next trade scheduled when they complete their phase of the project and have done their final walkthrough. For example, the electrician will notify the superintendent and the HTI when they are done. This communication creates smoother transitions and is a matter of consideration of others on the job. As the HTI finishes and completes the final check, the drywaller should be notified along with the superintendent.

New construction schedules the HTI as the last trade to perform work before the drywallers. You can request information on the new construction timeframe when the project is bid and accepted. Electrical cables are heavier than data and voice cables. If electrical cables are pulled in after data and voice cables, the electricians could damage the data cables. Also, the data and voice cables are low-voltage and require certain standards-compliant practices regarding routing around electrical power wires. Data cables have more stringent rules regarding issues such as bend radius and kinks. Also, plumbers often use torches when soldering pipes. Heat from the torches can easily damage voice and data cables. Pulling data and voice cables after the plumbing and electrical have been roughed-in allows the HTI to avoid these issues. Finally, the installation timeframe for the HTI is typically much shorter than that of most of the other trades.

The construction superintendent is responsible for maintaining the construction schedule. Modern construction companies use Microsoft Project or other software packages that are specific to construction to manage schedules. They can also update schedules remotely from the job site. Typically, schedules are updated and then posted on the job site, and updates are sent to each trade via e-mail. Trades are also called when there has been a scheduling change, especially when it affects their start time. Monitoring the overall construction schedule goes a long way toward enabling the HTI to do a proper scheduling of the rough-in and eventual trim-out and finish of the cabling infrastructure for residential networking.

Working Around the Electrical Rough-in

As previously mentioned, it is important for the HTI to be scheduled to start after the electrician is finished. The major factor of influence is the location of the outlets. Outlets for data, telecommunications, and other subsystems are usually located near an electrical outlet. Installing data outlets too close to the electrical outlets can cause interference in the lines. These outlets should not be mounted on the same stud as an electrical outlet, and the cables feeding the outlets should not share the same wall cavity as the electrical outlet. The only way to ensure that these conditions are actually met for a new construction project is to install the low-voltage cabling *after* the electrician has completed their rough-in work. Figure 12–2 illustrates the electrical rough-in.

Figure 12–2
Electrical Rough-In

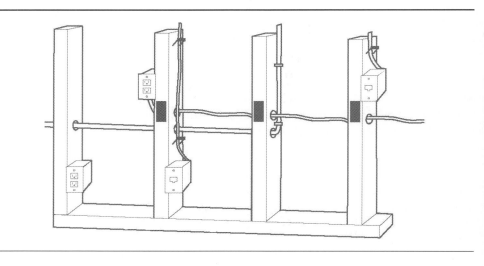

Cables are less expensive and more flexible than electrical wires. There are typically fewer cables than electrical lines. Therefore, it is easier to install the cabling after the electrician's installation.

Checking Local Codes

Local codes affect the cable installation, and they can affect the construction schedule. Following are a number of code-related issues that may arise:

- In many communities, electrical cables cannot be installed until there are shingles on the roof. This introduces another trade into the mix and can affect the construction schedule and the HTI schedule.
- In cold-weather states, drywall cannot be installed if the interior of the house is below freezing, so the general contractor often must provide temporary heat or delay the drywall installation. This weather restriction can seriously affect a construction schedule.
- Many local codes dictate whether you can drill through a truss, or they can limit the size of a hole drilled through a truss.
- Some of the most common codes define the rating of the wire used in the home. They may also require that the distribution device be UL-listed.

Verifying these code issues in advance can greatly help with scheduling and reduce unnecessary costs that could otherwise be incurred. It is the responsibility of the HTI to understand the building codes that affect the installation of the home network. Most cities and municipalities publish their codes and make them available for a fee. Check the Yellow Pages or search the

Internet for building codes for your location. There are also web sites that specialize in helping contractors and offer codes online for a monthly fee.

Web links:
http://www.nfpa.org/catalog/home/index.asp
http://www.bocai.org/codes.asp

Installing Nail Plates

Local codes also define when and where nail plates are required. *Nail plates*
are steel plates mounted to the face of a stud. These plates are installed if a hole
is drilled through a stud and wires are run through the hole. The purpose of the
nail plate is to protect the cable from nails or screws that the drywall installer
may use in the sheetrock installation. Nail plates also protect the wire from
other nails or screws used by the homeowner to hang pictures, for example.
Figure 12–3 illustrates the placement of a nail plate in new construction.

Nail plates

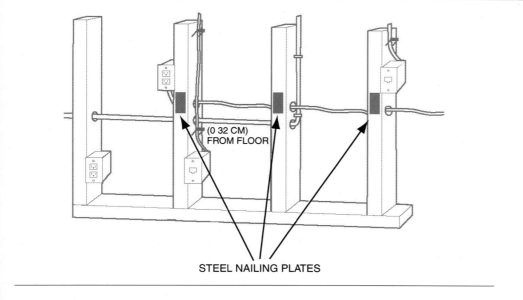

Figure 12–3
Placement of Nail Plate

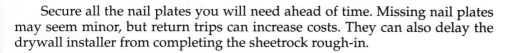

(0 32 CM)
FROM FLOOR

STEEL NAILING PLATES

Secure all the nail plates you will need ahead of time. Missing nail plates
may seem minor, but return trips can increase costs. They can also delay the
drywall installer from completing the sheetrock rough-in.

CABLE PLACEMENT FOR NEW HOME CONSTRUCTION

New construction has an obvious advantage over cable installation in an existing home because access is easier. This section focuses on the placement of
cable in new construction. Rules developed by the NEC must be adhered to so
as to protect cabling systems. Placing mud rings on outlet boxes, running cables, predrilling, staging cables, and pulling cable are covered in the following
sections.

Determining the Placement of Mud Rings

A *mud ring* is a square mounting bracket onto which the outlet faceplate can
be attached. In home networking, mud rings or drywall rings are preferred
over the use of outlet boxes where the local code permits. Cables using mud

Mud ring

rings have a larger bend radius than cables coiled inside an outlet box. The wall cavity can be used to store a small service loop without the worry of exceeding the bend radius. Figure 12–4 illustrates the placement of a mud ring in new construction.

Figure 12–4
Placement of Mud Ring

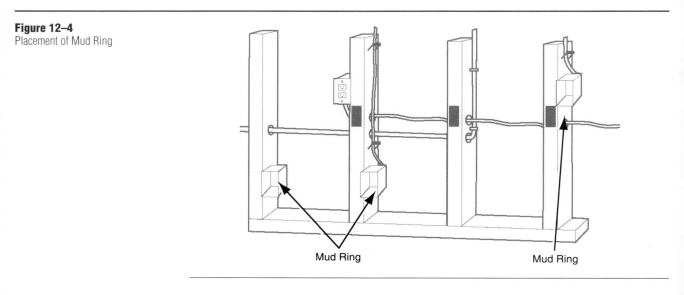

Mud Ring   Mud Ring

The first thing done on the job site is the placement of mud rings, which locate all the outlet locations and clearly identify locations when cable pulling begins. Care should be taken to measure the outlet height of the electrical outlets. Communication outlets should match the height of the electrical outlets.

Under-counter outlets for telephone, data, or video should match the height of the electrical wall switches. The installer should look for notes on the blueprints that designate special outlets. Handicapped outlets for telephone are generally located 48 inches above the floor, and standard wall phone heights are usually about 60 inches above the floor. However, it may be necessary to locate some outlets at other heights, particularly in the kitchen and home office, to fit within openings between cabinetry and counter tops. The height varies for mud rings or outlets intended for wall-mounted speakers and security cameras.

Mud rings should not be mounted to the same stud or in the same wall cavity as an electrical outlet box. The mud rings or the stud they are mounted on should be labeled by using a permanent marker as the rings are mounted. These markings are covered when the sheetrock is installed, but they are valuable in the pulling process.

> **NOTE** Always check applicable codes before performing any installation. Local codes may designate outlet heights. Some local codes may forbid the use of mud rings and require outlet boxes.

Mounting Mud Rings

The HTI should strive to maintain consistency in the orientation of mud rings compared to the electrical outlets, regardless of whether they are mounted horizontally or vertically.

reel. When the pull is complete, cut the cable at the box or reel and place the label near the end of the cable. This process is repeated for each cable.

CABLE MANAGEMENT FOR NEW HOME CONSTRUCTION

Cable management is the process of keeping cable bundled, so that it is easier to pull, and organized, so that the SMC can be connected neatly. Proper labeling ensures that the right cable will be terminated to the right connection. The final step is installing nail plates to protect the cable during the rest of the construction phase and from nails and screws put into the wall by the homeowner.

Cable Management
Cables should not be stapled into place because staples can pinch the cable and cause additional loss in the cable. It is generally acceptable to allow cables in a residential environment to lie above the joists. Joists are generally placed on 16-inch centers, which provides adequate support for the cables. If it is necessary to fasten cables, Velcro cable ties are preferred over nylon cable ties. Figure 12–5 shows different ways to provide cable support.

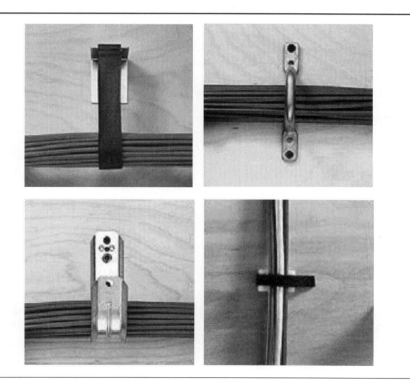

Figure 12–5
Cable Support

If it is absolutely necessary to staple a cable, do not use conventional staples. Special staples are available that bend for the diameter of the cable and have an insulator that will prevent stress on the cable jacket. These staples require a special staple gun designed for this application.

Installing the Structured Media Center (SMC)
The SMC should be mounted with the main work area at eye level. The SMC is available in a variety of sizes (sizes generally define the length of the cabinet). Almost all SMCs are designed to fit between 2 × 4 studs as shown in Figure 12–6. In extremely large installations, more than one SMC may be required.

Figure 12–6
Installing the SMC

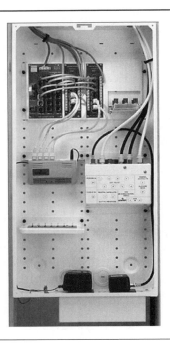

Multiple SMCs should be mounted next to each other at the same height. Screw the cabinet mounting brackets securely to the 2 × 4 studs. Make sure to allow for the thickness of the drywall that will be applied. A completed cabinet installation should be flush with the finished wall as shown in Figure 12–7.

Figure 12–7
Securing the SMC

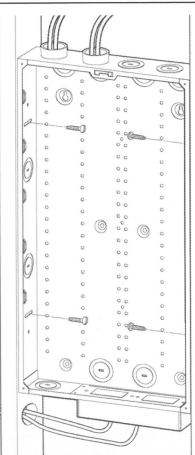

Many SMCs have some sort of knockouts at the top of the cabinet. Remove the proper size and number of knockouts. Prior to pulling the cables into the cabinet, install grommets in the holes to protect the cables from chafing. Figure 12–8 shows the top of the SMC.

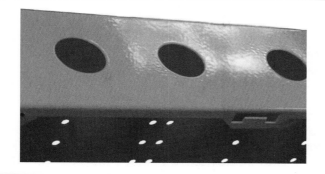

Figure 12–8
Top View of the SMC

Some cabinets have precut holes with rubber or plastic plugs sealing the holes. Cut an "x" into the center of the plug and pull the cables through. If a larger hole is required, cut out the center of the plug but leave the outer edges, which act as a grommet and protect the cables from chafing. Figure 12–9 shows the preparing of the SMC for cable.

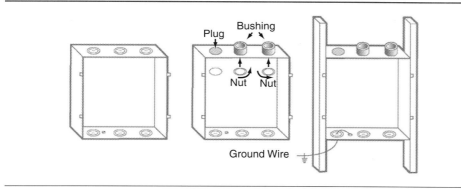

Figure 12–9
SMC Prep

Pulling Cables into the SMC

The top plate, the double 2 × 4s at the top of the wall that the SMC is mounted on, can now be drilled for the cables. Depending on the number of cables, this may require more than one hole. Cables should be loose in the entry holes. It should not be necessary to exert a large amount of force to pull the cables into the distribution device.

In pulling cable, always avoid any sort of excessive tension or stress, keep within standards-compliant bend radius, and avoid kinks. These are all things that can adversely affect the transmission characteristics of low-voltage cabling.

Relabelling and Dressing Cables

Cables entering the SMC will be at various different lengths. Relabel the cables at a point approximately four to six inches after they enter the cabinet. The cables must be stored in the cabinet until the finish phase of the project. Select a convenient length and cut all of the cables to length. *Dress* the cables by forming them into manageable bundles and taping the ends of the bundles. The cables can now be coiled up and stored inside the cabinet.

Dress

Protecting the SMC

Ideally, the door for the SMC is not secured until after other trades are finished in the area because the finish could be marred. Also, a flanged door can interfere with the drywaller's work. However, the interior of the cabinet and the cables must be protected from dirt or harm. One of the last phases of the drywaller's finish work is to spray the walls. The interior of the cabinet must be protected from the drywaller's overspray.

If the door will not affect the completion of the drywall, it can be installed and covered with plastic to protect it. If the door installation affects the drywall, use a sheet of cardboard the same size as the opening of the cabinet. Duct tape can be used to tape the cardboard to the opening of the cabinet to provide inexpensive protection to the interior of the cabinet and the cables.

Dressing and Securing Cables
at Outlet Locations

Cables at outlet locations should be relabelled, and labels should be placed approximately six inches from where the cable protrudes from the wall. Cables should be cut so that approximately 12 inches protrudes from the wall.

Care should be taken to properly store the cables within the wall cavity. Drywallers may be working with power tools to make the cut-out in the sheetrock for the outlet. If cables are not dressed and stored carefully, drywallers or other trades can damage them.

In addition to drywallers, insulation contractors will be working near cables on outside walls. Insulation contractors often staple insulation into place. Move any cables that might be damaged by an insulation contractor's staples.

Keeping Wires Organized

Keeping the wires organized can be done with strings, zip ties, tie-wraps, or Velcro. Velcro is particularly effective for this job. Be careful not to use any ties that have metal in them. By bundling cables, subsequent workers are less likely to accidentally cut into a cable because a bundled cable is much easier to see than a single cable. You may also find it useful to use J-hooks to suspend your cable bundles in the ceiling or attic to provide protection for low-voltage wiring from the high-voltage power lines that share the same space. Figure 12–10 shows bundled cables.

Figure 12–10
Bundled Cables

Another important aspect of keeping your wiring organized is labelling them. Whenever you install cable, it is important that you document where the cables are laid and what types of cables they are. You can do this by using a cut sheet as you install the cable. A *cut sheet* is a rough diagram that shows the locations of the cable runs and indicates the numbers of the rooms to which the cables have been run. Later, you can refer to this cut sheet to place corresponding numbers on all telecommunication outlets and at the distribution panel.

In the data networking, generally the TIA/EIA standard of labeling is applied. It specifies that each hardware termination unit must have some sort of unique identifier marked on the unit or on its label. All labels must meet legibility, defacement, and adhesion requirements. In residential wiring installations, a simple numbering scheme is sufficient in most cases. A three-digit system on each cable works for most projects. In such a scheme, for example, the first digit may represent the floor level, whereas the next two numbers are just sequential numbers representing different types of wire. Specifically, you can have basement speaker wires (0/00-20), first-floor speaker wires (1/00-20), and first-floor coaxial cables (1/21-30). You can also use color-coding for different types of wires or rooms.

Finally, you should always allow a couple of extra feet at both ends of a cable run. The extra length you include needs to be coiled into a service loop and secured in the ceiling or in the termination box, which can prove helpful in the future if any remodeling or upgrading job causes the movement of equipment that goes beyond the previous point of termination. It also helps release any potential harmful tension in the cable such as that caused by a repair worker tripping on the original wire in the ceiling or attic, for example.

Installing Nail Plates Where Required

The last phase of the rough-in is the installation of nail plates. Go through all the cable routes and take a close look at locations in which cables pass through holes drilled in studs, top or bottom plates, joists, rafters, or trusses. Install nail plates at all locations, as mentioned earlier, where cables could be damaged.

The top plate above the distribution device invariably requires more than one nail plate because of the number of holes and the number of cables. The NEC and local codes typically define locations where nail plates are required. Nail plates are inexpensive, normally less than 25 cents (U.S.) each. It is wise for the installer to use nail plates at all locations in which cables pass through the house framing. Make sure to look at both sides of the wall; interior walls typically require a nail plate on both sides.

Walking through the job site and installing nail plates is also a good opportunity to pick up debris and scrap wire. It is the responsibility of every trade to clean up after their work is completed.

CHALLENGES WITH RETROFITS

Retrofitting poses certain challenges that must be overcome. Unlike new construction, retrofitting involves existing and typically inhabited homes. The work must be planned to present the least inconvenience to the occupants and to minimize physical obstructions.

Gathering Information

Gathering information is the first step to overcoming potential challenges in a retrofit. If you will be presenting a proposal to install a home network, there is some information that must be known in order to accurately bid the job. Of key importance is what the homeowner wants to accomplish with the home network, as discussed in Chapter 11. After the job is secure, some of the areas that you must gather information on and preplan relate to the following:

- Homeowner's schedule
- Blueprints and as-built drawings, if available
- Verification of local codes that affect the project
- Walls that can be fished and walls that cannot

Handling Scheduling Issues

Determining the homeowner's schedule goes beyond selecting a completion date. Keep in mind that installing cabling in an existing home can be intrusive. Depending on the difficulty and complexity of the installation, there can be a significant amount of dirt and noise. Certain days of the week or certain times of day may be better for some of the harsher aspects of the work. Other schedule-related considerations are children and pets. It may be wiser to perform certain tasks when children are away at school, which can alter workday hours.

Blueprints versus As-Built Drawings

As-built drawings

Blueprints are an invaluable tool for a residential retrofit. Secure any drawings that are available. *As-built drawings* reflect the home when construction has been completed and include any changes made after the initial blueprint. Keep in mind that there may have been changes since the house was constructed. Verify everything before beginning construction. Make sure to have a contingency clause in the installation contract. Blueprints provide the basis for the installation, but keep in mind that there can be many unforeseen situations when dealing with an older home.

Verifying Code Requirements

Verify local code requirements before the installation. The requirements for a new installation can be different from those for a retrofit. It is the responsibility of the HTI to know and understand the building codes that relate to the home network. It is also important that the HTI keep current on any changes to local codes. Determine which codes affect the project during the planning stage and make sure all phases are in compliance.

Fishing Wires

Fished

When presurveying the installation, identify walls that can have cables run through them, or can be *fished.* Look for vertical alignment when cable has to be run through more than one floor. Outside walls are generally more difficult to fish because of the insulation present. Some walls have horizontal 2 × 4s that are designed to act as a firebreak; these walls are nearly impossible to fish. In general, pulling wire between floors (in a multilevel house) after construction can be tricky business. Following are some basic steps to take:

1. While on the lower floor, you can use a sensor (a stud sensor, for example), to find the joists between which you will place a device.
2. Next, cut a hole about two inches in diameter between the joists that you located in step 1 and measure the distance from the hole to the external (outside) wall parallel to the joists.
3. Go upstairs and measure the same distance from the same outside wall. This measurement enables you to find the exact location of the same joists.
4. Drill a hole between the tack strip and the molding of the floor. Using a stop ring on the fishing chain, drop it so that it piles on top of the ceiling below.
5. Go back downstairs, open the ceiling, and locate the pile of fishing chain.
6. Attach the located fishing chain to about 10 feet of the pull cord and attach the cable to the cord.

7. In the final step, return upstairs and use the chain and pull cord to pull the cable or wire into position.

As is typical of installation projects, there may be additional steps to be carried out, depending on what specific challenges are involved with different projects. It is important to plan ahead of time.

Special Situations

Older telecommunications cable within a building can sometimes be reused. Technologies such as Home Phoneline Network Alliance (HomePNA) allow network signals to travel over existing telephone lines. Speed is a limiting factor for high performance, however. Most technologies that use existing wires or cable limit data speed to 10 Mbps, which may not be sufficient for large file transfers or video servers on a network. If a switch is used, old cable using HomePNA technology and newer cable using 100-megabit Ethernet can be mixed on the same network. Each connection operates at its designated speed. If a hub is used, all the connections to the hub are reduced to the speed of the slowest device connected to the hub.

When planning a retrofit installation, the installer may discover that some walls just cannot be fished. Surface mount raceway provides a means of cabling these areas. A *raceway* is a baseboard plate with a cover that allows cable to be concealed. It comes in various sizes to accommodate any number of cables. Fittings are available to make practically any offset or change in direction. Most surface mount raceway is plastic.

Raceway

There is a newer decorative style of raceway available. Decorative raceway can replace a baseboard. Raceway that simulates a cornice molding or chair rail is available. Although it tends to be expensive, raceway provides an attractive alternative to fishing in walls.

Some owners object to surface mount raceway of any kind. When walls can't be fished, the other alternative is to run cables outdoors. When cabling to a second floor is required, it may be necessary to go outside the house at the basement level, run cable up the outside of the house to the second floor, and re-enter the house at the second-floor outlet location. When running cables horizontally outdoors, it is usually best to run the cable under the eaves of the house, which protects it from the weather and helps disguise the cable run. Always remember that indoor cable is not rated for outdoor use. Besides the effect of water, the ultraviolet rays of the sunlight can damage the cable jacket material.

Retrofit Installation

As mentioned earlier, retrofit installations can pose various challenges to the installer. For the sake of efficiency, it is best to perform all similar tasks at the same time.

- When fishing walls, cut all the holes for outlets at the same time.
- Fish all the walls and leave a pull string in place. This string will be used when the cable pulling begins. Use the steps outlined earlier in the chapter to fish wires.

Walls that cannot be fished usually require surface mount raceway.

- Mount all the raceway to these locations. Special outlet boxes are available for surface mount raceway. Mount these boxes where required.

Some SMCs are designed to be a surface mount box; this is common for a retrofit application. Surface mount cabinets generally use some type of surface mount raceway to create a neat and professional appearance for the cable entering the SMC.

Some SMCs require the installer to cut a hole in the wall to mount the cabinet flush with the wall. Flush mount cabinets are designed to be mounted between the 2 × 4 studs. Cabinets are screwed to the studs from the inside of the cabinet. When using a flush mount cabinet, it may be necessary to fish all the cables entering the cabinet.

After all the walls have been fished and surface mount raceway has been mounted, cable pulling can begin. Just as in new construction, cables should be labeled as they are pulled. All cables to an individual outlet should be pulled at the same time. Space is limited in retrofits and there may not be access to a path for a second pull. Retrofits are a good place to consider using the newer composite cables or the so-called bundled structure cabling systems. Composite cables have a variety of different cables within the same cable sheath or bundle. Having two CAT 5 cables and two coax cables inside a single cable sheath can ease the pull in a retrofit installation that has limited space.

Cleaning Up

The last step in the rough-in process is cleanup. Each individual trade is required to clean up their work area in new construction and the final cleanup is completed once the project is finished. In a retrofit, the cleanup is extremely important and should be handled on an ongoing basis. In addition to wire scrapes, there is drywall debris and dust from cutting holes in walls and fishing. Cleanup should be completed promptly after cables are pulled and dressed.

SUMMARY

- New construction installation requires coordination with the construction superintendent and the electrician in order to gain access after the electrician but before the drywall installation.
- Securing blueprints for an installation in a new home enables the HTI to accurately determine the placement of cables to be run.
- Verifying local codes that affect the cable installation helps with scheduling issues and can avoid unnecessary costs that could be incurred.
- In new construction, mud rings or drywall rings are preferred over outlet boxes. They should not be mounted to the same stud as an electrical outlet box.
- Communication cables should be run no closer than six inches from power cables and all intersections should be at 90 degrees.
- The NEC defines the rules for structured cabling systems. Cables through framing members must be 1-1/4 inch from the edge or protected by a nailing plate. Cable run through metal must be protected by a bushing or grommet unless it is inside a conduit.

- Predrilling holes and staging the cables before actually pulling the cable provides for a more efficient installation. Once pulled, cables should be labeled according to TIA/EIA standards.
- The challenges of a retrofit installation can be eased with proper planning. Use blueprints or as-built drawings when available, and verify all applicable installation codes.
- Fishing cable through existing walls requires preplanning to avoid obstructions. When fishing is not an option, raceways can be used. Newer raceways provide updated styles to add a decorative touch. HomePNA can also be incorporated into the network.
- New construction cleanup is typically done after all trades have completed their work and is handled by the construction superintendent. Retrofit installations must be cleaned up by the HTI upon completion. This includes removing all debris left by cutting wire and drilling holes.

GLOSSARY

As-built drawings Refers to the final version of the blueprint.

Blueprints Detailed drawing used to determine the location of network outlets.

Communication cables Bundled cable.

Dressing cable The practice of forming cables into manageable bundles and taping the ends for better management.

Fishing Running cable through existing walls.

Floor joist Cross beam laid to support the floor.

Mud ring Low-voltage mounting bracket.

Nail plate Metal plate installed on studs to protect cables from damage after the drywall has gone up.

Pulling cable Running cable from the staging area to the location of the mud rings.

Raceway A unit used to hide cable, usually run along the baseboard.

Staging cable Setting up cable reels or boxes of cable in order to pull from a central location.

CHECK YOUR UNDERSTANDING

1. What is a document that shows the details of the construction of a house?
   a. Blueprint
   b. Construction schedule
   c. Construction proposal
   d. Bid document

2. The HTI should begin work after the:
   a. drywall installer.
   b. plumber.
   c. electrician.
   d. HVAC crew.

3. How far apart should communication cables be run from electric cables?
   a. No closer than ten inches
   b. No closer than six inches
   c. It doesn't matter.
   d. At least three inches

4. What defines when and where nail plates are required?
   a. Homeowner needs
   b. Electrician
   c. NEC
   d. They are not defined.

5. What is the advantage of using mud rings?
   a. Larger bend radius
   b. Installed at eye level
   c. Can be installed at any level
   d. Local codes will always approve.

6. Mud rings should not be mounted on the same stud as the electrical outlet box.
   a. True
   b. False

7. As each cable is pulled, it should be:
   a. as tight as possible.
   b. terminated.
   c. secured with tape.
   d. labeled.

8. What is the key to pulling cables quickly?
   a. Predrilling holes in the framing
   b. Staging cables
   c. Determining cable routes
   d. All of the above

9. What is the optimal height for a SMC or distribution device?
   a. At eye level
   b. Under a cabinet
   c. In the garage
   d. Under a window

10. Fishing cable refers to:
    a. cables that are tangled.
    b. running cables through existing walls.
    c. attaching cable to a wall stud.
    d. drilling cable holes.

11. What is a cable raceway?
    a. A container used to store cable
    b. Installing cable as fast as possible
    c. A unit used to hide cable
    d. A device to cut cable

12. All cable intersections should be:
    a. at 90 degrees.
    b. at 45 degrees.
    c. at 10 degrees.
    d. bent to fit in the space.

13. What does the TIA/EIA standard of labeling specify?
    a. All hardware must be labeled with the installer's name.
    b. All hardware termination units must have a unique identifier.
    c. All hardware must be labeled with the project name.
    d. All hardware must be labeled alphabetically.

14. What is the double 2 × 4 at the top of the wall above each mud ring known as?
    a. Top plate
    b. Outlet plate
    c. Pull plate
    d. Drill plate

15. What is the last phase of the rough-in?
    a. Installing the SMC
    b. Placing mud rings
    c. Pulling cable
    d. Installing nail plates

16. What is the purpose of the nail plate?

17. What are the advantage and the disadvantage of using technologies like Home Phoneline Network Alliance (HomePNA)?

18. Where can an HTI find information on local building codes?

19. Who is responsible for scheduling changes for a new construction project?

20. What is the main concern when determining cable routes in home network installation?

13

Trim-Out

 **OBJECTIVES**

Upon completion of this chapter, the HTI will be able to complete the following tasks:

- Identify trim-out tools

- Properly terminate CAT 5, coaxial, and audio cables

- Properly connect audio, video, antenna, and security components

- Understand wireless components

- Develop a home control portal

 **INTRODUCTION**

In this chapter, you learn about the trim phase of the installation process. In the trim phase, the media are terminated, which means to connect the wire to the connectors or outlets. There are different ways to terminate the different kinds of media in the different subsystems. After all the terminations have been completed, you learn how to connect the subsystems to the Structured Media Center. Wireless components and how they relate to the integrated home network are discussed (for example, security, computer subsystems, and entertainment). The chapter wraps up with an explanation of how to develop the home control portal to provide the homeowner access to automated components on a personalized website.

TRIM CABLES AND INSTALL OUTLETS

The process involved in the trim phase of an installation includes terminating the cables, connecting the subsystems in the distribution panel, and testing to ensure connectivity. This section deals with trimming cables and installing outlets per the design document. Prior to installing the outlets, they should be marked and the cables should be run. Having the right tools for the job allows the installation to proceed smoothly.

Trim-Out Tools

The tools that an HTI uses to install a home network were detailed in Chapter 1. As a review, the following tools are specifically needed for the trim-out phase:

- Labels (for labeling cable)
- Drill
- Drill bits (one-inch extension or auger bit)
- Extension power cords
- Screwdriver set (regular and Phillips)
- Pliers
- Hammer
- Keyhole saw
- Wire snips or wire cutters
- Category 5 UTP stripper
- Impact/punchdown tool with 110-bit and blade
- RG-6 quad-shield coax stripper
- RG-6 quad-shield coax F connector crimper
- Six- and eight-position telephone and Category 5 plug crimp tool
- Leviton coaster for Category 5 jack termination
- Utility knife
- Fish tape
- Level (6 inches)
- Extension power cords
- Flashlight
- Ladder
- Broom and dustpan
- Velcro straps

Specific tools needed for testing include the following:

- Multimeter or volt-ohm meter
- Category 5 field tester
- Modular plug breakout adapter
- Tone test set
- Inductive probe

Before proceeding with the trim and termination phase of the installation, review the following terms, all of which refer to problems with electrical circuits:

Open

- *Open*—The circuit is not complete or the cable is broken as shown in Figure 13–1.

Figure 13–1
Illustration of an Open Circuit

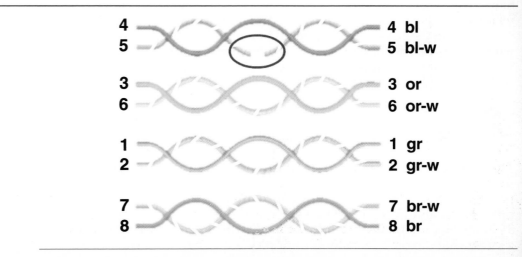

- *Short*—An incomplete circuit caused by a hot conductor coming into contact with a ground or metal component as shown in Figure 13–2.

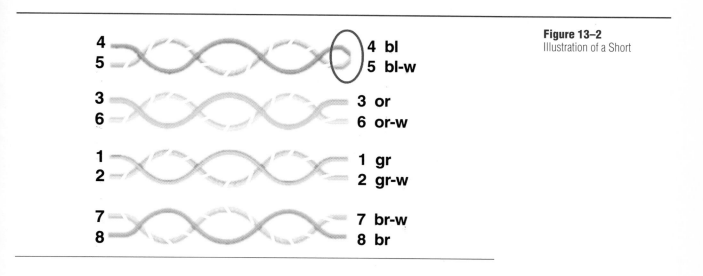

Figure 13–2
Illustration of a Short

- Reversal—Conductors on one end reversed on the other end as shown in Figure 13–3.

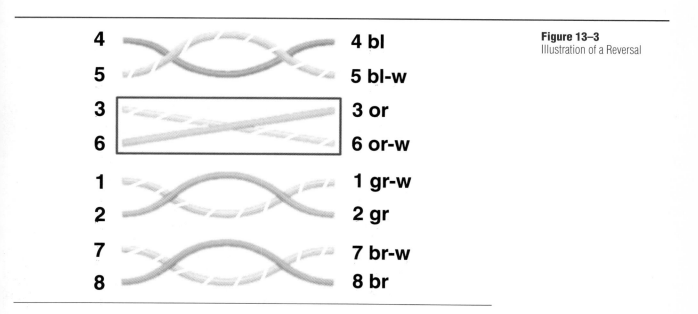

Figure 13–3
Illustration of a Reversal

- Transposed—Conductors are not in the same order at both ends of the cable as shown in Figure 13–4.

Figure 13–4
Illustration of Transposed Cables

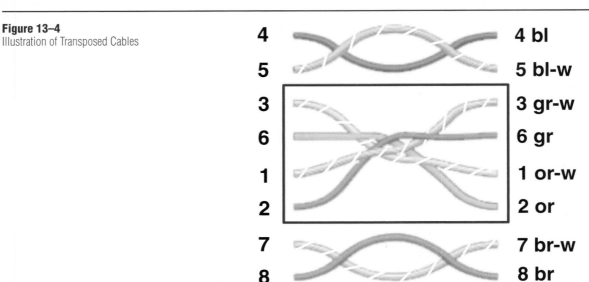

- **Split pair**—Caused by connecting wires from different pairs to "paired" pins (see Figure 13–5). This effectively results in an untwisted pair, which can cause a high level of cross-talk.

Figure 13–5
Illustration of Split Pair

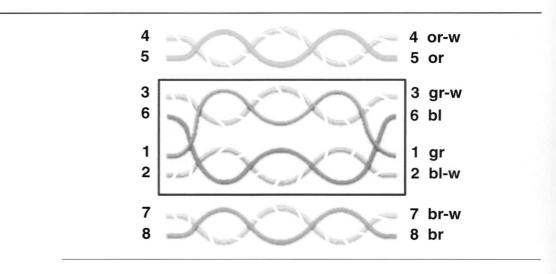

Twisted-Pair Cables

The specification for home networking (ANSI/TIA/EIA-570-A) allows the use of CAT 3 cables, but recommends the use of CAT 5 or CAT 5e cables. All twisted-pair cables should be terminated using the T568A wiring scheme. If two jacks are installed in a single faceplate, do not split the pairs of a four-pair cable to feed both jacks. Each jack should have its own four-pair cable run to it.

All terminations—whether they are in RJ-45 (8P8C) plugs, 110-type termination panels, or wall jacks—should use the standard T568A termination scheme. Care must be taken to minimize the untwist on the wires, and with RJ-45 you must ensure that the outer jacket of the cable is captured by the small strain relief tab inside the plug. Figure 13–6 shows the process to terminate twisted-pair cable.

Figure 13–6
Twisted-Pair Termination

Outlet cables should be trimmed so that 8 to 12 inches are left hanging from the outlet. Remember to relabel the cable if the label is cut off with the excess cable. Cables that are terminated in outlet boxes should be cut to eight inches. Cables that will be terminated in mud rings may be left longer. The larger wall cavity in this case accommodates longer cables without the risk of exceeding the maximum bend radius of one inch. It is difficult to coil excess cable into a standard outlet box.

Remove two to three inches of the outer jacket to expose the twisted-pairs. Always use the proper tool for removing the cable jacket to ensure that the insulation on the pairs is not nicked or damaged. Follow the manufacturer's instructions for the jack when terminating the pairs.

Jacks have the color code scheme imprinted. Both color codes, T568A and T568B, are identified. The installer has the option to use either code, but only the T568A scheme is compliant with the ANSI/TIA/EIA-570-A standard. Figure 13–7 shows a Leviton jack with imprinted color code.

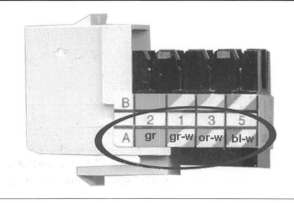

Figure 13–7
Leviton Jack with Imprinted Color Code

When terminating jacks, it is essential not to have more than one-half inch of untwist on the individual pairs. Less is better. Too much untwist can cause crosstalk, which is a major source of data errors for computer networks (as discussed in Chapter 3).

Terminating Category 5 Cable

To terminate Category 5 cable, you should follow the EIA/TIA 568-A standard. This standard defines line configuration when terminating cable.

* Use the 568-A standard on each end of the cable. The 568-A standard determines the order in which the wires are placed in the RJ-45 connectors, and the order in which they are punched down at the wall outlets.
* Strip back only as much cable jacket as is required to terminate the connecting hardware.
* Do not untwist the wires more than one-half inch at the connector or wall outlet.
* Be careful not to nick or cut the wires when stripping the outer sheathing.
* Make sure the plastic sheathing is crimped in RJ-45 connectors. Wire pairs may become exposed and damaged if the sheathing is not properly crimped.

The process of terminating wire pairs is illustrated in Figure 13–8 through Figure 13-11.

Figure 13–8
Terminate the Blue Pair

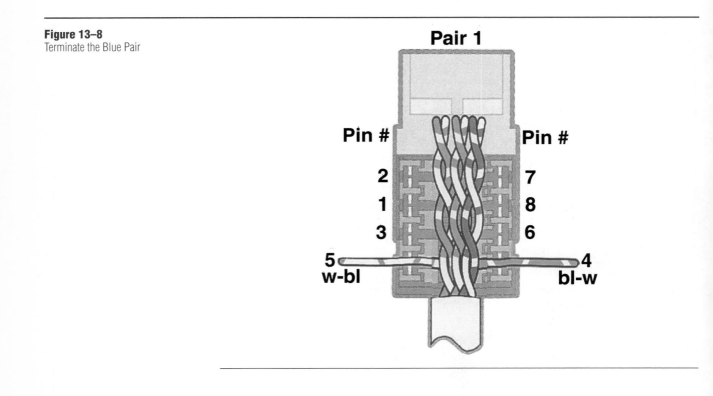

Termination Equipment for Category 5 Cable

Removing the cable jacket properly and terminating securely ensures that every wire connection is made with full contact between the wire end and its termination point. This technique results in a high-quality connection that cannot be obtained when a wire is nicked or poorly stripped. Leviton offers

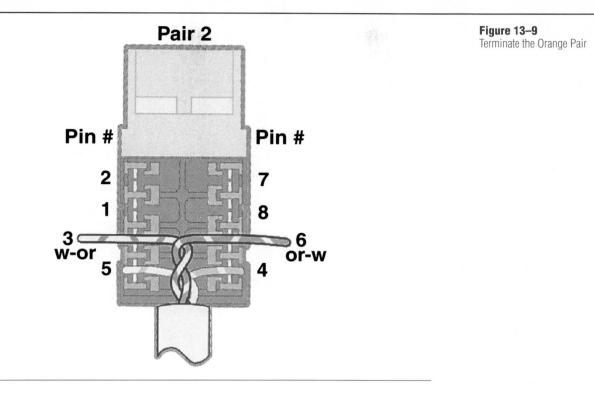

Pair 2

Pin #   **Pin #**

2   7
1   8
3   6
w-or   or-w
5   4

Figure 13–9
Terminate the Orange Pair

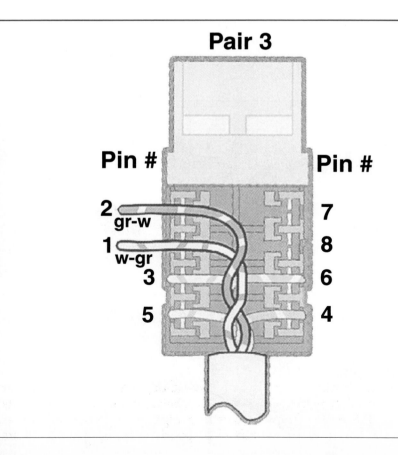

Pair 3

Pin #   **Pin #**

2   7
gr-w
1   8
w-gr
3   6
5   4

Figure 13–10
Terminate the Green Pair

Figure 13–11
Terminate the Brown Pair

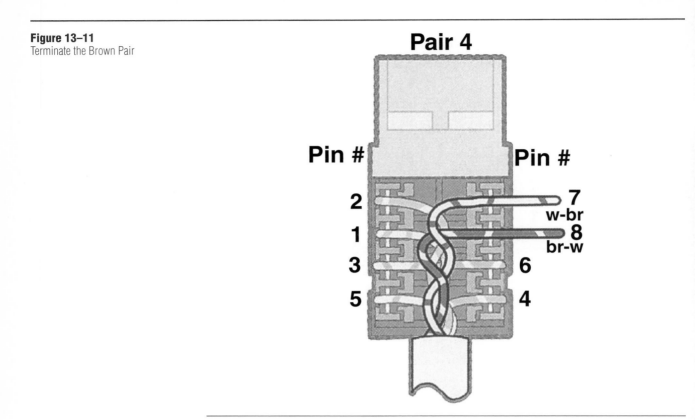

several termination tools for Category 5 cable including:

- UTP jacket-stripping tool shown in Figure 13–12
- Jack termination tool shown in Figure 13–13
- D814 wire punchdown/termination tool (and D814 tool blades) shown in Figure 13–14

Figure 13–12
UTP Jacket-Stripping Tool

Figure 13–13
Jack Termination Tool

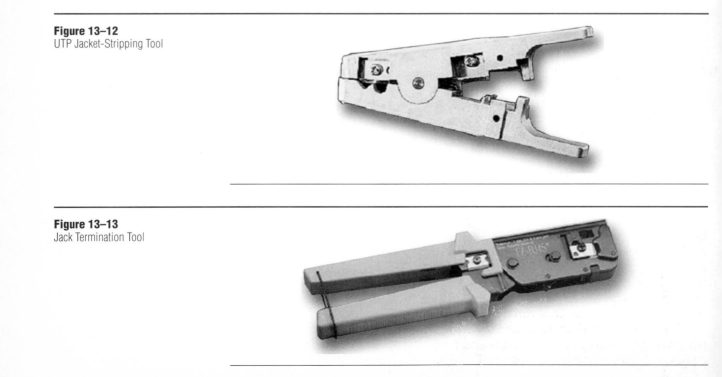

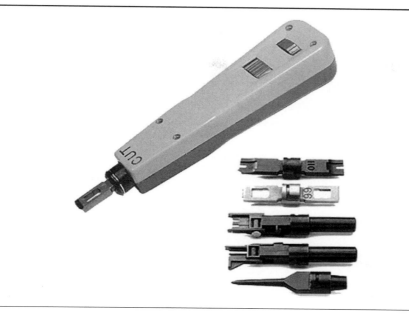

Figure 13–14
D814 Wire Punchdown/
Termination Tool

The *wire punchdown/termination tool* presses wire conductors into place in insulation displacement connectors (IDCs) and in QuickPort modular snap-in jacks. It also trims off the ends of the wire conductors. The wire punchdown/termination tool assures solid connections on an array of wire termination blocks. A push on the tool's handle and one of its five interchangeable blades easily terminates 22-, 24-, or 26-gauge solid wire into 110-style IDC punchdowns, as found on Leviton Category 5 jacks, Category 5 modules, and other Leviton modules.

**Wire punchdown/
termination tool**

NOTE Category 5 cable requires precise termination. Removing too much insulation can adversely affect the efficiency of the cable. Tools manufactured specifically for trimming low-voltage cable work more precisely than knives or diagonal cutters.

EIA/TIA-570 standards recommend the use of eight-conductor jacks only with a T568A/B wiring pattern on the outlet end. Leviton recommends one of the following three terminations on the distribution end of Category 5 cable:

- Category 5 module or patch block (or patch panel)
- Individual Category 5 jacks mounted in a housing, bracket, or panel with T568A/B wiring
- Unbridged patching module

Color Coding of Wires   Each twisted-pair is color coded to correspond with the color coding on the termination jack and the distribution module. In a given cable, each pair has a twisting at a different twist rate than the other pairs in the cable or bundle. The different twist rate gives the cable greater immunity from interference that can result in distortion. For a better installation, always untwist the least amount of cable necessary to make a connection.

Each twisted-pair must attach to its own color-coded location on the eight positions of a Category 5 jack. Standard four-pair UTP color coding uses four colors in a distinctive combination. The color combination identifies the pair number (1 through 4) as well as the first or tip wire and the ring wire within the pair. Table 13–1 shows the wiring color code for Category 5 cable.

Table 13–1
Wiring Color Code for
Category 5

Wire Pair# and Lead Functions	Color Code
1 Tip 1 Ring	White—Blue Blue
2 Tip 2 Ring	White—Orange Orange
3 Tip 3 Ring	White—Green Green
4 Tip 4 Ring	White—Brown Brown

Low-voltage cable is typically 24-gauge. It must be handled carefully in order for it to carry a signal with the least amount of loss or distortion. Use the tools and methods recommended in this manual in order to maintain the integrity of the system and keep callbacks to a minimum. Two higher grades of cable are also available:

- Category 5e
- Category 6

Category 5e meets the standards set out in Addendum 5 of EIA/TIA-568-A; standards for Category 6 cable are still in development.

Terminating Category 5 Jacks   All voice and data Category 5 cable are terminated in a Category 5 jack with 110 IDC connectors using the 568A wiring pattern. EIA/TIA-570 recommends the use of eight-conductor Category 5-compliant jacks only.

To terminate a Category 5 jack, you need the following:

- UTP stripping tool
- D814 wire punchdown/termination tool with 110 blade
- Leviton coaster with jack holder

The following steps are guidelines for terminating a cable at a Category 5 jack:

1.   Allow a 24-inch service loop at the wall plate.

2.   Remove as little of the cable jacket as possible from the cable, keeping the jacket intact as close to the termination point as possible.

3.   Use a Leviton UTP stripping tool or equivalent to strip the jacket.

4.   Follow the color code, matching up similar colored wires to their counterpart locations on the jack. Use *only* the T568A wiring pattern, as shown on the lower portion of the jack label, when wiring the jack.

5.   Untwist each cable pair a maximum of one-half inch from the termination point.

6.   Use a Leviton D814 wire punchdown/termination tool or equivalent for inserting the cable into the jack and trimming off the ends, one pair at a time, starting at the rear of the jacks with the blue pair.

NOTE Leviton's plastic coaster for holding jacks during punchdown provides a flat surface for punching down cable and cutting off the ends on the jacks. The coaster fits in your pocket and is readily available when you do not have a level work surface.

Installing RJ-45 Connectors

The following are instructions for attaching the RJ-45 connectors to a Category 5 cable.

1. Strip back two inches of sheathing from the cable.
2. Untwist the pairs; flatten and arrange them in the proper color-code order following the 568-A standard.
3. Trim the wires to one-half inch from the sheathing.
4. Insert the wires into the RJ-45 connector.
5. Make sure that the wires remain in the proper order.
6. Make sure that each wire slides under the crimping "teeth" in the connector.
7. Make sure that the sheathing is far enough inside the connector to be crimped.
8. Crimp the connector using an RJ-45 ratchet-style crimper.
9. Hold the connector and lightly pull on the cable to make sure all the wires and sheathing are securely crimped.

Installing RJ-45 Wall Receptacles

When installing the RJ-45 wall receptacle, make sure to use a receptacle configured for the 568-A standard. In this example, the receptacle is labeled for use with 568-A and 568-B configurations.

1. Strip back two inches of sheathing from the cable.

2. Place the wires in color-coded slots for the 568-A standard. Maintain the original twist ratios as much as possible. Remember that the pairs should not be untwisted more than one-half inch between the punchdown point and the cable sheathing.

3. Use the punchdown tool to punch the wires into place. Make sure that the cut side of the punchdown tool is facing outward.

4. The wires should be cleanly cut during punchdown. Place the plastic cup over the exposed wires and snap it into place.

5. The Category 5 receptacle snaps into the wall plate and can be used in conjunction with other snap-in receptacles.

NOTE Category 5 certification testing, also known as *compliance testing*, is highly recommended for verifying your cable installation. For certification, wiring should be tested for continuity, attenuation, crosstalk, and cable quality.

Audio Cable

Audio cables are selected by their wire gauge. Generally, they are not twisted-pair. Audio cables are usually two-stranded conductors, 16-gauge or larger, that are not uniformly color-coded, as shown in Figure 13–15. Because one conductor in an audio cable can be distinguished from the other, special attention should be paid to polarity. Sometimes the conductor itself may be differently colored. One may have a copper color, whereas the other has a silver color. The insulation on the conductors may also be colored differently. Other times, the insulation on one conductor may have a rib in the plastic insulation that is not present on the other conductor. Audio wire does not use color coding in the same way as twisted-pair copper wires.

Figure 13–15
Audio Cable

Audio cables are most commonly terminated on screw terminals or binding posts. A screw terminal uses a screw that tightens into a plastic base. A binding post is a stud that protrudes from a plastic base. The wire is wrapped around the stud and a nut is tightened to secure it. Split the conductors of the audio approximately one inch. Strip one-half inch of insulation from the conductors. Be extremely careful not to nick the conductors in the stripping process. Wrap the conductors around the screw terminal or binding post and then tighten them. Always wrap the wire around the terminal in a clockwise direction to keep it from slipping out as the screw or nut is tightened. If the wire is wrapped counterclockwise, the tightening motion tends to push the wire out. Never wrap the wire more than 180 degrees. Wrapping the wire so that it overlaps itself (more than 180 degrees) can cause the wire to be cut when tightening.

Coaxial Cable
Care must be taken when handling coaxial cable (coaxial cable is shown in Figure 13–16). Kinks and tight bends can change the impedance or introduce loss into the cable. Improperly installed connecters can cause signal leakage and can provide a way for interference to enter the cable. When RF interference or noise enters the cable and is transmitted down the cable system, it can degrade the signal.

Figure 13–16
Coaxial Cable

Coax cables should be cut to length with six to eight inches left sticking out at the outlet. Make sure to replace the label on the cable if it is cut off with the excess cable. As with twisted-pair cable, longer lengths can be left at the outlet if the outlet is mounted on a mud ring.

Special stripping tools are used on coaxial cable. The tool has two blades set at different depths. One blade slits the outer jacket only; the other blade slits the outer jacket, the braid, and a portion of the center dielectric insulation.

The blade should not cut or nick the center conductor. Remove excess materials, leaving the proper amount of braid and center conductor exposed. The proper connector is pushed onto the end of the cable. Single-braid and quad-shield cables require different connectors.

The new generation of coaxial connectors, shown in Figure 13–17, does not require a crimp around the circumference. They are compressed longitudinally, forcing a compression ring up into the connector. This creates a uniform compression around the entire connector, and this type of connector requires a compression tool.

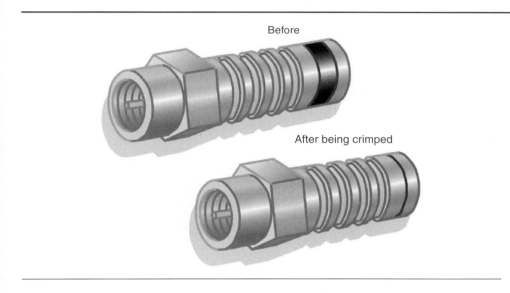

Before

After being crimped

Figure 13–17
New Generation of Coaxial Connectors

After the connector is installed on the end of the cable, the connector is screwed to a *bulkhead connector*, sometimes called a *feed-through connector*. To ensure that the connector is tightened correctly, a torque wrench should be used. Loose connectors can be a source of signal leakage and can be subject to interference. Overtightening a connector can sometimes strip threads on the bulkhead connector.

Terminating RG-6 Cable

Terminate RG-6 cable by means of professional-quality F connectors. Figure 13–18 shows RG-6 cable with an F connector. Do not use twist or push-on connectors. Weatherproof connectors are required for any terminations made on the exterior of the house.

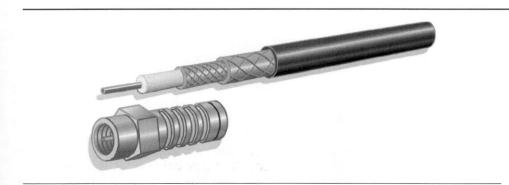

Figure 13–18
RG-6 Cable with an F Connector

The recommended terminations for video wiring are compression lock or threaded male F-type connectors at both ends of the cable. A high-quality push-on F-fitting can be used as well, but if you have any questions about its integrity, use a compression lock or threaded style instead. The small amount of extra installation effort will pay off in the end.

Leviton recommends a high-quality, commercial-grade coax stripping tool (for example, a Leviton C5914) for stripping all RG-6 quad-shield cable and a comparable crimping or compression tool for assembling the connector. The type of tool depends on the style of F connector used.

Leviton offers coaxial stripping tools for one-step stripping of RG-6 quad-shield cable as shown in Figure 13–19.

Figure 13–19
Coaxial Stripping Tool

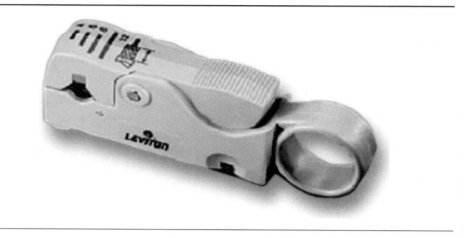

Terminating Coax Connectors

Leviton recommends RG-6 quad-shield coax cable for your television and video installations. You need the following tools to terminate coax cabling:

- Coax cable crimping tool
- Coax cable stripping tool

Coaxial cable can be terminated using one of the following:

- Compression F connector
- Crimp-on F connector
- Twist-on F connector (does not require a crimping tool)

The following are guidelines when terminating connectors for coaxial cables:

- Use a high-quality and properly adjusted stripping tool to strip back approximately three-quarters of an inch of the coax cable's outer jacket. Do not use the stripper to pull off the stripped end.
- Approximately one-half inch of the center conductor should be exposed, and approximately one-quarter inch of the braiding and foil should be exposed.
- Use your fingers to separate the outer braid from the white dielectric insulation. The outer braid should not be folded back over the shielding; it should be raised so that it will fit within the outer channel of the F connector.
- Slide the F connector over the cable so that the dielectric fits within the center channel of the connector and the outer braid fits within the outer channel of the connector.

- Make sure the white dielectric is flush with the back of the F connector.
- The center conductor should protrude one-quarter inch from the F connector.
- Use a quality coax crimper to crimp the connector.
- Double-check the center conductor and trim it to one-sixteenth to one-quarter inch from the connector at a 45-degree angle.
- Make sure that the connector is securely on the cable.
- Allow for a 24-inch service loop at the wall plate.

Installing F Connector Wall Taps   Use the following steps when installing F connector wall taps:

1. Screw the F connector at the wall tap into the wall plate.
2. Use a barrel-type connector in the wall plate.
3. Terminate all unused ports to prevent signal distortion.

Connecting Audio Components

Speaker wire is run in pairs to the appropriate locations in different rooms to serve the left and right speakers. Figure 13–20 illustrates audio speakers connected to a 1 × 6 passive audio module.

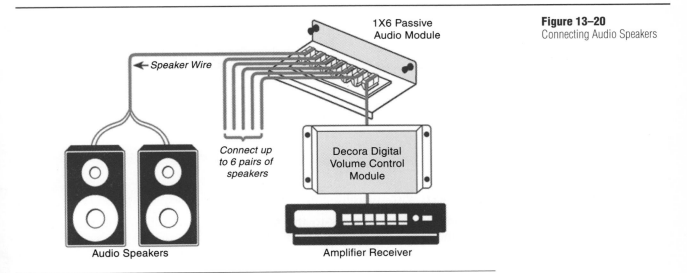

Figure 13–20
Connecting Audio Speakers

The speaker wire should be terminated in one of two ways:

- For stand-alone speakers, connect the ends of speaker wire to a speaker-connector wall plate (the type with banana jacks or the five-way binding posts is preferable).
- For in-wall speakers, just connect the ends of the speaker wires directly to each speaker's inputs.

Follow these steps to install an audio network:

1. Prewire speaker wire in pairs throughout the home. Source devices serve as a starting point.
2. Connect devices to the control amplifier (preamplifier).
3. Connect source devices to the receiver (integrated amplifier).
4. Connect the audio output interfaces of the receiver to a connecting block or speaker selector device to distribute the audio signal to multiple locations.
5. Connect the connecting block/speaker selector device to the speaker wiring.

6. In each room, terminate the speaker wire at the volume control device.
7. In each room, connect the volume control device to the speakers.

It is advisable to start with a small-scale audio network and have only selected rooms integrated into the audio network. However, to ensure scalability, the speaker cable should be run to every room in the house.

Connecting Video Components

Connecting video components requires prewired video cable between the distribution panel and the rooms. Figure 13–21 illustrates coaxial connection to a 3 × 8 bi-directional video amplifier.

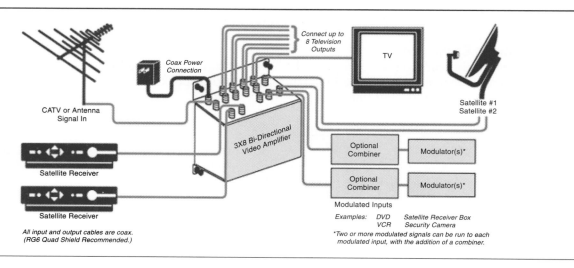

Figure 13–21
Connecting Video Components

Proceed with the following steps:

1. Connect source devices to the modulator (to distribute the video signal around the house by using unused television channels).
2. In each room, connect the modulator to the coaxial wall jack that sends to the distribution panel.
3. In each room, connect the television to the coaxial wall jack that receives from the distribution panel.

To terminate video components, it is necessary to connect devices in the rooms to the coaxial network and then connect the coaxial network to the distribution panel to enable the distribution of the video signals.

To connect devices, follow these steps:

1. Prewire video cable between distribution panel and rooms.
2. Connect source devices to the modulator (to distribute video signals around the house by using unused television channels).
3. In each room, connect the modulator to the coaxial wall jack that sends to the distribution panel.
4. In each room, connect the television to the coaxial wall jack that receives from the distribution panel.

To connect the RG-6 cables at the distribution panel:

1. All prewired RG-6 cables should end at the distribution panel.
2. All cables should be labeled (source, length).

3. Terminate each receiving RG-6 cable to an output with amplification at the distribution panel.
4. Terminate each sending RG-6 cable to an input modulator at the distribution panel.
5. Terminate the RG-6 cable from the antenna or cable television feed to the antenna/television input on the distribution panel.
6. Program the devices' channels and assign unused channels to the modulators.

Connecting Antenna Components

There are specific rules for installing antennas. Instructions should be supplied when you buy these products. When installing antennas and lead-in wires, watch for the following:

- Avoid crossing outdoor antennas and the lead-in conductor over light and power wires. Also avoid attaching them to the service mast or on electric poles carrying voltages over 250 volts.
- The lead-in wires must be securely attached to the antenna, and both the antenna and lead-in wires need to be securely supported.
- Unless permitted, antenna conductors shall not be run under open light and power conductors.
- On the outside of the building, position and secure antennas so that they cannot swing too close to any high-voltage power and electric lines.

For additional information regarding antenna installation, consult Article 810 of the NEC. The best test of antenna installation is to check the signal received on the television after it is hooked up.

Proper installation of a satellite dish includes ensuring that it is pointed in the right direction. This is determined by location. If necessary, use a certified installer for the actual dish installation. Coordinate with the satellite installer to make sure that the cable is run using the best route to the SMC. When connecting DSS (Digital Satellite System) or DBS (Direct Broadcast Satellite) to the input interface at the distribution panel, you cannot simply plug the DSS output into the distribution panel input. These methods use an RF-modulated video signal to transmit, which is different from the regular video signal that a coaxial cable transmits. The RF video signal should be handled like another source device—like a VCR, for example. This way the signal can be distributed by utilizing unused channels over modulators.

Alarm Systems

Alarm cable is generally a two-conductor cable. For security reasons, alarm cable should not be exposed. Intruders could cut exposed cables. Alarm cables usually connect contact devices, glass breakage sensors, and motion detectors to a central controller.

Most alarm devices use a screw terminal-type termination. Wires are separated and the ends of the wire are stripped approximately one-half inch. Wires are wrapped 180 degrees around the screw terminal. Just as with audio wire, security wires should not be wrapped 360 degrees around the screw terminal; this can cause the wire to be cut when the terminal screw is tightened.

Avoiding Common Termination Mistakes

Prior to terminating cables, perform a visual inspection of cables, check that all cables are labeled on both ends for easy identification, look for obvious damage, check that the cable is not bent, and ensure that each cable is connected to the intended device.

To ensure a good connection for cable terminations, follow these guidelines:

- Terminate all unused internal and external ports. This is critical for maintaining a strong and clean signal.

- Install DC blockers with terminators on all unused camera ports.
- When stripping RG-6 cable, be careful not to nick the solid center conductor. The outer edge carries much of the signal. Also, avoid cutting through the shielding layers. Fine-tune your stripper settings and use the same RG-6 cable type to maintain consistent quality cuts.
- Expect poor performance from any cable run that has been:
  - kinked. Maintain sweeping 90 degrees turns or less. No sharp turns.
  - forced. Excessive pulling force degrades performance. Maintain wire geometry.
  - spliced. Never insert a splice between wall outlets and the distribution panel.
  - nicked. Nicked or cut wires affect signal quality. Watch for attic truss splice plates.
  - compressed. Avoid staples or anything else that crimps.
  - overrun. Never run more than 90 meters from connection center to wall outlets.
  - induced. AC wire, fluorescent lights, and dimmer switches induce interference.
  - untwisted. The jacket must crimp into the connector to prevent untwisting.
  - bent or folded. A slight bend (around 15 degrees) in the cable is acceptable; any more bend in the cable can cause problems. Avoid folding cable, which can cause breakage.

STRUCTURED MEDIA CENTER

Before you trim and terminate the cable for the Structured Media Center, be sure that all cables are long enough to connect to the modules. The RG-6 cables should reach about halfway down the distribution panel box, and the Category 5 cables should almost touch the bottom of the distribution panel box.

Cables that enter the distribution panel should be grouped together by type. It is highly recommended that you use Velcro to group the cables before you install the distribution panel box in the wall. Cables that are neat and organized not only make the connection process easier, but also make the installation look more professional. Figure 13–22 shows an SMC 100 and an SMC 140.

Figure 13–22
SMC 100 and SMC 140

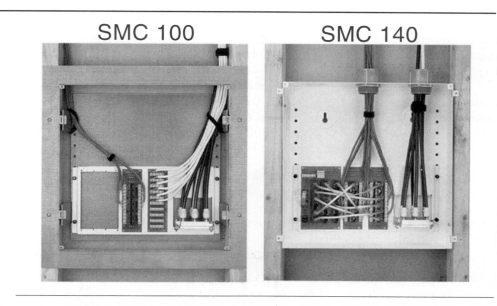

SMC 100   SMC 140

Depending on the number of cables that are required, you may need to slightly adjust cable groupings. Pull the cables through and let them hang in the distribution panel box. The configuration of the SMC should provide openings for each of the groupings of cables (CAT 5, coax, and so on). You may need to use the openings in both the top and the bottom of the distribution panel box.

> **NOTE** For a surface-mounted distribution panel, the cables should enter the distribution panel box in the same order, except that the group of cables should be routed through the openings in the back of the distribution panel box, combining the Category 5 cable groups in the center opening.

The modules installed in the SMC are dependent upon the requirements of the home network. For example, if computer networking is part of the system, a computer module will be installed. A power module must be installed for modules that require a power source. The following are SMC modules:

- Computer Module
- Entertainment Module (sometimes separated into Audio Module and Video Module)
- Telephone Module
- Home Automation Control Module

Installing Terminal Equipment
Pre-made distribution panels usually come with ready-to-implement modules. Each module has its space and preset connections in the panel, which makes the assembly easy. Figure 13–23 shows an SMC with modules installed.

Figure 13–23
SMC with Modules

Commonly, there is one module per subsystem:

- Audio/video unit
- Computer unit

- Telecom unit
- Security unit
- Home automation control unit/Home control supersystem unit

These modules simply need to be installed in the preset spaces after planning the internal layout of the SMC and adding divider plates, tray guides, and cable combs as necessary to ensure orderly installation.

Terminal equipment in the Structured Media Center (SMC) should be installed in a logical and orderly manner. All terminations should be grouped by services delivered. All telecommunications should be grouped together; data terminations and video terminations should be within their respective groups.

Items requiring power should be given special consideration. Hubs, switches, gateways, amplifiers, and modulators require a power source in addition to cable termination.

Equipment layout must also accommodate cable routing within the distribution device. Cables must be routed neatly without exceeding the minimum bend radius.

Dress and Terminate Cables

In many applications, such as in a multi-dwelling unit telecommunications room (MDU-TR), 110-style IDC termination hardware is used. 110-type terminations should follow the standard color code: white/blue, white/orange, white/green, and white/brown of the T568A wiring scheme mentioned earlier. Figure 13–24 shows 110-style termination blocks. Most 110-type terminations for home networking are on printed circuit boards. The printed circuit board creates the T568A wiring scheme for the different services provided through the termination panel. Terminations on a 110-type panel must have a minimum of jacket material removed and must have no more than one-half inch of untwists.

Figure 13–24
110-Style Termination
Blocks

If it is necessary to bundle cables, Velcro fasteners should be used. Velcro fasteners do not damage the cable and are easily adjusted when needed.

Coaxial Cables Coaxial cables in the distribution device should be routed so that the minimum bend radius is not exceeded. All coax cables should be grouped together and terminated in a location reserved for video.

Terminating Phone Wiring

Electricians commonly terminate phone wire with faceplates and modular jacks. This is found mainly in new construction. For a typical home, the electrician will have roughed-in wall boxes or plaster rings at all the potential telephone outlets. The type of wall box determines the type of faceplates that will be used.

Patch panel

In the Structured Media Center, a *patch panel* allows organized phone wiring terminations. A patch panel is also referred to as a punchdown block or terminal block. Using the patch panel, you can patch one line to another. The patch panel also acts as a central hub for the telephone network from which cable runs to wall outlets in various rooms throughout the home. This is an important part of the phone network.

The following are patch panel interfaces:

- **Punchdown interface**—Use a punchdown tool to push the wire down into the wire receptacle, which strips and connects the wires.
- **RJ-style interface**—Directly plug each phone wire run into modular jacks.

The *Network Interface Device* (*NID*) is the demarcation point between the home phone network and the telephone service provider's network. Only the service provider has access to the NID. The patch panel can reside in the main distribution panel or it can be wall-mounted. It is designed to facilitate future changes to the home phone network. The patch panel consists of a plastic frame with several evenly spaced wire receptacles, eliminating any potential need to reroute cables through the wall. The patch panel can have any of the three types of interfaces listed previously, depending on complexity, price, and even the installer.

Network Interface Device (NID)

Terminating the Computer Subsystem

The patch panel constitutes the central wiring node of the home data network or LAN. The LAN uses a patch panel or punchdown block for reasons similar to the home telephone network—that is, for data distribution. For the computer network, patch panels should be rated Category 5 at a minimum, to ensure high-speed data transmission. Cable runs from the different wall outlets can be terminated quickly and easily at one point, without a lot of splicing and crimping, thanks to the punchdown block. When the cables terminate on one side of the patch panel, wires called cross-connects are used to connect the terminated cables to a series of RJ-45 connectors on the other side of the patch panel.

Connecting Alarm to Phone Line

The security subsystem is usually connected to other subsystems. The phone subsystem is used by the security-monitoring service to notify them that the alarm, smoke detector, or carbon monoxide detector has detected an emergency situation. The phone line runs through the security unit in the distribution panel. You and the homeowner should consider dedicating a phone line to the security system.

Another option is having a wireless security-monitoring service. This adds extra security because a security alert would not be affected if the phone lines were cut by an intruder.

Connecting Alarm to Other Interfaces

In addition to the phone subsystem, the security system can be connected to the home automation control subsystem as well. With a web interface, the security system can be controlled from anywhere in the world.

The home automation control subsystem is easily connected to the security system because different security control panels support connections to the X-10 home control modules. These modules can control lighting via timers and programming, for example. The security system can also be connected to the intercom system.

Connecting the security subsystem to the World Wide Web provides for programming that enables the monitoring of the house via cameras and web cameras. If the alarm is tripped, the homeowner can determine the entry point.

Terminating Home Automation Control Wires

Terminating the cable for a home automation control subsystem is similar to terminating the wire for a light switch. Lighting is installed by connecting the hot wire to one side of a light switch. The neutral and ground wires are run directly to the light, bypassing the switch. The neutral wire serves as the return path for current. When you terminate the wires this way, the switch has only one possible way of controlling the light: by interrupting current flow in the hot wire.

X-10, the standard for *home automation control* (*HAC*), can limit what a device can do based on the termination. For instance, incandescent light fixtures, in which the light bulb has a filament, can be controlled because the filament provides a path for X-10 signals. However, low-voltage and fluorescent lights cannot take advantage of X-10 control in this wiring and termination scheme because they have no filament to ensure X-10 signal flow. For complete home-lighting control using X-10, you have to use special appliance or dimmer switches that connect all the wires, including the hot, neutral, and ground wires. This means coordinating with the electrician to make sure that the ground wire is also run to the switch's junction box, instead of bypassing it. This way, all three wires are available and terminated at the junction box.

Home Automation Control Processor

Home control supersystem devices require programming. They provide a GUI (graphical user interface) displayed on a screen that facilitates programming. In reality, each subsystem has its own control interface, which can be viewed via the GUI by a simple press on a button, touch on a screen, or click on an icon. Each subsystem can also have a configuration file that can be accessed remotely, which also presents an opportunity to "flash" upgrade the files. However, in order to control all the systems within a home, that is, the *supersystem* (everything tied together in one CPU) around the clock, customer support is preferred. Only a few high-end products do actually tie into one GUI control interface. Some products in this category include Echelon and Crestron.

Once programmed, adding new devices into a supersystem control, or upgrading, can be very challenging. Most additions require a reprogramming of the entire supersystem CPU. Even with the high-end products, it is still not possible to add all the functions of each subsystem to the same CPU. Incorporate only the essential functions from each subsystem into the supersystem control.

> *Web links:*
> http://www.echelon.com/
> http://www.crestron.com/default2.asp

Installing and Configuring the Home Automation Control Processor

There are two types of home automation control processors available with the X-10 home network system: a simple version and a sophisticated version.

- The *simple* version needs no programming. It uses a minicontroller and requires appliance modules to control the appliances' on/off state and an X-10 unit to control the lighting.
- The *sophisticated* version requires programming. It uses either a programmable controller per room or wireless controllers.

  Programmable controllers can be programmed directly or via a connected PC.

Home automation control (HAC)

Supersystem

When programming via a PC, a set of consecutive commands, known as macros, can be created and reused for building more complicated home control programs.

Downloadable macros are available at http://www.x10.com.

Programming kits, which are available from IBM, include control software setup, X-10 interface hardware, a lamp module, an appliance module, and a wireless control system.

Installing and Configuring Subsystem Components

Home automation subsystem components consist mostly of different types of sensors that trigger some event, such as an on/off event for an appliance. For example, you can make an X-10 home automation control network interactive by adding sensors to the programming of the network. Sensors can be used for security when set to trigger a random event, so that the home seems inhabited, or a specific input can trigger a specific programmed event.

Various sensors are available for home automation subsystem applications, including the following:

- *Photocells:* Trigger events based on the amount of sunlight hitting the cell. Commonly used to turn lights on at dusk.

Photocells

- *Motion detectors:* Trigger events when movement in a specified range is detected. Commonly used to turn on lights in a hallway when somebody enters.

Motion detectors

- *Motion-detector floodlights:* Trigger events when movement in the detection field occurs. Can be used to trigger a chain of events such as turning on the porch light or turning on the hallway light.

Motion-detector floodlights

- *PowerFlash modules:* Can be implemented in almost any stand-alone sensor. An event is triggered when the sensor is activated.

PowerFlash modules

- **Water sensor module:** Sensors relay to shutoff device installed at the water source. When water is detected at the sensor, it sends a signal to turn the main valve off.

In addition, sensors are available to monitor the temperature in the house, when mail is delivered, or when a car enters a driveway. There are also time sensors to trigger specific events at certain times during the day, week, or month. New technology continues to develop wired and wireless solutions to further automate a homeowner's life.

To extend the capabilities of the home automation control components, the sensors need to be combined with a home automation controller. In this setup, one event detected by a sensor can trigger multiple events and control several appliances at the same time. Most home automation control systems on the market are made up of customized sensors and a central controller or processor with preset functions. The actual configuration of the components is usually described in great detail in the installation instructions or user manual of the specific product.

Labeling Terminal Devices

All cables and devices in the distribution device should be clearly and neatly labeled. A unique identifier is assigned to each cable within the distribution device. Twisted-pair termination blocks should be labeled so that termination locations can be clearly identified. Devices such as modulators should be labeled with information including the type of device and the channels or frequency at which it operates.

Labels should not be exposed. A common practice with the distribution device is to attach the labeling to the inside of the enclosure. Figure 13–25 shows an SMC with the components labeled.

Figure 13–25
Labeled SMC

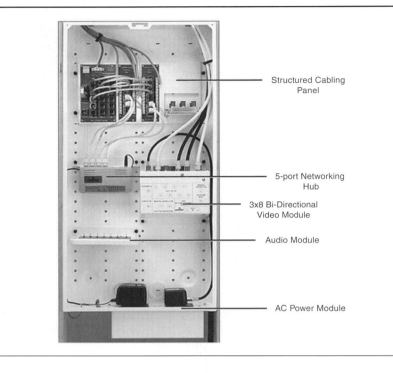

Structured Cabling Panel

5-port Networking Hub

3x8 Bi-Directional Video Module

Audio Module

AC Power Module

After all the labeling is complete (both faceplates and distribution devices), you should provide a separate listing of the cables, their locations, and other system features to the homeowner or builder. This documentation should include all cable listings by number or a floor plan showing the exact locations and numbering sequence for the cabling, if possible. This documentation is provided in the package of information that is presented to the customer upon completion of the job.

Running Cross-Connects or Patch Cords

The last step for the cable installation is to run cross-connects or patch cords. The patch cord delivers the actual services to the user outlets. It is common for voice outlets to use silver satin patch cords, whereas data patch cords will be CAT 5 or CAT 5e. CAT 5 patch cords can be used for voice; however, the distribution device can be easier to manage if different color or style patch cords are used for voice and data. Figure 13–26 illustrates where patch cords connect in the SMC.

Coax cables for video may terminate directly onto a video splitter. Sophisticated services such as modulating a channel may require the use of patch cords.

All patch cords should be run neatly, with no excessive bends. Care should be taken so that patch cords are not damaged when the door of the distribution device is closed.

Installing the Power Supply

Most distribution panels are designed as a distribution device for low-voltage wiring only; they are not designed for the installation of standard AC cable. When power is required for distribution modules, power modules should be installed. Most of them have the additional benefit of providing surge protection for the distribution modules.

Depending on the distribution panel, the location of the AC power module might vary. Generally, it is located at the bottom of the panel to allow easy access to a power line. An AC power module is shown in Figure 13–27.

Figure 13–26
Patch Cords in the SMC

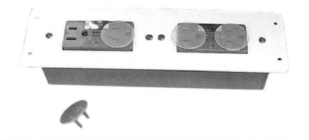

Figure 13–27
AC Power Module

The following items are important when working on the distribution panel:

- Never wire live AC cable. The circuit must be disconnected at the power source and confirmed with a tester.
- Communications wiring and components should not be installed during lightning storms or in wet locations.
- Never touch wires that are not insulated or terminals unless they have been disconnected.

Installing the AC Power Module Like any piece of electrical equipment, the AC power module should be installed following all appropriate electrical codes and safety precautions, including the following:

- Running at least one dedicated 15-amp circuit to the distribution panel
- Using 14/2 or 12/2 Romex house wire
- Making sure that the power to the circuit is off before wiring the AC power module
- Using only copper or copper-clad wire from the power source
- Understanding that the AC power module cannot act as a lightning arrestor

• Do not use a power strip in place of the power module. This can void the UL listing of the distribution panel.

Figure 13–28 illustrates how an AC power module is installed.

Figure 13–28
AC Power Module Installation

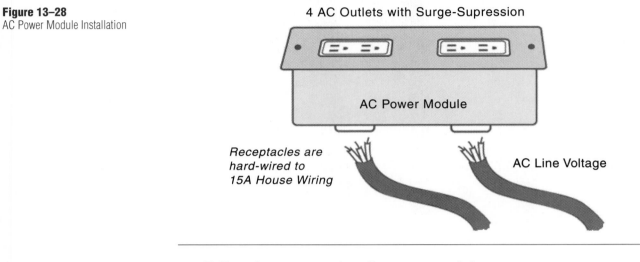

Follow these steps to install a power module:

1.   Turn off the circuit breaker before you begin working with the distribution panel.

2.   Remove the large rectangular knockout at the bottom of the distribution panel. Place the power supply to the left of the junction box on the bottom ledge of the enclosure so that the AC power cord comes out from the right of the power supply and the DC power cord comes out from the left.

3.   Punch out one or both of the cable entry knockouts that are located at the bottom of the module housing and pull the cable(s) through.

4.   Use the clamps provided to route the DC power cord for the power supply behind each tray you install. Maintain at least a one-quarter-inch space from any other cables.

5.   Drop the housing into its opening.

6.   Connect the incoming power cable to the surge suppressor receptacles. The wiring connections depend on whether one or two dedicated circuits are being run to the AC power module. Neatly wrap the extra cord with cable ties as shown in Figure 13–29.

Figure 13–29
Module Plugged In

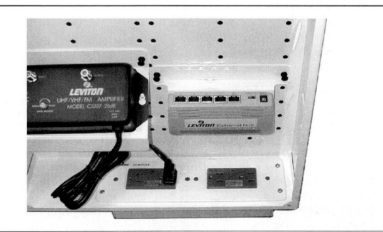

Follow these steps to install receptacles using the single cable of one dedicated circuit:

1. Attach a "pigtail," or short piece of white neutral (N) conductor, to the silver terminal screws on each receptacle. These screws are located on the sections labeled white on the back of the receptacles.

2. Twist the two white pigtails together with the white or neutral conductor from the incoming power cable, securing the ends with an appropriate wire connector.

3. Attach a black or hot pigtail to each brass-colored terminal screw on the back of each receptacle. These screws are located on the sections labeled black on the back of the receptacles.

4. Twist the green ground leads from the receptacles with the green or bare copper grounding conductor from the incoming power cable, securing the ends of the wires with an appropriate wire connector.

5. Carefully place the receptacles into the housing, making sure that all wiring is safely tucked inside the housing.

6. Secure the AC power module to the distribution panel using the four #6 X 3/8-inch screws that come with the module and tighten the module securely; these screws also secure the module's housing to the distribution panel.

7. Test the module by turning the power on at the service panel and checking the voltage with a meter.

8. Until the work is completed, install safety plugs into the receptacles to protect them from construction debris.

9. If you are installing two separate dedicated circuits, each incoming cable connects directly to a single receptacle without the use of pigtails or wire connectors.

Grounding the Distribution Panel   A grounding lug can be found at the bottom of the distribution panel. An appropriate ground wire should be secured between the grounding lug and an earth ground point, such as a grounding rod or a grounded metallic water pipe. Insert the ground wire into the lug and tighten with a screwdriver. A minimum No. 10, solid conductor, bare copper wire is recommended to ground the unit.

WIRELESS COMPONENTS

There are many wireless solutions that can be integrated into the home network. Ideally, the wireless technologies should be used as an extension of the cabled network.

The advantages include the following:

- Portability
- Ease of installation
- Capability to access unreachable areas

There are some disadvantages to using wireless in the home network, including the following:

- Limited bandwidth
- Security
- Interference
- Distance limitations
- Proprietary protocols

There are several things to consider with wireless technology:

• Interference is a major problem with wireless components. For example, a home near an airport is subject to constant interference. An "interference study" can be done to determine whether a specific area is prone to this problem.

• Interference can be caused by having too many wireless devices on the network.

The following sections discuss wireless solutions in security, computer subsystem components, and home entertainment.

Wireless Security Subsystem Components

As discussed in Chapter 6, security systems use the zoned control approach. These zones are secure cells that are monitored independently by different types of sensors. Security breaches are detected by zone and then transmitted to a central controller to confirm an actual breach. The concept of supervision describes the monitoring of the security zones by a central controller. Most systems monitor power failure, telephone-line problems, loss of internal time measurement, breaches or attempts to tamper with the system, or low batteries. Wired and wireless security supervisions are basically the same, except for the type of media used to transmit signals. Figure 13–30 illustrates the different wireless components for a security system.

Figure 13–30
Wireless Components for a
Security System

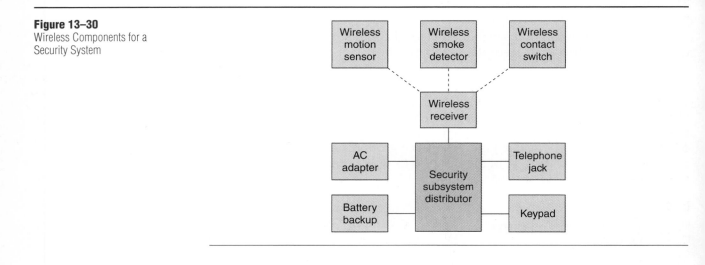

Most wireless systems require some parts to be wired. Although there are several wireless security sensors available that emit high-frequency coded signals to the base station, you may still need to have a phone line to call out to the monitoring service in case of a security breach. A wired phone line is usually required for security systems because wireless security systems are prone to interference from other signals (for example, radio, television, other wireless phones, and so on).

Wireless Computer Subsystem Components

Wireless Ethernet solutions based on the 802.11b standard are the common denominator to ensure a future proof wireless LAN. These wireless networks commonly consist of radio transceivers and special-purpose NICs. There are various solutions available from several manufacturers. Figure 13–31 illustrates wireless components in a home network.

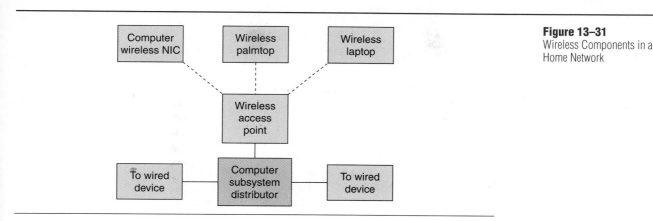

Figure 13–31
Wireless Components in a Home Network

Wireless computer subsystem components are usually more expensive than their wired counterparts, and their data transfer rate is commonly slower. Wireless technology is relatively new, but it is expected that wireless LANs will become more reliable, faster, and affordable over time.

Another wireless computer subsystem option is using HomeRF products. These products were developed especially for the residential market and are priced accordingly.

Wireless Entertainment Subsystem Components

Not only can your computer subsystem be wireless, but so can your entertainment subsystem. More audio and video products are being made that can be connected via IR or RF technology.

Audio   Your audio system can be distributed around the home without any wires by utilizing RF technology. There are wireless speakers that receive signals from wireless line-level distribution systems, which connect to audio source devices and send their signal to the receiver that in turn connects to the amplifier.

Video   Just as audio components can be wireless, so can video components. A wireless video network commonly consists of source devices such as VCRs, DVD players, or DBS. You also need a wireless transmitter and a wireless receiver. The transmitter connects to the video source via standard RCA-type connectors, whereas the receiver has an RCA and an RF output interface that connects to the TV. These wireless baseband video distributors transmit only one signal at a time. You also need to make sure that the system can receive IR signals, so that the users can use remote controls to pause, rewind, and control the volume of the video devices.

The most important aspect of a wireless video solution is the frequency spectrum on which the signals are distributed. There are systems that distribute over 900 MHz and others that use the 2.9 GHz band. The higher the frequency, the better the signals pass through physical obstacles and the less susceptible they are to interference. In addition, high frequency offers more bandwidth than lower frequency solutions.

Wireless Phones

Although wireless phone distribution is limited to two or three lines, when compared to wired phone distribution, it has the definite advantage of ease of installation. First, you need at least one conventional phone line to hook into. There are wireless phones on the market that offer one- or two-line solutions without having sophisticated call features such as intercom or transfer functions. Newer wireless phone stations offer up to eight extension phones that communicate with one base station, which also functions as answering machine and system controller. These systems also have intercom, call transfer, caller ID, conference calling, and headset features. These devices, however, are prone to interference, and some of them do not offer options to install fax machines or additional modems.

IR Control Components

Infrared components utilize light instead of radio waves to transmit signals. That means that the two devices must be near each other and facing each other in order to communicate. If the light beam is obstructed by an object, inanimate or animate, the two devices cannot communicate. Many computer and A/V devices have IR interfaces, such as printers, laptops, PCs, PDAs, VCRs, DVDs, video cameras, and so on.

As a backup in case the IR signals fail, you may also want to connect infrared components with additional runs of Category 5 cable to each IR target and IR emitter location.

Wireless Home Automation Control Components

Wireless home automation control components may be sensors that utilize RF or IR to send signals back to their base stations. These components include wireless security components, wireless entertainment subsystem components, or wireless sensors for photoreception, motion detection, or temperature measurement. Like other wireless components, they are prone to interference or might be intercepted. An intercepted signal could interfere with privacy or pose a security risk.

 **HOME CONTROL**

With the Internet becoming more and more a part of the everyday lives of consumers, it comes as no surprise that the home networking market has also discovered new ways to embrace the power of the Internet for home control purposes. There are three emerging home control concepts that the HTI should understand:

- Home control portals
- Remote home control
- Remote network management

Home control portals

Home control portals are personalized websites that the HTI creates for the homeowner. They can function as family calendars, as a collection of favorite links, as web cam supervision, and as home control interfaces.

One example of *remote home control* is a web-based home control portal, which gives the homeowner the ability to control and check on his home when on the road or traveling. It is possible, for example, to turn on the heating or air conditioning before leaving for a trip to a summer house or to check current temperatures in the main house.

For the HTI, the Internet can be a valuable tool to save time and money. A problem with a home network would normally require a service call to check the equipment. With the help of specialized software that enables *remote network management*, problems can be checked and fixed remotely from your office. Also, most software updates can be completed automatically without the necessity of a house visit.

Web link: http://www.cybermanor.com/Services/Process/Family_ Web_ Center/ family_web_center.html

Choosing Remote Network Management Software

Several remote computer and network management software packages are available today. For the consumer market with its relatively small networks, PCAnywhere and WinVNC are common products. Figure 13–32 shows an example of a home portal using a network management software package.

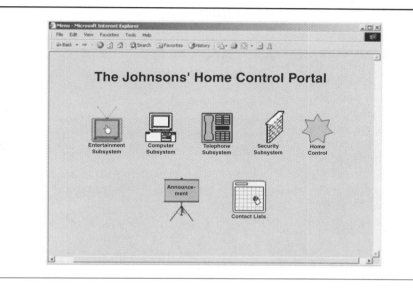

Figure 13–32
Example of Home Portal Software

PCAnywhere needs to be installed on both computers—on the main computer in the home and the control computer in the HTI's office. After installation and setup, it will emulate the home computer's interface and thus enable the remote user to take advantage of all programs and functions installed on the home computer. This solution works only if the home computer is running and if the home computer has software-based home control functions installed.

WinVNC (Virtual Network Computing), on the other hand, does not need a host component to be installed on the home computer, but it can't transfer files for software updates like PCAnywhere does.

Web link:
http://www.symantec.com/pcanywhere/Consumer/
http://www.dcs.ed.ac.uk/home/vnc/

Creating and Customizing a Home Control Portal

Web-based home control is commonly achieved by computer-based proprietary interfaces that are distributed with certain home controls. These interfaces usually consist of self-installing software that is web-based. Once installed, the web-based interface can help the residents control their thermostats or lighting by means of any Internet-based computer.

To create a home portal, the HTI creates a page in a word processing program, inserts links to favorites and web-based home controls, and simply saves it as an HTML page. This page is then transferred to a server, from which it can be accessed over the Internet.

You can either use an ISP to host these portals or host them on your own server. The advantages of using an ISP are as follows:

- Reliable and good Internet connection, having usually in the neighborhood of 99% uptime reliability
- Management of services (web server)
- Greater security through data encryption

SUMMARY

In this chapter, you learned how to terminate the different types of media both at the wall outlet and for the devices. You also learned the following:

- Trim is an important step that requires organization so that all cables can be terminated at the appropriate location. Use of the proper tools enables the HTI to trim and terminate efficiently.
- CAT 5 cables require precise termination. Each twisted-pair is color-coded to correspond with the color coding on the termination jack. Handle carefully to reduce the loss of signal and distortion.
- Special stripping tools are used with coaxial cable. New connectors compress instead of crimp. A loose connection to the bulkhead or feed-through connector can cause leakage.
- Audio cables are run in pairs in order to provide stereo sound. These cables are terminated on screw terminals or binding posts. Wire is wrapped clockwise and never more than 180 degrees.
- DSS (Digital Satellite Systems) or DBS (Direct Broadcast Satellite) uses an RF-modulated video signal to transmit. This should be handled like a VCR when connecting to the distribution device.

- Terminal equipment in the Structured Media Center (SMC) should be installed in a logical and orderly manner. All terminations should be grouped by services delivered. Items requiring power (hubs, switches, amplifiers, and so on) should be given special consideration.
- The home automation control processor is a computer that provides a GUI interface. There are two types, a simple version and a sophisticated version. They allow each subsystem to be controlled via the control panel and over the Internet.
- The use of wireless components can extend the security system and the entertainment system. Wireless components are prone to interference.
- Remote network management software allows for a customized portal that provides an interface for the components installed in a home network.

Now that you have the network connected and have checked the viability of all the cables, you will learn about the finish phase in the next chapter.

GLOSSARY

Home automation control (HAC)   With a web interface, HAC allows the security system to be controlled from anywhere in the world.

Home control portal   A personalized website created for an integrated home network.

Motion detectors   Trigger events when movement in a specified range is detected. Commonly used to turn on lights in a hallway when somebody enters.

Motion-detector floodlight   Triggers events when movement in the detection field occurs. Can be used to trigger a chain of events such as turning on porch or hallway lights.

Network interface device (NID)   Demarcation point between the home phone network and the telephone service provider network.

Open   The circuit is not complete or the cable is broken.

Patch panel   Enables organized phone wiring terminations.

Photocell   Triggers events based on the amount of sunlight hitting the cell. Commonly used to turn lights on at dusk.

PowerFlash module   Can be implemented in almost any stand-alone sensor. An event is triggered when the sensor is activated.

Short   An incomplete circuit caused by a hot conductor coming into contact with a ground or metal component.

Supersystem   Refers to all subsystems tied together in one CPU.

Wire punchdown/termination tool   Presses wire conductors into place in the insulation displacement connector (IDC) and in QuickPort modular snap-in jacks.

CHECK YOUR UNDERSTANDING

1. What process is involved during the trim phase of installation?
   a. Terminating the media
   b. Connecting the subsystems in the distribution panel
   c. Testing each cable to determine for connectivity and any problems
   d. All of the above

2. What is another term for patch panel?
   a. Network interface device
   b. Punchdown block
   c. Distribution panel
   d. Face plate

3. What allows organized phone wiring terminations back at the NID?
   a. Patch panel
   b. Distribution panel
   c. Outlet
   d. Modular jack

4. What acts as the central hub for the phone network?
   a. NID
   b. Distribution panel
   c. Patch panel
   d. None of the above

5. The patch panel has what type of interface?
   a. Punchdown interface
   b. Pole and nut terminal
   c. RJ-style interface
   d. All of the above

6. What is the demarcation point between your home phone network and your telephone service provider's network?
   a. NID
   b. Distribution panel
   c. Patch panel
   d. Residential gateway

7. Who has access to the NID?
   a. The telephone service provider and the homeowner
   b. Only the HTI
   c. Both the telephone service provider and the HTI
   d. Anyone

8. What is the preferred method to bundle cable that goes to the distribution panel?
   a. Twist ties
   b. Strapping tape
   c. Velcro
   d. String

9. Before working on the distribution panel, which of the following rules should you follow?
   a. Never wire live AC cable.
   b. Never install wiring during lightning storms.
   c. Never touch wires that are not insulated or terminals unless they have been disconnected at the other end.
   d. All of the above

10. In order for wiring to be certified, which of the following do you need to test for?
   a. Continuity, attenuation, impedance, and reactance
   b. Continuity, attenuation, crosstalk, and cable quality
   c. Attenuation only
   d. Crosstalk only

11. Which of the following should be done prior to termination to avoid common mistakes ?
   a. Label all cables
   b. Look for obvious damage

   c. Check for bends
   d. All of the above

12. Cables routed within the distribution device must be neat without:
   a. exceeding the minimum bend radius.
   b. exceeding the length specification.
   c. exceeding the width radius.
   d. exceeding the maximum folding radius.

13. Which of the following terms describes a short between two terminals?
   a. Open
   b. Short
   c. Cross
   d. Split

14. When connecting satellite television, it should be handled like a:
   a. telephone cable.
   b. VCR.
   c. coaxial for video.
   d. wireless component.

15. What is a personalized website created for the homeowner that functions as a home control interface called?
   a. Home Control Portal
   b. Management Control Portal
   c. Home Control Window
   d. Home Management Control

16. Why is the patch panel an important part of the telephone network?

17. How do open, short, and cross circuits differ?

18. What are some of the testing tools you need to test cables and connections?

19. When performing a visual inspection of cables, what should you look for?

20. What role does the telephone service provider play in your home network? Where is the demarcation point between the telephone service provider's network and the homeowner's network?

Finish Phase: Testing and Troubleshooting

 **OBJECTIVES**

Upon completion of this chapter, the HTI will be able to complete the following tasks:

- Identify cable testing instruments

- Match cable testing instruments to the type of cable on which they operate

- Understand the sorts of problems identified by cable testing equipment

- Understand the basic procedures of repairing cable problems

- Provide homeowners with a comprehensive cable test result report

INTRODUCTION

In this chapter, you learn about the finish phase of the installation process. After connectivity has been established, you install wall plates over exposed cables. Afterward, you need to test each subsystem and the entire network as a whole. If there are any problems, you will need to troubleshoot the network.

In this chapter, you learn also about the types of faults common to residential structured wiring systems, the tools used to identify fault types and locations, and how to repair those faults. You identify common testing tools and the cable types on which they work. You learn how testing tools help installers implement fully functional residential networks and how to deliver reports that offer assurance to the homeowner of the network's functionality and reliability.

FINISH PHASE TASKS

Wall outlets constitute the point of termination of the phone network wiring runs in various rooms in a house. For today's modern networked homes, it is useful to consider what other home network outlets you may need in a room when you choose phone outlets. There are all kinds of combination outlets in the market. It is possible to find many modular outlets that require double- or triple-gang-sized junction boxes, creating the opportunity to connect (in addition to phone jacks) data, audio, and/or video networks all in one large outlet. Jacks and plugs are used to make electrical contact between communication circuits. Jacks are therefore used to connect cords or lines to telephone systems. You can buy faceplates equipped with RJ-11 modular jacks.

Faceplates and modular jacks are the common way that electricians terminate homed or phone wire drops in the home. You will find these mostly in new constructions. Jacks and plugs can generally be described in two ways:

1. Physical size (number of conductor positions)
2. Physical configuration (number of wires actually connected to them)

Structured Wiring Faceplate Configurations

Structured wire wall plates must accommodate a variety of snap-in modules and devices. Figure 14–1 shows some examples of Leviton wall plates. Home networking companies offer modular wall plates that can accommodate different structured wiring solutions. Leviton, Inc., for example, offers both single-piece wall plates already configured for a set number of modules and one- and two-gang wall plates that can accept two-, three-, four-, or six-port inserts.

Figure 14–1
Leviton Wall Plates

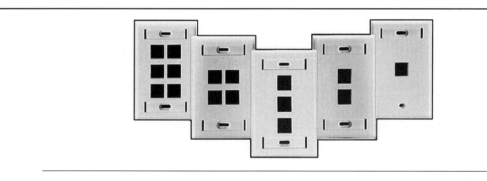

The Leviton wall plates described in the following sections can accommodate many devices, including the following:

- Telephone, fax, and data ports (Category 5 jacks)
- Standard video ports (F connectors)
- DSS with video and telephone (F connectors and Category 5 jacks)
- Speaker terminals (binding posts)
- Audio/video jacks (RCAs and BNCs)
- Infrared remote control emitters (Category 5 jacks)

Flush-Mount Plates   These wall plates come preconfigured in a single-gang housing. Make sure that all wall plates are UL-listed and conform to NEMA and ANSI standards. The following flush-mount configurations are available in several neutral colors:

- one-port
- two-port

- three-port
- four-port
- six-port

Wall-Mount Plate for Telephones Use when wall mounting a telephone. Special wall-mounting plates are available that include an RJ-11 jack and mounting hardware.

Wall Plate Inserts You can also purchase wall plates that are not preconfigured. They are available with two-, three-, four-, five-, or six-port inserts that can be mounted in multigang wall plates for applications requiring higher density terminations than are possible in a single wall plate. A wall plate insert is shown in Figure 14–2. The HTI can use a four-port insert to mount two video (TV and VCR), one RJ-11 (phone), and one RJ-45 (computer) in a single outlet box.

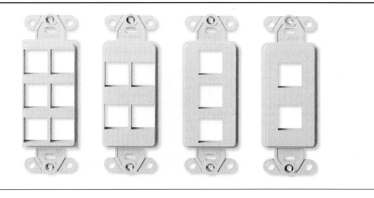

Figure 14–2
Wall Plate Insert

Installing Faceplates

The proper faceplate should be used for each individual outlet. Audio cables will have different connectors than video or telephone cables.

Faceplates should be installed neatly and tight to the wall. They must be straight and plumb. When tightening faceplate attachment screws, note that overtightening can cause the faceplate to crack.

Inserting Connectors and Jacks into Faceplates

Cable terminations at connectors, modules, and Category 5 jacks are critical to the integrity of your structured cable installation. Figure 14–3 shows the 10Base-T, modem, Data 2, and video jacks installed on the faceplate.

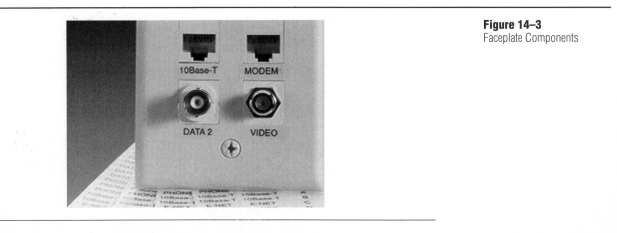

Figure 14–3
Faceplate Components

To install connectors and jacks into the wall plates, do the following:

1. Match "UP" on the wall plate to "UP" marked on the jack or other device.
2. Seat the lower catch into the plate retention feature.
3. Roll the jack or other device into place until the latch snaps in place.

Labeling Faceplates

Faceplates should be labeled in the same manner that the cable is labeled; it is not acceptable to hand-write labels. Professional installations have printed labels applied to the faceplates. Each label should identify the location and the service provided by the outlet. If it is a telephone jack, it is a good idea to incorporate the letter T in the label. If it is a data jack, the letter D incorporated into the label identifies it as such.

When labeling coaxial cable in a Grade 2 (Level 2) installation, one cable should be identified as a video service delivery cable and the other cable should be identified as a video source cable. There are many labeling tools that can be used to achieve a neat job as shown in Figure 14–4.

Figure 14–4
Labeling Tool

Note that labeling telecommunications outlets and connectors may sometimes depend on the preference of the homeowner or the builder. Most customers do not want writing to be instantly visible on the face of the outlet. You may need to discuss some aspects of labeling with the appropriate stakeholders before the process is completed.

Labeling Cables

The importance of properly labeling your cables becomes readily apparent during the testing phase. Accurate test results come, in part, from accurate labeling of the cables both at the distribution point in the distribution panel and at the point of termination. Label all your cable ends before pulling them to their termination points.

After completing the prewiring stage of an installation, there could be dozens of cables to deal with. If you can't identify the cables (because of poor, incorrect, or nonexistent labeling), you can use a pocket toner to solve the problem. A *pocket toner* sends a tone down the cable from the wall jack so the cable can be identified at the SMC. To use the pocket toner, insert the sender portion of the pocket toner into a wall jack. At the SMC, connect one cable at a time to the speaker until a tone is heard; then label that cable accordingly.

Cleaning Up

Cleanup should be completed at each outlet location immediately after the label is installed. Drywall dust, bits of wire, and bits of insulation must all be cleaned up and disposed of according to applicable environmental laws. A small, battery-operated vacuum can remove the drywall dust quickly and efficiently.

Testing, Troubleshooting, and Certification

The final steps in the finish phase are to test the cable and document the results. Testing ensures connectivity. If a problem is encountered, you need to determine the problem by troubleshooting. After the problem is solved, test the cable again. If no more problems are found on the cable, it can then be tested to see if it meets certification standards. There are no requirements that the cable be certified, but it is a good way to protect you in the future if the homeowner contends that there are problems with the network. By certifying the cabling system, you show the homeowner that your cabling job meets or beats the industry recommendations.

 **TESTING CABLE**

Low-voltage cable is not as durable as other types of cable, but it has been specifically designed to meet strict requirements. Low-voltage cable follows its own industry standard and requires specific tools and care to maintain the integrity of a structured wire system. It is more easily damaged during rough-in and drywall installation than other electrical cable, and damage can go unnoticed. For these reasons, it is important that low-voltage cable be tested thoroughly. This is a relatively simple process. Testing verifies that no faults occurred during the installation. Problems that occur later will be due to other reasons such as system abuse or damage from individuals other than the installer.

Problems with low-voltage wiring installations do not always show up as readily as those with AC cable. Testing is critical to confirm that the system is working properly. In most cases, you will be performing your tests before dial tone is available from the telephone service provider, signal is available from the CATV service provider, or a TV or DSS antenna has been installed. Therefore, the testing described as follows assumes that no signal or attached consumer electronics equipment is available. It is important that all tests be recorded and documented.

Test Rules

The following are rules that must be followed during the test to make it valid:

- Cabling and components cannot be moved during testing.
- Both pass/fail indications and the actual measured values (frequencies) should be recorded.
- Reconfiguration requires retesting.
- Qualified adapter cords should be used to attach the test instruments to the link under inspection.

> **NOTE** Basic link testing is used to verify the performance of permanently installed cable. Channel testing is used to verify the performance of the overall channel.

Most companies that manufacture distribution panels and elements require the HTI to provide testing documentation for guarantee purposes. It is crucial that you use the forms that come with the product and keep a copy of the test documents that the testers can print out.

Visual Inspection

The first type of test is a visual inspection. Many problems can be found quickly before spending too much time doing instrument testing. Figure 14–5 shows some examples of damaged cables.

Figure 14–5
Damaged Cables

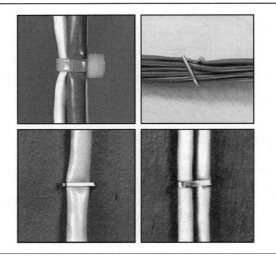

When visually inspecting cables, check for obvious damage such as cuts to the sheath, cables that are bent beyond the recommended bend radius, or cables that are stretched or twisted.

Table 14–1 shows the minimum bend radius for twisted-pair, coaxial, and optical fiber cables.

Table 14–1
Minimum Bend Radius

Cable Type	Minimum Bend Radius
Twisted-pair (100-ohm UTP)	Not less than 4 times the cable diameter.
Coaxial (75-ohm)	Not less than 20 times the diameter of the cable when pulling. Not less than 10 times the diameter of the cable when placing or dressing the cable.
Optical fiber (2- and 4-fiber)	Not less than 10 times the diameter when there is no tension. Not less than 15 times the diameter when cable is subjected to tension.

Basic Testing

The second type of testing is considered basic. Basic testing covers all but CAT 5 testing and can be used for the following cables:

- Audio cable (for speakers and volume controls)
- RG-6 quad-shield cable
- Inbound cable from the telephone service provider demarcation point
- CAT 5 cable that does not terminate in CAT 5 jacks at both ends of the run
- CAT 5 cable used for powering video cameras or IR targets and IR emitters
- CAT 5 cable for telephones that terminate on the 1×9 bridged telephone module

WARNING: Do not do any testing on energized wiring or circuits. Disconnect all sources of electric current before starting your tests. To eliminate the risk of potentially fatal electric shock, be sure that all distribution panels and modules are disconnected from either the AC or DC modules.

Testing for Cable Faults

In basic testing, you test cables for the following faults:

- *Open:* Also known as an *open fault*. This means that the circuit is not complete or the cable/fiber is broken. An illustration of an open fault is shown in Figure 14–6.

Open

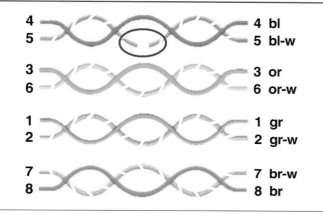

Figure 14–6
Open Fault

- *Short:* An incomplete circuit caused by a hot conductor coming into contact with a ground or metal component, where the current does not follow its intended path. This type of fault provides a bridge across the intended load and creates a low-resistance path, or *short*, in parallel with the true load. This causes the current, which seeks the path of least resistance, to bypass the intended device, rendering the device inoperable. An illustration of a short is shown in Figure 14–7.

Short

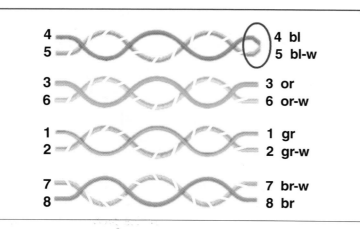

Figure 14–7
Short

- *Cross:* The occurrence of a short between two terminals; usually occurs when too much bare copper conductor is stripped and then the end is not trimmed after connection. An illustration of a cross is shown in Figure 14–8.

Cross

Figure 14–8
Cross

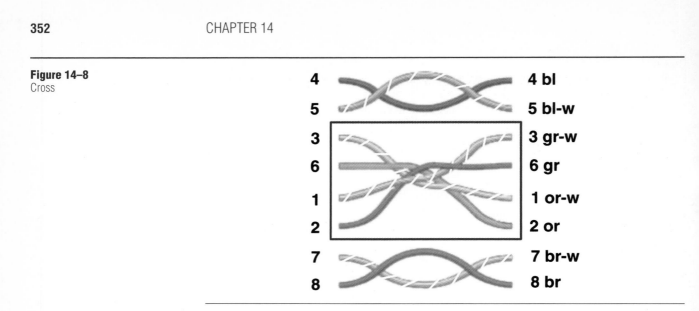

4	4 bl
5	5 bl-w
3	3 gr-w
6	6 gr
1	1 or-w
2	2 or
7	7 br-w
8	8 br

Split

- *Split:* This occurs when two wires of a pair are split or separated and improperly matched with wires from another pair. An illustration of a split is shown in Figure 14–9.

Figure 14–9
Split

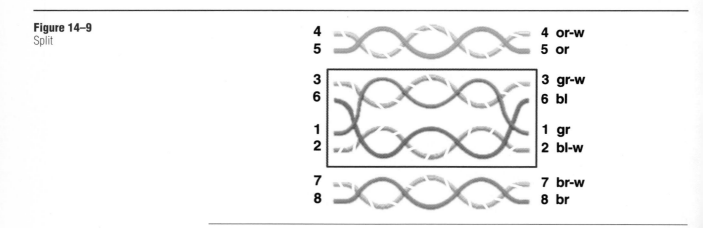

4	4 or-w
5	5 or
3	3 gr-w
6	6 bl
1	1 gr
2	2 bl-w
7	7 br-w
8	8 br

Reversed polarity
or rolled pair

- *Reversed Polarity or Rolled Pair:* A rolled pair results when the tip and ring leads are reversed in connecting to the network. An illustration of reversed polarity is shown in Figure 14–10.

Figure 14–10
Reversed Polarity

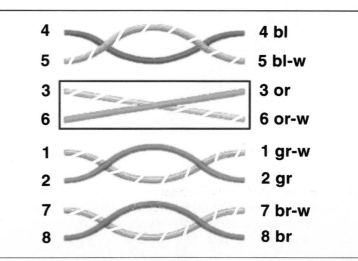

4	4 bl
5	5 bl-w
3	3 or
6	6 or-w
1	1 gr-w
2	2 gr
7	7 br-w
8	8 br

Testing Audio Cable

Testing audio cable includes checking the resistance reading of the cable for both terminated and unterminated audio cable:

- **Terminated audio cable:** Check the resistance reading of the cable from the audio module in the distribution panel. Make sure it is close to the measurement of the resistance of the terminals on the volume control or speaker to which the audio cable run is terminated. This measurement indicates that the cable is continuous and without shorts.
- **Unterminated audio cable:** Check for an open fault between the wires in the speaker or volume control audio cable.

Testing Speaker Cable

Look for damage in the cable run, and perform a standard continuity test. Check all connections for loose strands, and be sure that the terminal screws are secure. After all audio module connections (that is, volume control and speaker) are complete, turn the system ON with the volume turned all the way down and slowly raise it to test performance.

Testing Coaxial Cable

Coaxial cable can be tested for continuity between the distribution panel and the termination point. It can also be tested for breaks in the cable itself and for cable resistance. Breaks can be checked using a multimeter or coax cable tester. The multimeter also checks for resistance. Be sure that connections at coax splitters are secure and that all connections are hand tight.

Connecting a small, portable television to each video F-connector jack indicates cable continuity and whether signal level is adequate. A snowy picture is an indication that signal level is too low. This problem may be caused by a faulty cable, or you may need to add a video amplifier. If your picture shows several wavy lines and "ghost images," check to make sure that all terminations without equipment attached have the 75-ohm termination caps in place. Hand-held coaxial cable testers are available from several manufacturers to simplify this testing. Some cable testers for telephone applications, such as the TelScout TS100 from Textronix, Inc., offer twisted-pair and coax testing in a single unit, enabling you to locate faults in coaxial cables. Prior to testing with a TV set, test with an ohmmeter or tester.

Testing RG-6 Cable

The link length of RG-6 cable should not exceed 200 feet. On RG-6 cable, a cable tester searches for opens and shorts. RG-6 cable attenuates between 3 dB and 7 dB per 100 feet, depending on frequency.

For unterminated RG-6 quad-shield cable, use a multimeter or volt-ohm meter to check the resistance between the shield and center conductor on RG-6 quad-shield cable. If the meter shows a finite resistance reading (below 100 kilohms), the cable has a short and that cable run fails the test.

For RG-6 coax that is terminated in a splitter, or in a run with a 75-ohm termination cap, the resistance measurement should be about 75 ohms to pass the test. A coax cable tester indicates faults and breaks in coax cable. Prior to the installation of the endpoint device (for example, television, VCR, and so on), the cable can be tested by plugging it into the back of a portable device such as a small television or radio.

Testing Telephone Cable

With the dial tone installed and all initial testing completed, final verification testing can be done with either a buttset or a telephone test set. As shown in Figure 14–11, a *buttset* is a testing handset with touch-tone pad, alligator clips, and various features to help you verify that the telephone circuits are functioning properly.

Buttset

Figure 14–11
Buttset

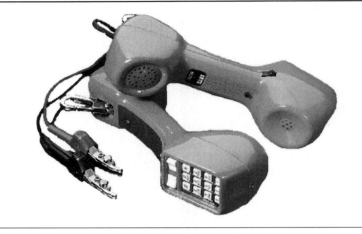

If a failure is encountered during the final verification test, check the wiring to the jack for proper connections, and check the wiring (including the distribution device) to see if any obvious disturbance has occurred since rough-in. Some phones might not work with polarity reversed, so be particularly careful to check the wire colors for proper polarity. If no fault is obvious, retest as for rough-in.

> **NOTE** If dial tone is not available, perform the same tests as for rough-in before the jacks were installed. A splitting adapter may be necessary to test each line of each jack. A toner can be used with a normal telephone or buttset to test for dial tone. Most toners provide an audible tone to the phone as well as sufficient voltage to power the phone to generate touch-tone or rotary dial digits.

Testing CAT 5 Cable
When field-testing Category 5 cable installed for data transmission, do the following:

- Use a Category 5 field test set, as shown in Figure 14–12, following the manufacturer's instructions.
- Use a Category 5 field test set for testing all Category 5 cable installed for data transmission.
- Only cable runs with Category 5 jacks available on both ends of the run (wall plate Category 5 jack and Category 5 module at the SMC end) can be tested for Category 5 compliance.

Certifying the Wiring
It is important to test the media for continuity, attenuation, crosstalk, and cable quality; this is called certifying the wiring. Certified wiring demonstrates to the customer that the cabling is working well. Not all home integrators certify their wiring, but you should. Doing so distinguishes you from the competition.

The TIA Certification test measures the following:

- **Length:** The link length does not exceed 90 meters (295 feet).
- *Attenuation:* How much signal degradation occurs from one end of the link to the other.
- *Near-end crosstalk (NEXT):* A signal from one pair of wires that is picked up by other pairs.

Attenuation

Near-end crosstalk (NEXT)

Figure 14–12
Category 5 Field Test Set

Both the Continuity/Wire Map test and the Category 5 certification test should yield a Pass or Fail result. If you get a Fail rating, use a cable tester to pinpoint the location of the problem. After the problem is fixed, it must be retested.

> **NOTE**   A non-TIA/EIA-570-A system that passes Category 5 testing is not necessarily TIA/EIA-570-A-compliant.

 **TEST EQUIPMENT**

Many different test instruments are available for testing the infrastructure of residential control systems. These instruments include simple tools that measure whether an electrical circuit exists across two ends of a cable, instruments that perform tests that certify a cable for adherence to specific standards, and instruments that can complete a wide variety of testing functions on many different cable types. Good testing equipment does the following:

- Help keep your installation CAT 5 (or another cable) compliant with standards
- Find any errors or disruptions
- Point out problems in your installation techniques so that they can be corrected

- Add value to your installation, to the builder, and to the homeowner
- Make it possible to provide the manufacturer extended warranty on installation

Matching Test Equipment to Cable Type

When an installer understands which tools can help track down particular problems, there is the opportunity to create properly functioning wiring plants within the constraints of project deadlines and budgets.

The first distinction in cable-testing equipment is along the lines of which type of cable is to be tested. Coax, unshielded twisted-pair, serial, and fiber-optic cables each have scanners and testers devoted to their analysis. Within each of these cable-type groupings, there are devices designed to find short and open circuits, determine proper termination, and hold cables accountable to published standards. It's important to know which instruments are capable of showing desired information about a particular cable, and why one or another might reasonably find a place in the installer's toolbox.

Digital Multimeters

Digital multimeters are a type of test equipment that locates shorted or open wires, and identifies slave and master wires easily. Digital multimeters come with different feature sets, depending on the manufacturer and the range of functions or tests they are intended to perform. Figure 14–13 shows a Fluke 110.

Figure 14–13
Fluke 110

When installing three-way switches, it is crucial to locate the wires in the wall that have power. It is often easy to connect the slave and master wires incorrectly, resulting in a faulty switch. A digital multimeter or similar tester resolves this problem. It can be used to test alarm system wiring, install three-way switches, and test voltage and resistance on cables and wires. Most digital multimeters can measure AC and DC voltage, AC and DC current, resistance, diodes, continuity, and transistors. The DC voltage function can even be used to test 9-volt or 1.5-volt batteries.

Cable Testers

A *cable tester* is a device that can test opens, shorts, and other wiring problems for one to 25 pairs of wires or any combination of 25-pair and modular jack terminations. A cable tester should be able to accommodate cable runs of up to 762 m (2,500 ft.) accurately. One such cable tester is the MicroNetBlink, made by Fluke Networks. This device tests individual conductors rather than wire pairs, which allows for testing of all wiring configurations including USOC, T568A, and T568B.

> **NOTE** The Universal Service Ordering Code (USOC) system was developed to connect customer premises equipment to the public network. These codes are a series of Registered Jack (RJ) wiring configurations for telephone jacks that remain in use today. Developed by the Bell System and introduced by AT&T in the 1970s, USOC was adopted in part by the FCC, Part 68, Subpart F, Section 68.502.

Test Equipment for Unshielded Twisted Pair

Before a cable can be tested for adherence to a specification, the installer must be confident that the cable is complete and correct. It is quite possible to use CAT 5 cable and CAT 5 connectors to create a finished cable run that doesn't come up to the standards of CAT 3. When this happens, network devices may very well indicate that a link exists, but performance will suffer and some applications may not run at all.

Small hand-held units may be all that are required to verify that cables are correctly constructed and complete from end to end. This type of instrument typically uses preconfigured tests to verify the cable and reports the results via multicolored LEDs on the front or top of the device. These testers frequently include a tone generator as well, which can be used along with a second unit to determine which cable end at the distribution panel matches a particular remote end. These test instruments are useful for validating that a link exists, but they cannot help the installer with either the location of any problem or the quality of the link.

Finding the location of a problem requires a test instrument that contains a *time-domain reflectometer*, or *TDR*. The TDR, as shown in Figure 14–14, sends an electronic signal down a cable and measures both the time and quality of the signal that returns from the other end.

Time-domain reflectometer (TDR)

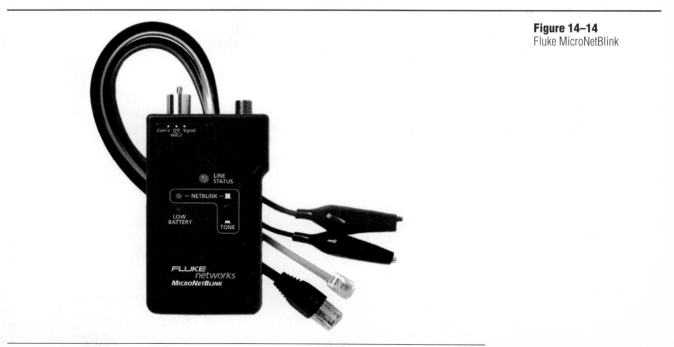

Figure 14–14
Fluke MicroNetBlink

Measuring the time from signal-sending to reflection allows the TDR to calculate the distance to the other end of the cable. Monitoring the quality of the returned echo—whether it is a single clear echo or a series of weak echoes, for example—can tell the operator whether there is damage to the cable that, although not completely interrupting the circuit, could cause performance problems for hardware and applications.

Some applications and network devices require the availability of network links having at least a certain minimum quality. Some customers require that the network cable meet a particular standard—Category 5e or Category 6, for example—and want to be assured that the cable plant ordered is the cable plant installed.

Test Equipment for Coax Cable

Test equipment for coax cable is primarily used for ensuring that the cable jacket has not been compromised and that the center conductor has not broken. The primary tool for testing both places a signal of known strength and quality into one end of the cable and measures the quality of the signal on the other end of the cable.

Analog Sound Level Meter

An analog sound level meter, as shown in Figure 14–15, tests the power of a home theater system. This highly accurate tester displays the precise decibel level of any sound source (for example, an entertainment system, a security system siren, a smoke detector, or some other safety alarm) and verifies that the devices listed are in proper working order.

Figure 14–15
Analog Sound Level Meter

The built-in microphone typically measures sound from 54 dB to 126 dB in different selectable ranges. Select fast or slow response time for measuring sudden noise spikes or average sound levels over time. A maximum hold function locks in the highest reading on the display.

Digital Sound Level Meter

The digital sound level meter, as shown in Figure 14–16, is primarily used in home networks for the entertainment system. It is more precise than an analog sound level meter because it can get readings to one-tenth of a decibel. This tool is used to balance the home theater system to match the room acoustics with the A/V equipment's internal test tones.

It offers a wide 40- to 130-decibel range, and features a $3\frac{1}{2}$-digit LCD display and a 30-segment analog bar graph. Plus, its built-in one-inch microphone allows for easy one-hand operation.

Figure 14–16
Digital Sound Level Meter

Test Equipment for Serial Cable

There are two primary concerns with serial cables. The first is the question of whether a mechanical connection has been made on each necessary wire within the cable. This testing is typically accomplished with a simple volt-ohm meter (VOM) that delivers a tone when a complete circuit is measured. The second concern is that the cable has been constructed properly, with pins in the connector on one end connected to the correct pins of the connector on the other end. A VOM can be used for this testing, although touching the probes of the VOM to each pin on the serial connector can become labor-intensive, especially if RS-232 25-pin connectors are required. For complex cables, a breakout box can indicate the pin assignment of each conductor in the cable and whether the voltage present on the conductor is high, low, or zero. A breakout box, as shown in Figure 14–17, indicates which pins are active through illuminated LEDs.

Figure 14–17
Breakout Box

When considering a breakout box, it's important to make sure that the box is equipped with connectors that match the connectors used for the particular installation. This is especially important because there is no electronic communications method with more variation in interface and standards than serial communications. 9-pin, 25-pin, RJ-11, 2-wire, 4-wire, and 25-wire are just a few of the types of cable encountered in different building automation systems. Although most breakout boxes are available with test probes, using a single pair of test probes on a 25-pin connector negates the advantages of using the breakout box in the first place.

It is possible and mechanically simple to create a series of adapters for the various connectors, but care should be taken to ensure that the connectors in such a case don't interfere with the testing process by introducing errors or injecting unnecessary noise.

Test Equipment for Telephone Installations

Many companies manufacture equipment for testing low-voltage wiring. Leviton, for example, manufactures the Tone Test Set and the Inductive Speaker Probe. Used together, these instruments do the following:

- Locate individual wires in a horizontal run along with any breaks or terminations that might be present
- Test for continuity
- Check for shorts and opens
- Identify tip and ring polarity
- Identify the line condition for clear line with dial tone, busy line, and ringing line

The low-cost Tone Test Set, as shown in Figure 14–18, can test for all types of wire applications, including telephone, data, CATV, HVAC systems, and security/fire alarms. Results are shown on an easy-to-read LED display. Telco standard 6A-type alligator clips with piercing pins securely grip 66-clips, screw heads, screw bodies, and wire-wrapped threaded terminals, allowing testing of all types of configurations. A six-pin, two-conductor plug lead is also provided for connection to modular jacks.

Figure 14–18
Tone Test Set

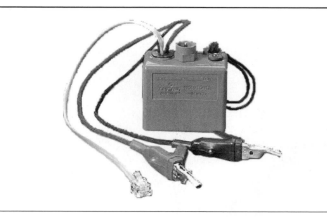

The Inductive Speaker Probe, as shown in Figure 14–19, is used in combination with the Tone Test Set. It readily detects audible frequency tones so that wires, cables, and metallic circuits can be traced and identified without damage to their insulation. The Inductive Speaker Probe's duckbill and needlepoint tips provide great flexibility for inspecting wiring in tight spots, cables under tension, or larger cable bundles. A built-in speaker eliminates the need for a buttset (in noisy environments, a buttset can be attached to the probe's connecting tabs).

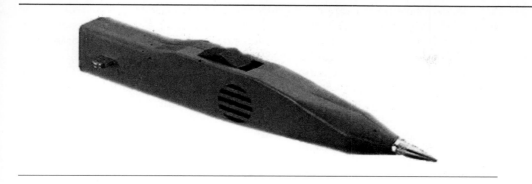

Figure 14–19
Inductive Speaker Probe

Testing Documentation

All testing should be recorded and documented. Some manufacturers require documentation to honor the warranty. The test documentation proves to the homeowner and any regulatory agency that you have completed your installation satisfactorily. Some manufacturers provide you with forms to record the cable number, the module in the distribution panel to which the cable run is terminated, and the results of the test (pass or fail). Always keep copies of your test documentation.

TEST THE SYSTEM

There is no reason to wait until servers, controllers and other components are installed to discover that a cable has been improperly terminated. Each cable should be tested for correct termination, mechanical soundness, and proper operation (or adherence to specification) when it is installed. If this testing is performed, you can concentrate on the setup and testing of components, rather than cable, when components are installed.

Total System Operation

A complete cable infrastructure test prior to component installation allows you to concentrate on the components when problems arise during later stages of system installation. You should note that successful early cable testing does not provide an absolute guarantee that no cable problems will arise. Issues such as crosstalk (signals being broadcast between conductors) appear only as the cables are placed into use. On the other hand, thorough early testing does minimize such problems, allowing for other systems to be isolated and checked for errors with less concern that signals are simply not getting through.

Verifying Component and System Functionality

There are a number of test devices that are useful for verifying the operation of the total system and its components. These devices include network sniffers, dial-tone simulators, and serial analysis devices.

Although many of them are expensive, understanding the functionality of these devices can help you determine whether your budget should include one or more for an individual installation project.

A network packet sniffer can provide detailed information on the way information packets are flowing across the network, allowing the installer to debug extremely subtle problems. It shows the originating point, destination, and configuration of network packets flowing through the system. Network sniffers can be stand-alone units or can be software running on either a laptop or a desktop computer equipped with a network interface card (NIC). Because the network sniffer goes far beyond simple cable testing, it is a crucial tool when chasing down performance or operational problems in complex networks.

Dial-tone simulators can be used to place a dial tone on residential telephone lines before they are connected to the telephone service lines. Dial-tone simulators can help verify correct operations for call distribution devices, residential PBX systems, and other telephone applications.

Like network packet sniffers, serial analysis tools go beyond verifying connectivity, providing signal level and bit flow information to the technician. When serial devices—sensors, command panels, and control processors—don't seem to be communicating with each other, the serial analysis tool can help determine whether the problem lies with a signal not moving from point to point along the network or with an incorrect or corrupted signal being received.

Presenting Test Results to the Owner

As part of the documentation, you should include reports of the cable plant tests performed. This is especially important if your total system testing includes certification tests for any CAT 5, CAT 5e, or CAT 6 cabling run through the residence. Some test equipment vendors, such as Fluke and Agilent, offer software that automates the process of creating reports from test results. These applications present varying combinations of raw test results, along with analysis and (in the case of a certification test) verification that test results indicate compliance with the standards governing the certificate.

If automated report-generation software is not used, a full report listing all tests performed, test results, and any remedial action taken in the case of errors found should be presented to the customer as part of the final installation documentation. Complete test effort and result documentation will reassure the customer that all systems are built on an infrastructure that will support current and future needs. It will also provide a documentation basis for future updates, upgrades, and repairs to the installed system.

TROUBLESHOOTING

For complete home networking systems, the manufacturers usually offer testing equipment and troubleshooting tips with every device.

For self-assembled equipment, it is always advisable to follow these general rules for troubleshooting:

- Follow the proper sequence for troubleshooting device and network-related problems.
- Become familiar with some of the more common hardware and software problems.

The ability to effectively troubleshoot computer-related problems is an important skill. The process of identifying the problems and trying to solve them requires a systematic step-by-step approach. Some of the suggestions here are more than what will be required to solve basic hardware and software problems, but they help provide a framework and guidelines when more complex problems arise.

Step 1—Gather Information

The first step in troubleshooting is to gather information such as the following:

- What are the symptoms?
- What is happening or not happening?

- Is it hardware-related (check for lights and noises) or software-related (errors onscreen)?
- Is it local (this workstation only) or remote (possible network-wide problem)?
- Does it affect only this application?
- Is this the first time it has happened or has it happened before?
- Was anything changed recently?
- If another component is substituted, will it work?

Step 2—Locate the Problem

After determining that a problem does indeed exist on a cable, the next issue for the installer is to locate the problem. There are three major steps to finding the location of the error:

1. Examine and test connectors at both wiring closet and remote jack ends of the cable. Cable termination and connection points are prime points for cable problems because of the inherent difficulties in making properly configured, mechanically sound connections.

2. Visually inspect cable runs where they make sharp bends, pass through constricted areas, or are in close proximity to sharp metal edges and surfaces. Cable jacket materials are tough, but nails, screws, junction box edges, and other sharp metal objects can pierce the cable covering, causing a short circuit or breaking the conductor entirely. In addition, all cables have minimum bending radius limits—limits on just how sharply the cable can be bent without damaging the conductors inside the jacket.

3. Test the cable run with a time-domain reflectometer (TDR) to determine the distance from the end of the cable tested to the point at which the signal stops. The readout from a time-domain reflectometer (TDR) can indicate the presence of open circuits and partial shorts, among other cable faults. Visually inspecting a troubled cable is important, but a conductor broken by bending during installation may not be visible, and sheetrock or other finishing work done after cable installation may make visual inspection impossible. In these cases, technicians are left with two choices: Use a TDR to isolate the location of the problem or pull an entirely new cable. In most cases, locating the trouble spot will be less expensive and less disruptive to the project's work flow.

How a TDR Operates A TDR operates by injecting a signal into one end of a cable and then measuring the delay until the reflected signal returns to the device. The time measured is used to calculate the length of the cable. If the returned length is different from the total length of the wire, there is a break in the conductor at the point indicated by the TDR. It's a simple concept, but one that requires high-speed electronics to execute.

Interpreting TDR Results Traditional TDRs also require a skilled operator to interpret the returned signal and its display on the testing unit. With these traditional units, the operator must have been trained to evaluate whether spikes mean shorts or open circuits, what may be assumed from the presence of multiple echoes, the importance of dips in the display, and so on. This training is specific to the unit and is an important part of using this type of TDR successfully.

More recent digital TDR units remove much of the interpretation from the user's responsibility—providing readouts on error type, circuit length, and other information discovered through testing. The key piece of information that the user must bring to using either type of TDR is the knowledge of the tested cable's length—comparing this number to the circuit length returned by the TDR tells the operator whether the cable has a fatal error along its length.

Step 3—Identify Potential Solutions

After gathering information and locating the problem(s), the next major step is to identify potential solutions. Rank possible solutions in order of most-likely to least-likely cause.

- Check the simplest possible causes first. (Is the power turned on?)
- Check the easiest-to-check problems first. (For example, try a system reboot.)
- Verify hardware first and then software. (Do any lights come on?)
- Start with the physical layer of the OSI model; then move to the data link and network layers (check the NIC before the IP address).
- Troubleshoot from the bottom up. Always check OSI Layer 1 first (the power cord layer).

Use Table 14–2 as a template for a troubleshooting log.

Table 14–2
Troubleshooting Log

Problem	Symptom observed	Problem identified	Solution
The Internet cannot be reached by the PC in the kitchen.	The web browser displays an error message.	The CAT 5e cable connected to the PC was not fully inserted into the wall jack.	The cable was firmly and securely inserted into the wall jack.

Test the most likely solution based on your best guess and then check the results.

Step 4—Correct Problems

By far, connectors and cable endpoints are the most common locations in which errors occur. Visually and electronically checking the connectors for mechanically correct connections is an important step of correcting wiring problems.

In order to correct connector problems, you must understand the nature of a correct connection and be aware of the requirements put in place by the various standards organizations for connectors that meet certification levels.

Visual Inspection   Some connectors and cable types lend themselves to visual inspection. RJ-11 and RJ-45 modular plugs, for example, are often clear, providing a way for technicians to look at the plug and see whether all conductors are in contact with the plug's contact pins. If not, an obvious error has been discovered. Transparent plugs also make it easy to see whether other installers have adhered to specifications regarding untwisted wires; In Category 5 cables, the untwisted portion can be no longer than one-half inch.

Termination Problems   Coax cables reveal less to visual inspection, although they should be checked for neat termination and the presence of the center conductor. Serial cables are also not as open to problem resolution from visual inspection, although the plug shell can be disassembled and connection of the conductor to connector pins can be verified in many cases.

When a cable doesn't work, proper error-correction methodology can help ensure that repairs are made efficiently and effectively. The correction can range from reterminating the cable, to splicing in a new section of cable, to re-pulling an entirely new cable. One procedure for determining the sequence of corrective actions to follow resembles the sequence of diagnostic efforts: Identified problems, connectors and terminators, and replacement.

NOTE Splicing should not be considered an option for most residential networking and entertainment cables. Although some special situations may seem to call for a connectors splice—cutting out the damaged section of cable, placing connectors on the new cable ends, and then placing a short jumper cable in-between—it should be remembered that each mechanical connection provides an opportunity for mechanical failure and the certainty of signal strength attenuation.

If visual inspection or testing with a TDR pinpoints a problem spot, fix the problem. In reality, repairing connections is the most common repair to be made; in most cable types used for residential networking, cable replacement is a far more reliable and ultimately cost-effective solution than trying to splice a section of new cable into the run. Wrapping with electrician's tape should not be considered a suitable repair for cable jacket damage that is severe enough to affect the function of the cable.

If testing does not pinpoint the problem source, connector replacement remains the repair method with the greatest likelihood of bringing the cable back to full function. From a material and time-cost basis, terminator replacement is a reasonable course of action whenever cables are not working as they should.

Step 5—Double-Check

When you think you have found the problem and corrected it, double-check to make sure everything still works.

Step 6—Document the Corrections

When corrections are made (whether to the physical cable plant or to software), the changes should be well-documented. This is especially important when the changes are made or when problems were caused by environmental factors that could not be completely corrected. This documentation helps to track down future problems, and it will be invaluable to you when it is time to make changes to the physical plant.

Change notes should be complete, including the following information:

1. Change made. Were connectors replaced, entire cables run, or other repairs made? If the problem was environmental (that is, the cable jacket was cut by a screw point), state how the environment was changed to ensure that the problem will not recur. Be sure to indicate which patch-panel ports were affected by the change.
2. Date of the change.
3. Reason for change. Include symptoms, diagnostic tools used, and final diagnosis of the problem.
4. Person making change. Each technician should take responsibility for the work performed.

This may seem like a great deal of trouble for each change made, especially if the change involves simply reterminating a cable. When a copy of the change record is presented to the customer with the final documentation bundle, however, it will provide persuasive evidence of a thorough testing effort for the infrastructure of the system.

SUMMARY

The finishing stage of the installation process involves testing the individual subsystems as well as the entire network. In this chapter, you learned testing techniques, how to troubleshoot, and how to fix problems if a device or network fails.

- Testing the infrastructure, or cable installation, of the home automation system is a crucial step in ensuring the proper function of the completed system. Installers have a number of tools and techniques available to them to assist in making sure that the cable system will support the applications to be installed.
- A multimeter, which incorporates a voltmeter, an ohmmeter, and an ammeter, is an important first-line tool in testing cables and wiring. With a multimeter, the installer can ensure that a given conductor is intact from end to end.
- A time-domain reflectometer, or TDR, indicates the electrical length of the cable. It shows the point at which the cable's conductor ends and provides an indication of the point of any break in the cable. A TDR also shows the location of any shorts or other non-catastrophic damage in the cable.
- Breakout boxes are used to guarantee the correct pin assignment of serial cables. These boxes can show whether a given

pin is held with high voltage, low voltage, or no voltage, indicating whether the proper signals are being received at the end of the cable.

- Cable certification test instruments conduct a series of tests on a cable, indicating through the results whether the cable meets a particular standard that allows it to be certified to a given level of performance. This information can be crucial for homeowners and installers implementing high-bandwidth applications.
- When problems are found on the cable, termination replacement and cable replacement are normal remediation techniques. Splicing is not appropriate for coaxial or Category 5 cable—for either of these cables, cable replacement is the preferred remediation step.
- A full report of all tests, test results, and remediation steps should be presented to the homeowner. This report provides assurance to the homeowner that the cable plant is capable of supporting all current and future applications; and it provides crucial information to any technician called upon to repair, update, or upgrade the home automation system.

GLOSSARY

Attenuation   The amount of signal degradation that occurs from one end of the link to the other.

Buttset   A device that can listen to a telephone wire. Also known as a telephone test set.

Cross   Two wires are crossed.

Near-end crosstalk (NEXT)   A signal from one pair of wires that is picked up by other pairs.

Open   The circuit is not complete or the cable is broken.

Reversed polarity or rolled pair   The tip and ring leads are reversed.

Short   An incomplete circuit caused by a hot conductor coming into contact with a ground or metal component.

Split   Two wires of a pair are improperly matched with wires from another pair.

Time domain reflectometer (TDR)   Device that injects a signal into one end of a cable; then measures the delay until the reflected signal returns to the device.

CHECK YOUR UNDERSTANDING

1. Which of the following is *not* part of the home automation infrastructure?
   a. Home networking cable
   b. Residential gas lines
   c. Telephone wiring
   d. Serial automation wiring

2. What is the most common point of failure for home automation cable?
   a. Jacket
   b. Conductor

   c. Termination
   d. Insulator

3. When should cable ends be labeled?
   a. Before pulling them to termination points
   b. After pulling them to termination points
   c. Labeling is not necessary.
   d. When connecting them to the DD

4. Most companies that manufacture distribution panels and elements will require _____ for guarantee purposes.
   a. store receipt
   b. testing documentation
   c. usage logs
   d. pictures of installation

5. The minimum bend radius for coaxial cable is not less than 20 times the diameter of the cable when pulling. It is not less than _____ when placing or dressing the cable.
   a. 10 times the diameter of the cable
   b. 20 times the diameter of the cable
   c. 30 times the diameter of the cable
   d. 40 times the diameter of the cable

6. Basic testing covers all but:
   a. fiber optic.
   b. coaxial.
   c. CAT 5.
   d. speaker wire.

7. A "cross" is defined as a(n):
   a. circuit not completed.
   b. incomplete circuit.
   c. occurrence of a short between two terminals.
   d. two wires split or separated and improperly matched.

8. TDR is the acronym for what testing device?
   a. Tool for determining resonance
   b. Time-delay radiometry
   c. Test determination of readiness
   d. Time-domain reflectometer

9. Which tool is preferred for determining proper serial cable pin assignments?
   a. VOM
   b. Breakout box
   c. TDR
   d. Oscilloscope

10. A TDR's primary return is what measurement?
   a. Length of the cable
   b. Category of cable
   c. Type of connector
   d. Signal quality

11. What is the tool used to show the contents of network traffic?
   a. TDR
   b. Oscilloscope
   c. Network packet sniffer
   d. Dial-tone simulator

12. The first step in determining the source of network cable trouble is:
   a. TDR.
   b. feeling the jacket for warmth.
   c. removing the cable from the building.
   d. visual inspection of connectors.

13. Which of the following is *not* an appropriate remedial technique for residential automation cable problems?
   a. Replacing the connectors
   b. Splicing a replacement section into the cable
   c. Replacing the cable
   d. Replacing a jack

14. What information should be included in the testing report to the homeowner?
   a. List of tests performed
   b. Complete list of test results
   c. List of remedial actions taken
   d. All of the above

15. What does the TIA certification test measure?
   a. Speed, attenuation, resistance, and impedance
   b. Speed, attenuation, resistance, and voltage
   c. Speed, attenuation, length of link, and NEXT
   d. Speed, attenuation, length of link, and reactance

16. What are the ways to connect to jacks when running four-pair wiring?
   - You can connect four pairs of UTP wire into four separate RJ-11 jacks.
   - You can connect two pairs into a single jack (for dual-line phones or key phone systems), leaving two pairs for spares.
   - You can use a single-line RJ-11 jack, which leaves three pairs for any future upgrades or needs.

17. What is basic link testing used for?

18. What is channel testing used for?

19. What is the difference between an analog sound level meter and a digital sound level meter?

20. Why is test documentation important?

Multiple-Dwelling Unit Wiring

OBJECTIVES

Upon completion of this chapter, the HTI will be able to complete the following tasks:

- Identify minimum points of entry (MPOE) and demarcation points for service providers

- Identify physical, power, and connectivity requirements for telecommunications, cable, and data communications service providers

- Plan for backbone and redundant cable installations

- Understand the requirements for fiber optics

- Select proper routing for various cable types

- Understand building code requirements for cable runs

OVERVIEW

In this chapter, you learn about structural wiring for communications and automation in multiple-dwelling units such as apartments and condominiums. You identify the requirements for points of entry provided to, and by, telecommunications, entertainment, and data communications utilities. You learn about requirements for providing electrical service to telecommunications and data communications closets, and for running cable assemblies through dwelling units.

BUILDING ENTRY

Building entry focuses on the *minimum point of entry* or *MPOE* (pronounced "em-poe") for services brought into a multiple dwelling unit (*MDU*). The cables from service providers are brought in for distribution to individual units from the telecommunications rack. Understanding

Minimum point of entry (MPOE)

MDU

369

how to total the electrical power required for the telecommunications rack will ensure an adequate power supply.

Identify Minimum Point of Entry and Demarcation Points

In a building containing MDUs, the point where the telephone company cable meets customer premises equipment is called the MPOE or the demarcation point. Logically, the MPOE is the structure at which the telecommunications transmission cables join with the internal telecommunications distribution wires. Legally, the MPOE is the point at which the wiring shifts from telephone company ownership and responsibility (outside the MPOE) to building management ownership and responsibility (inside the MPOE). Figure 15–1 shows an example of the MPOE.

Figure 15–1
Minimum Point of Entry
(MPOE)

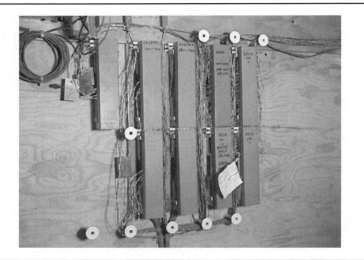

> **NOTE**   This legal separation may not occur if the building's owners have allowed the telecommunication service provider to retain ownership of the internal distribution wires and provide service under direct contract with each occupant. It is important that you know, and verify, ownership of any customer premises wiring and equipment before beginning any work within the building.

In single-family dwellings or small office buildings, the MPOE may be called the Network Interface Device (NID) or demarc. In all of these settings, the function of the device is the same—to provide a mechanism for bringing the communication provider and customer equipment together in a safe, structured manner. Traditionally, the MPOE was only concerned with wiring for Plain Old Telephone Service (POTS), but the advent of broadband Internet access has brought new emphasis to these demarcation points as a central component in both voice and data communications.

When looking for the MPOE, the first step is searching in the right location. In all new construction, the MPOE will be located at the point closest to where the telephone cable crosses the property line, or the closest practical point at which the telephone cable enters the building. Typically, the MPOE will be within 12 inches of the point at which the telephone company's cable enters the building.

In most modern apartments or multi-use buildings of any size, the MPOE will be in a separate room, usually the basement. A separate room allows for security and for the ability to maintain a consistent climate.

The Telecommunications Rack

This is typically enclosed in a telecommunications closet. The main piece of hardware in the telecommunications closet is the punchdown block, which allows multiple-input cable pairs to be distributed to units within the building.

Four kinds of punchdown blocks are available for use in the telecommunications closet, 110, 66, Bix, and Krone. 110 and 66 are the most popular with 110 being the most current.

- The Type 110 block, as shown in Figure 15–2, is the latest technology. It is able to support higher bandwidth communications and networking standards, including Category 5e and Category 6 networking.

110 Punchdown Block

Figure 15–2
110 Punchdown Block

- The Type 66 block, as shown in Figure 15–3, is used in many voice-only wiring plans, where high-speed data networking will not be an issue. It uses clips to bridge between connected pairs, and is able to support bandwidth only up to the Category 5 networking standard.

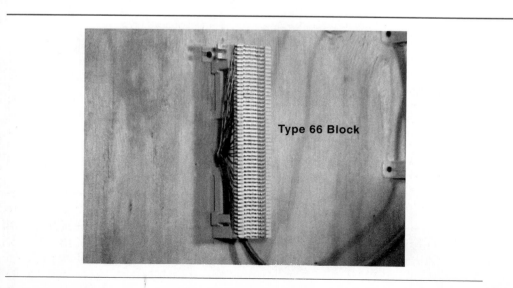

Type 66 Block

Figure 15–3
Type 66 Punchdown

DSL

In addition to punchdown blocks, filters for *DSL* (Digital Subscriber Line) broadband Internet connectivity may be deployed in the telecommunications rack. These filters allow DSL Internet connectivity and voice communications to share a single copper pair by preventing the high-frequency DSL carrier from interfering with voice communication. DSL filters are most frequently deployed within wall jacks between the wall jack and telephone instrument cord; but they may be placed in the telecommunications rack to allow separate voice and data outlets to be installed in the individual residential units.

Power Requirements

Total electrical power required for a telecommunications rack is simple to calculate, and a necessary figure to possess for installing safe, reliable telecom equipment. A spreadsheet that lists power requirements for all devices located in the telecom rack will allow for proper electrical service sizing. Table 15–1 is an example of a spreadsheet used to organize and total requirements so that important components are not missed when calculating total power load.

Table 15–1
Power Requirements
Spreadsheet

Quantity	Manufacturer	Model	Description	Volts	Amps
1	Compaq	2015	Monitor	110	2
1	Compaq	5100	System Unit	110	4
1	Sun	Ultra 10	System Unit	110	4
1	Cisco	Catalyst 5500	Switch	110	16

Power requirements for individual components can be found on labels or plates affixed to the rear or bottom of the equipment. A typical plate or label is affixed near the electrical cord's connection to a piece of equipment and shows the device's power requirements in volts, amps, watts, and frequency. An example is shown in Figure 15–4.

Figure 15–4
Power Requirement Plate or
Label

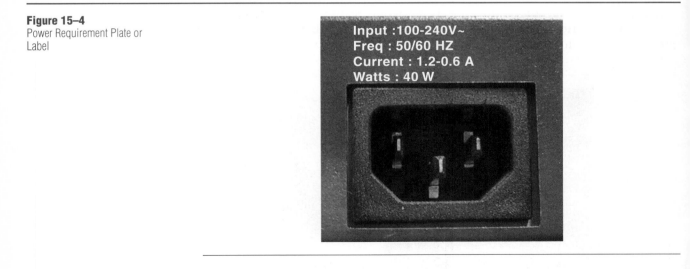

Power requirements for all equipment within the MPOE should be totaled. This includes climate control, security, automation, and other equipment that may not be part of the telecommunications rack, but may share the space within the MPOE. A safety margin, which varies according to the overall load, should be added, along with an allowance for anticipated future equipment. This margin will be a larger percentage for smaller overall loads. The result will be an electrical service that will meet communications needs.

In addition to basic electrical service, provisions should be made for two features:

1. Continued equipment operation when the electrical service is disrupted.
2. Immediate equipment shutdown in case of an emergency.

Continued operation is provided by an uninterruptible power supply (UPS), which has a bank of batteries to provide electricity when main electrical service is lost.

> **NOTE** A UPS using batteries is preferred over a standby power supply. While standby supplies may be less expensive, there will always be a switchover time between main power and the batteries. The switchover time is very short, but it may be enough to cause some components to reset or malfunction. In addition, the batteries of a UPS provide some protections against spikes, dips, surges, and other electrical disturbances. This protection is not provided by the standby power supply.

Instant equipment shutoff is provided by an *Emergency Power Off* (EPO) switch that cuts all power to circuits in the event of fire, flood, or other disaster. An EPO switch should be installed in every MPOE to protect life and property in the event of an emergency.

Emergency Power Off (EPO)

Service Provider Cable Requirements

Service provider cable enters the building at or near the MPOE, with multiple conduits carrying copper and fiber to one or more MPOEs within the building perimeter, as shown in Figure 15–5.

In general, you will have little control over how service provider cables are provisioned into the building. In most cases, service provider conduits are placed into walls or the foundation of the building as part of initial construction, and will not be moved or modified. For this reason, it is important to install conduits sufficient to handle any anticipated communications growth at the time of initial construction.

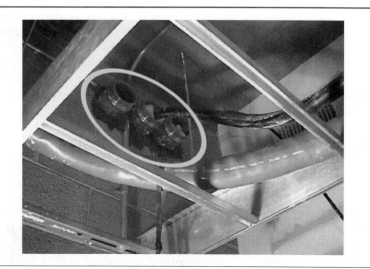

Figure 15–5
Conduit Entrance

MULTIPLE-DWELLING UNIT CABLING

Redundancy is usually referred to as two or more different services brought into an MDU, providing a choice of service to the tenants. All services are brought into the Main Distribution Frame (MDF) and then to the Intermediate Distribution Frame (IDF). This is what makes up the cabling backbone of a building. Fiber-optic cable is standard for backbone cabling.

Redundant Service Provider Cabling

Redundancy refers to the availability of a backup should one system fail or if a choice of two service providers is required. In cabling, redundancy is important for providing reliable transmission links into the building and easily-maintained wiring within the building. Redundancy in transmission links may be provided by a single service provider, but a growing number of building managers are electing to offer service from multiple communications service providers so that residents may benefit from competitive pricing and enhanced reliability.

When multiple service providers bring cables into the building, each will make use of an MPOE. All providers will be brought together in the Main Distribution Frame (*MDF*) where telecommunications, Internet, and other communications providers house their customer premises equipment. In a very large facility, the MDF may be a room separate from the MPOE. When that is the case, it will require the same sort of electrical service and climate control considerations as those given to the MPOE. If the MPOE and MDF share a room, then the electrical service and climate control must be sized to handle the needs of all equipment that will be in either rack.

MDF

Backbone Cabling

Interconnection between telecommunications closets, equipment rooms, and entrance facilities is provided by a backbone cabling system. A backbone can refer to one building or it can extend between multiple buildings. Cables for individual floors are led from the MDF to an Intermediate Distribution Frame (*IDF*) located within the building core on each floor. The IDF is where each telecommunications and Internet vendor can designate and split off cable pairs for individual customers.

IDF

In some instances, building layout or unit density may indicate the need for additional distribution frames. Depending on the communications service and provider involved, trunk cables may be bundled 50-pair copper or fiber-optic cable. Cables that run vertically between floors of the building are called *risers*—those that stretch horizontally within the confines of one floor are called *ties*. All riser and tie cables—and, for that matter, all cables that run inside ceiling or wall spaces—must be plenum rated. Plenum cables have special fire-resistant jackets. Since the cost of plenum cables is often only slightly more per foot than regular cables, some installers recommend plenum cables for all cable runs within buildings. You may make decisions based on economy and experience, but there is no choice on those cables that are within walls and ceilings—they must be rated for plenum installation.

Fiber Optics

Fiber-optic cables have become common pieces of infrastructure in offices and manufacturing and are becoming more frequently used in apartments and condominiums, as shown in Figure 15–6 and Figure 15–7. In many cases, fiber will be used as part of the backbone wiring plant, connecting the MDF to IDFs and on to SDFs. In these installations, there will be a switch or router that also performs a media conversion within the distribution frame.

Fiber-optic connections are moving beyond the backbone in some office and industrial applications. However, fiber-to-the-desktop is still far from common for the personal computer because of the cost of fiber-optic network adapters.

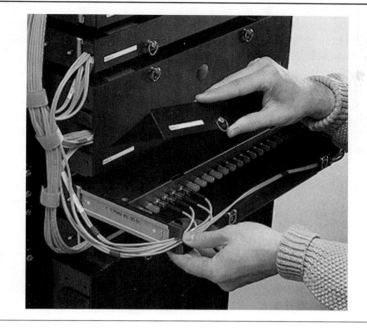

Figure 15–6
Fiber Optics

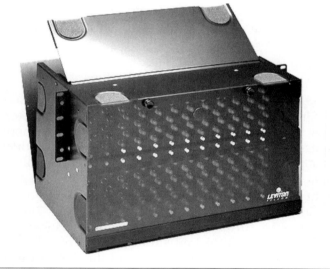

Figure 15–7
Fiber-Optic Panel

When fiber-optic cables are terminated in patch bays, the resulting racks look like the unit shown in Figure 15–8.

Figure 15–8
Fiber-Optic Patch Panel

There are several varieties of fiber-optic terminator in use; FC, SC, ST, and LC are the most common. Some of the different terminator types are shown in Figure 15–9. Terminators vary from vendor to vendor and application to application—you should be aware of the type of connector used by the vendor chosen to equip the distribution frames, MDF, and MPOE. You must then acquaint yourself with the splicing and termination techniques used for the type and brand of connector chosen. Fiber-optic splicing is very different from splicing building network or telephone cables. Although there are many different toolsets and procedures for splicing, most installers find it more comfortable and productive to learn and perfect one procedure and stick with it.

Figure 15–9
Fiber-Optic Connector and
Terminator Types

In addition to the varieties of connectors and terminators, there are two different modes used in fiber-optic communications.

1.   **Single-mode cable** is a single stand of glass fiber that has one mode of transmission. Single-mode fiber has a relatively narrow diameter, 8.3 to 10 microns, through which only one mode will travel. Single-mode provides a higher transmission rate and up to 50 times more distance than multimode, but it also costs more. Single-mode cable is ANSI/EIA/TIA-492CAAA.

2.   **Multimode cable** provides high bandwidth at high speeds over medium distances. Light waves are dispersed into numerous paths, or modes, as they travel through the cable's core, typically 850 or 1300 nm. Typical multimode fiber core diameters are 50, 62.5, and 100 micrometers. However, in long cable runs (greater than 3000 feet [914.4 ml) multiple paths of light can cause signal distortion at the receiving end, resulting in an unclear and incomplete data transmission. The jacket for 50/125 multimode cable will be marked ANSI/EIA/TIA-492AAAC, and for 62.5/125 multimode, the cable will be marked ANSI/EIA/TIA-492AAAA-A.

Most current equipment and applications are multimode fiber operating at either 50/125 or 62.5/125 micrometers in wavelength. Single-mode components and applications also exist, but are not as widespread as multimode.

NOTE   Fiber-optic cable is unique. Single-mode fiber-optic cable cannot be used in place of multimode, and multimode cable cannot be used in a single-mode application. Be sure that the proper cable is installed for the mode required by the equipment used in the installation.

UNIT DISTRIBUTION

Once cable has been brought into the building and distributed to the individual units, it is then routed within the residential unit or office. Central controllers should be located with the distribution frames nearest the individual residence. Of importance is the need to provide an auxiliary disconnect outlet. For consistency, all wall plates, switches, etc., for each individual unit should be the same.

Cable Routing for Apartment Outlets

Routing cables to and within individual residential units is governed by two concerns:

1. Safety.
2. Cable length required to maintain compliance with performance standards.

For safety, telecommunications and Internet access cables must not be routed in the same conduits or cableways as electrical service cables.

From the distribution frame to the telephone instrument or communications equipment, the distance should be no more than 100 meters (328 feet), as shown in Figure 15–10. Because the cable from the wall plate to the instrument should be no longer than 10 meters (32.8 feet), the distance from the distribution frame to the wall outlet should be no more than 90 meters (295 feet).

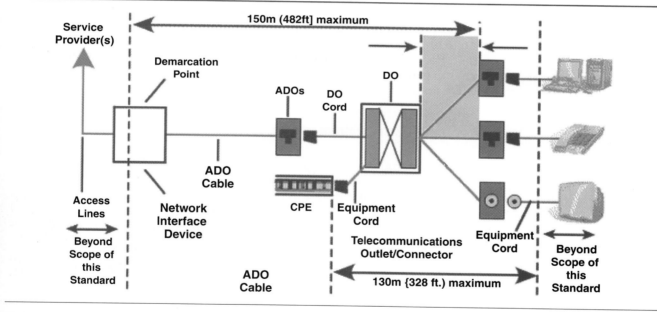

Figure 15–10
Path from the Distribution Frame to the End-User Device

The cable from each wall outlet to the intermediate or satellite distribution frame should be "home run" rather than daisy-chained or connected from one outlet to another. Each individual cable must conform to the distance specification. Cables must be run within walls and partitions in new or renovated apartments and condominiums.

Locating Local Central Controllers

Central controllers—whether telecommunications or Internet—should be located with the distribution frames nearest the individual residence. Obviously, the number of controllers and their locations will be governed by the physical size of the overall building. If overall cable runs can be kept within allowable parameters—160 meters (462 feet) from MDU to wall plate, and 100 meters (328 feet) from controller to instrument—then the controllers can be located in the same room as the MDF and the MPOE. In buildings too large for these cable run lengths to be observed, multiple controller locations (within intermediate or satellite distribution frames) may be required. In very large buildings, multiple MPOEs may be necessary to keep internal cable run lengths within standards.

Electrical service requirements to individual rooms housing central controllers will vary according to the equipment housed in each distribution frame.

The Auxiliary Disconnect Outlet

ADO

The Auxiliary Disconnect Outlet (*ADO*) provides an easy way of disconnecting or re-deploying lines in locations that are readily accessible to telephone and Internet service technicians. The ADO is not a requirement in installations where the MPOE and MDF share a room, but provisions for the ADO are part of most telecommunications equipment, and the ADO may be used with no detrimental effects even when it is not strictly necessary.

Wall Plate Locations in Residential Units

Wall plates for telecommunications and Internet service installed in the same position in all units creates a uniform, equal appearance. This uniformity allows for increased serviceability since technicians will not have to spend unnecessary time searching for wall plates in the event of service disruptions or change orders. Uniformity also eases calculations of overall cable length and wire order requirements since a standard length of cable within the walls of each unit can be calculated and added to the tie cables connecting each unit to the IDF on its floor.

GENERAL CONSIDERATIONS

Firestop

Firestop material in a building can help stop a fire from spreading in a building. Any penetration or coring, whether during installation or troubleshooting, should be repaired with firestop material to maintain the integrity of the ceilings and floors. The installation of an approved firestop material is required by the NEC. Another potential danger to the building is lightning strikes. Protection is provided within the architecture of the building, but cables must be well grounded where they enter the building. Each building will have its own policy regarding service technicians. It is important to understand what that policy is when a service call is required.

Firestops in Multiple-Dwelling Units

Firestop systems are designed to protect against fire, deadly gases, and toxic smoke through openings that exist in joints and gaps in fire-resistant walls, floors, and floor/ceiling assemblies. More information on firestop systems can be found at http://www.firestop.org. Firestop material was discussed in Chapter 9. Firestop systems are a crucial requirement of the fire-preventive capabilities of any building. The importance of firestop material cannot be overemphasized in a multiple-dwelling unit. Whenever firestops are punctured or penetrated by conduit or cable, installers must ensure that any voids are within fire code allowances. Spaces around conduit or cables should be well sealed with firestop material, special refractory putty that will preserve the integrity of the firestop even when passages for cable or conduit are necessary. In addition to putty, firestop material includes sealant, coatings, and a variety of firestop connectors and cable boxes. Figure 15–11 shows firestop material filled in the spaces around the cable running through the ceiling.

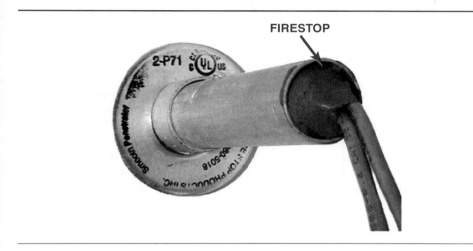

FIRESTOP

Figure 15–11
Firestop Material

Troubleshooting Terminal Access

After initial installation of the wiring plant has been completed and tested, additional verification is necessary whenever new or additional service is initiated to a new tenant or owner. Basic telecommunications testing equipment, including network tone generators and craftperson handsets, will help the technician ensure that the correct cable pair has been terminated at the intended wall plate, and that factors such as polarity are properly configured. Figure 15–12 shows a network tone generator that can assist you in tracing cable pairs across punchdown blocks and into residential units. Technicians working for the building owner or communications service providers will require access to both ends of interior distribution cables.

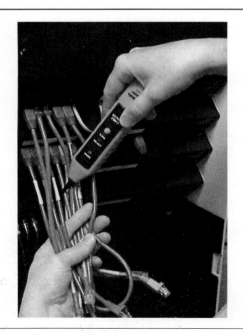

Figure 15–12
Network Tone Generator

Punchdown blocks and patch panels are often only casually labeled, if at all. Technicians, especially those who will be working frequently at a particular facility, will find that if blocks and panels are labeled neatly and professionally, they will spend less time sorting out cables. This will save time when performing service and maintenance.

Lightning Protection

General lightning protection for telephone and Internet service within the MDU will be provided within the architectural protection of the building. In order to take advantage of that protection, the cables must be well grounded where they enter the building. Precise grounding requirements vary according to local building codes, but all cables must be grounded within 1.6 meters (slightly over 5 feet) of entering the building at the MPOE. The effectiveness of the ground will depend on many variables, including local geology and service provider grounding in the service vaults. For this reason, it is best to consult with a professional engineer if there are any questions concerning the specifics of grounding provider cables in a given project.

Allowing Service Provider Access

There are a variety of reasons to house telecommunications and Internet service equipment in rooms dedicated to their needs. Security, climate control, and electrical service continuity are all considerations. Another crucial consideration is a mechanism for allowing technicians from service providers to have access to service provider premises equipment. Depending on the circumstances and policies of the building's owner, service providers may be provided with keys to the MPOE, MDF, and any IDF facilities where access is required. Alternatively, policies and procedures for having a building employee provide access when needed can be established and communicated to the service provider. In either case, access must be readily available so that no tenant or unit owner is left without communications for extended periods of time.

SUMMARY

There are differences in scale and complexity between telephone and Internet service provisioning of a single-family home and a Multiple Dwelling Unit (MDU). This chapter discussed the following topics:

- Service cable from the telecommunications provider enters the building at the Minimum Point of Entry (MPOE), which is located as close as possible to the point at which the cables cross the property line.
- From the MPOE, the cable is led to the Main Distribution Frame (MDF), where the main cable begins to split into individual areas or lines.
- Depending on the provider and the nature of the installation, the point of demarcation between service provider equipment and customer equipment may be at either the MPOE or the MDF.
- The MPOE and MDF may either be in separate physical rooms or share a room within the facility.
- Cable length between the distribution frame and the customer wall plate may be a maximum of 160 meters (525 feet). If this length can't be maintained from the MDF to the wall plate, Intermediate Distribution Frames (IDF) must be deployed on individual floors.

- Local controllers as required by the local telecommunications service provider are installed on the distribution frame; maximum cable distance between controller and telephone instrument is 100 meters (328 feet), of which no more than 10 meters (33 feet) can be between the wall plate and the instrument.
- When cables and conduits are run between distribution frames or between distribution frames and residential units, they may pierce firewalls either vertically, as risers, or horizontally, as ties. When they pass through the firewall, firestop putty should be used around the cable or conduit to ensure that the integrity of the firestop is maintained.
- Proper electrical grounding is a crucial part of lightning protection for telephone equipment. An electrical ground must be properly installed within 1.6 meters (slightly over 5 feet) of the MPOE to protect all systems within the building.
- The MPOE and the MDF must be accessible to technicians from the building owner and the telecommunications provider. Depending on the policies of both, the telecommunications provider should be given keys or access codes to the MPOE and MDF rooms, or owner representatives should have policies and procedures necessary to allow anytime access to the facilities.

GLOSSARY

ADO   Auxiliary Disconnect Outlet. Provides an easy way of disconnecting or re-deploying lines in locations that are readily accessible to telephone and Internet service technicians.

DSL   Digital Subscriber Line. Broadband Internet connectivity.

Emergency Power Off (EPO)   A switch that cuts off all power to circuits in the event of fire, flood, or other disaster.

Firestop   The material used in a firestop system that is designed to protect against fire, deadly gases, and toxic smoke through openings that exist in joints and gaps in fire-resistive walls, floors, and floor/ceiling assemblies.

IDF   Intermediate Distribution Frame. Where each telecommunication and Internet vendor can designate and split off cable pairs for individual customers.

MDF   Main Distribution Frame. Where telecommunication, Internet, and other communications providers house their customer premises equipment.

MDU   Multiple Dwelling Unit. This is a building designed for multiple residents; for example, an office or an apartment.

Minimum Point of Entry (MPOE)   The location where the telecommunications provider cables enter the building. Also known as the demarcation point or demarc.

CHECK YOUR UNDERSTANDING

1. MPOE is the acronym for:
   a. Median Point of Energy
   b. Minimum Point of Entry
   c. Maximal Point of Entry
   d. Minimal Price of Entropy

2. The MPOE is the location where:
   a. all residential lines are jumpered together.
   b. all DSL filters are installed.
   c. telecommunications provider cables enter the building.
   d. Internet Service Provider routers are housed.

3. What is the maximum length of the telephone cable from controller to instrument?
   a. 50 meters
   b. 200 meters
   c. 136 meters
   d. 100 meters

4. Where are the power requirements for individual components found?
   a. Power requirements are found in the manual only.
   b. On labels or plates on the bottom or rear of the component.
   c. Knowing the power requirement is not necessary.
   d. Inside the component

5. What does MDF stand for?
   a. Main Device Frame
   b. Managed Distribution Frame
   c. Main Distribution Frame
   d. Managed Device Frame

6. Cables for individual floors are led from the MDF to the:
   a. Intermediate Distribution Frames.
   b. Immediate Distribution Frames.
   c. Installed Distribution Frames.
   d. Installed Device Frames.

7. What is the preferred method of calculating power requirements?
   a. Lining up devices and adding the requirements on the labels
   b. A spreadsheet that lists power requirements
   c. It is not necessary to calculate power requirements.
   d. Only calculate power load if local codes require it.

8. How many fiber-optic modes are available?
   a. 2
   b. 3
   c. 16
   d. 7

9. Fiber-optic cable is not unique. A multimode fiber-optic cable can be used in place of a single-mode fiber-optic cable when needed.
   a. True
   b. False

10. Telephone cable run within a wall or ceiling space must have what sort of rating?
    a. Plenum
    b. Fire-resistant
    c. Wall
    d. Safety

11. Technicians may remotely disconnect or reconfigure a customer line at the:
    a. MDF
    b. MPOE
    c. ADO
    d. SDF

12. The punchdown block type supporting the highest bandwidth is:
    a. Type 66
    b. Type 110
    c. Type RJ-45
    d. There is no difference in bandwidth.

13. What is the maximum distance allowable between the MPOE and the ground point?
    a. 1 meter
    b. 1.6 meters
    c. 5 meters
    d. 3 meters

14. Which material should be used to maintain the integrity of a firestop when conduit or cables pass through?
   a. Asbestos tape
   b. Fiberglass batting
   c. Portland cement
   d. Firestop putty

15. Standards that describe the amount of smoke and the type of gases a burning cable can generate are developed by:
   a. TIA/EIA-570-A
   b. National Fire Protection Association

   c. National Electrical Code
   d. National Fire Code

16. Why should the MPOE be located in a separate room?

17. Why should a UPS be used over standby power?

18. What are the two concerns that govern cable routing to MDUs?

19. How should punchdown blocks and patch panels be labeled?

20. What are some of the reasons to house telecommunications and Internet service equipment in rooms dedicated to their needs?

Customer Service, Training, and Support

OBJECTIVES

Upon completion of this chapter, the HTI will be able to complete the following tasks:

- Walk through the finished project with the customer

- Obtain "sign-off" for the project

- Provide the customer with all documentation

- Provide training to the customer

- Offer maintenance contracts to the customer

- Identify upgrade opportunities

- Set up a customer support department

INTRODUCTION

Providing excellent service, support, and training is the best way to ensure that customers have a good feeling about the company and the work done. Customer support is often seen as a time-consuming activity that does not directly create revenue. Some even see it as just another cost. Small or mid-size businesses tend to underestimate the influence of good or poor customer support on business opportunities. Customer support should be viewed as a means to advance the business. When done correctly, providing customer support can translate into new opportunities for a business.

This chapter helps you deal with the customer. You learn the project completion tasks that need to be done in order to be paid. You also learn the importance of training the customer how to use the home automation system and subsystems. Next, you learn that upgrades and maintenance contracts are important additions to your bottom line. Finally, you learn

383

how to set up a customer support department and offer customer support using the World Wide Web. A web-based customer support department can be a cost-saving move.

PROJECT COMPLETION TASKS

You should give as much care to wrapping up a home integration project as you have shown throughout the entire process. The project-completion tasks described here are critical to ensuring that the customer feels that the work has been completed satisfactorily. This section provides guidance on how to perform the following tasks for the customer:

- Conducting the final walkthrough
- Getting final project acceptance or "sign-off"
- Providing warranties
- Providing documentation
- Supplying system test results
- Equipment and system orientation

Conducting the Final Walkthrough

Final walkthrough

When the work is done, it is time to perform the *final walkthrough* with the customer. This involves going through the entire project with the customers and verifying that each item was completed to their satisfaction. At the conclusion of the walkthrough, if nothing is found wrong with the installation, the customer signs off on the work. This clears the way for getting the final payment.

Figure 16–1
Punch List of Items that
Require Additional Work

XYZ Company
Punch List

		Passed	Failed	Comments
Rough-In	Frame Inspection	☒	☐	
	Device Locations	☒	☐	
	Wire Routes	☒	☐	
	Standards & Codes	☒	☐	
Trim	Terminations Outlets	☐	☐	
	Wire Labels	☐	☒	Rewire labels in the LV and master BR
	Equipment Installation	☒	☐	
Finish	Continuity Test	☐	☒	Retest in the LV and master BR
	System Configuration	☒	☐	
	Calibration Test	☒	☐	
	System Operation	☒	☐	

JP	5454122	2-24-2002
Inspector's Name	Project Log	Date

If there are problems discovered, a *punch list* will be generated. The punch list is a list of the items that are deficient or that will require additional work. Common items on a punch list are loose jacks, missing labels, or anything that might keep a customer from being fully satisfied with the job. Payment may be withheld based on the number and severity of the items on the punch list. Figure 16–1 shows a punch list.

The walkthrough ensures two things. First, it verifies that the work has been completed to the customer's specifications and is of acceptable quality. Second, it provides complete clarification about what is missing or left undone. Both parties benefit from doing a walkthrough at the end of a job: The customer knows that all the work has been done, and the installers know that they can expect payment as arranged.

Before taking the customer through the final walkthrough, make sure that the work areas have been cleaned up. If scrap material and trash are left behind and the areas are generally dusty and dirty, the customer may not be willing to sign off on the project. It will also give a better impression to the customer if the job site is clean and ready to be occupied.

Getting Final Project Acceptance or "Sign-Off"

The final step in the project cycle is the customer *sign-off,* which occurs after the walkthrough. The customer signs off when the customer agrees that the project has been completed. Figure 16–2 shows a sample sign-off.

Punch list

Sign-off

Figure 16–2
Sample Sign-Off Sheet

The XYZ Company
Customer Sign-off Sheet

THE XYZ Company shall not be liable for any special, incidental, indirect, or consequential damages or for the loss of profit, revenue, or data. The sole remedies for our liability of any kind with respect to the products of services provided by the XYZ Company or its subcontractors shall not exceed the total charges paid for these products or services.

The XYZ Company will, in no way, be responsible for the integrity of any data stored on computers, drives, tape backup systems, or any other product that contains data.

The XYZ Company reserves the right to use subcontractors to fulfill our obligations under this agreement.

If the attached proposal, along with its terms and conditions are acceptable, please sign below and return to the XYZ Company. The signed proposal along with the agreed upon payment shall constitute acceptance of this proposal and will be our authorization to continue to proceed with your project.

The XYZ Company is not responsible for modification made by the customer to the XYZ company system configuration during installation or after acceptance of the project. Any troubleshooting services required as a result of customer modifications will be billed at standard hourly rates of $125 an hour.

Accepted by:

_____
Customer Name   Date:

Submitted by: The XYZ Company

_____
XYZ Company Authorization   Date:

Please sign this acceptance page and fax back to the XYZ Company, 555-555-1313

The HTI then generates a document for the customer to sign. This document states that the customer has inspected the installation and that it is complete, minus any discrepancies noted on the punch list. Also, the customer accepts the installation and agrees to pay as arranged. Any hidden defects are covered under the warranty, which is discussed in the next section.

Providing Warranties

A *warranty* is a written statement provided to the customer. The warranty states that if any hidden problems develop within a clearly defined and reasonable period of time, it is the installer's responsibility to correct those problems. A sample warranty is shown in Figure 16–3.

If the problem was caused by an outside factor, such as another contractor damaging the cable when working on the HVAC system, the original installer is not held responsible. It is important to certify the cabling in order to protect the installer. Later, if any problems develop, the installer will have proof that the problem did not exist at the time of installation (if the problem was not a hidden one). It is often good business practice to fix a problem whether it falls under the warranty or not, especially if it might help secure future work from the client or recommendation for other potential customers.

Laws may require the installer to clearly state and conspicuously cite all the terms and conditions related to customer transactions in the documentation. If customers are aware of exactly what will be provided beforehand, they are more likely to be satisfied. It is acceptable to use standardized contracts, but make sure that they are written in language that the customer can understand.

Providing Documentation

Documentation related to any completed home integration project is the key to future follow-up and maintenance work on a residential network. Some of these are original documents the HTI received before the start of the project, whereas the rest are developed during the installation and at completion. Depending on the size of the home integration company, some of the documents might be created by a designer, an engineer, or the salespeople. Usually, it is good practice to document the network, generating a record of components installed and their locations, product information, special configuration issues, the general network structure, and a schematic showing how all the subsystems tie to each other at the central control processor. The following is a list of documentation generated during a project.

Sales Documentation:

Customer Requirements Checklist
Proposal
Scope of Work
Contract
Service Overview
Terms and Conditions

Design and Engineering Documentation:

Customer Requirements Checklist
High-Speed Internet Access Installation Checklist
Floor Plan
Schematic Block Diagrams
Signal Flow Diagrams
Bill of Materials
Pricing Summary
Proposal

Scope of Work
Contract
Service Overview
Terms and Conditions
Material and Devices Order
Wire Detail Chart

Figure 16–3
Sample Warranty

WARRANTY

Leviton Integrated Networks Structured Media™ Warranty

Leviton offers a two part warranty for its Leviton Integrated Networks Structured Media product offerings. The first part is a warranty on Leviton components used in the product line. The second part is an extended system warranty, *which is only offered for Leviton Integrated Networks Certified Installations that are installed, tested, documented, and registered to Leviton by Leviton certified installers (Program Installation Partners). Please ask your installer or builder if your installation is a Certified Installation.*

A. LIMITED TWO-YEAR LEVITON COMPONENT WARRANTY AND EXCLUSIONS

Leviton warrants to the original consumer purchaser and not for the benefit of anyone else that the Leviton labeled devices, plugs, jacks, connectors, splitters, cables, system media centers, and electronics included as part of a Leviton Integrated Networks residential installation, at the time of its sale by Leviton is free of defects in materials and workmanship under normal and proper use for Two Years from the purchase date. Leviton's only obligation is to correct such defects by repair or replacement, at its option. If within such two year period the product is returned prepaid, with proof of purchase date, and a description of the problem to Leviton Manufacturing Co., Inc., Attention: Quality Assurance Department, 59-25 Little Neck Parkway, Little Neck, New York 11362-2591. This warranty excludes and there is disclaimed liability for labor for removal of this product or reinstallation. The warranty is void if this product is installed improperly or in an improper environment, overloaded, misused, opened, abused, or altered in any manner, or is not used under normal operating condition or not in accordance with any labels or instructions. There are no other or implied warranties of any kind, including merchantability and fitness for a particular purpose, but if any implied warranty is required by the applicable jurisdiction, the duration of any such implied warranty, including merchantability and fitness for a particular purpose, is limited to two years. Leviton is not liable for incidental, indirect, special, or consequential damages, including without limitation, damage to, or loss of use of, any equipment, lost sales or profits or delay or failure to perform this warranty obligation. The remedies provided herein are the exclusive remedies under this warranty, whether based on contract, tort or otherwise.

B. LEVITON INTEGRATED NETWORKS CERTIFIED SYSTEM TEN-YEAR APPLICATIONS ASSURANCE AND TEN-YEAR EXTENDED WARRANTY

Leviton's approved Certification Program passive structured cabling products, when properly installed in the Structured Media Center and throughout the residence by Certified Program Installation Partners with the appropriate category rated cable in strict compliance with Leviton practices and procedures and the electrical performance criteria of the TIA/EIA-570-A standard, will support and conform to EIA/TIA-570-A specifications covering ANY CURRENT OR FUTURE APPLICATION which supports transmission over the category 5 cabling links for a period of ten years from the the date of certification provided the installation remains as originally installed and certified. In addition, these same Leviton-approved Certification Program products will be free from defects in material and workmanship for a period of Ten Years from the date of certification, as long as the product remains installed in the Certification Program System.

The specific terms and conditions of the Ten-Year Applications Assurance and Ten-Year Extended Warranty are set forth in the Leviton Integrated Networks Certified Systems Program Partner Agreement.

Important: The warranty does not cover abuse, misuse, or damage to the installation arising from causes beyond the control of Leviton or the certified installer.

YOUR RESPONSIBILITY TO MAINTAIN YOUR WARRANTY

To maintain the warranty for your wiring system, you may only change certain connectorized jumpers in the SMC, and add or remove equipment at the wallplates, but not behind the wallplates.
You may not make any other changes. Note: Only a certified installer may change any fixed wiring and/or add certified equipment (with appropriate documentation and certification to and from Leviton).
Under your warranty, you may comfortably enjoy the freedom to add and rearrange equipment at the wallplate jacks provided that you do not alter the jacks or the wiring to them in any way. Additionally, though you may choose to move patch cords in your SMC, Leviton recommends that you make these changes sparingly, and under the advisement of a certified installer or a Leviton customer service representative. Do not make any changes other than patch cord placement within the SMC.

Installation Documentation:

Punch List
Change Order
Material and Devices Order
Wire Detail Chart
Installation Instructions
Test Documentation Forms
Completion of Sign-Off and Certification Form

Customer Support Documentation:

Completion of Sign-Off and Certification Form
Warranties
User Manuals
Help Manual/Self-Troubleshooting
Cheat Sheets
Passwords
FAQ
Glossary
Customer Support Contact Information
Service History Records
Revised Documentation

Typically, copies of essential documentation can be left with the home-owner/user. Copies should also be kept at the company office for future reference and maintenance or upgrade work.

> **NOTE**   It is very important to retain all modified blueprints because they represent the final outcome of the installations.

Supplying System Test Results

As part of the total project documentation you should include reports of the cable plant tests performed. This is especially important if your total system testing includes certification tests for any CAT 5, CAT 5e, or CAT 6 cabling run through the residence. Some test equipment vendors, such as Fluke and Agilent, offer software that automates the process of creating reports from test results as shown in Figure 16–4. These applications present varying combinations of raw test results, along with analysis and, in the case of certification testing, verification that test results indicate compliance with the standards governing the certificate.

If automated report-generation software is not used, a full report listing all tests performed, test results, and any remedial action taken in case of errors found should be presented to the customer as part of the final installation documentation. Complete test effort and result documentation reassure the customer that all systems are built on an infrastructure that will support current and future needs and provide a documentation basis for updates, upgrades, and repairs to the installed system.

Equipment and System Orientation

The walkthrough also presents the opportunity to introduce your client to the new equipment and general structure of the network. Perform a demonstration of the way the integrated system works. Also, give the customer basic training and orientation on the handling and use of the system. If the HTI connected

Leviton Integrated Networks
HOME SYSTEM

Installation Number ___**16542**___

Installer System Certification

We certify that this structured media system installation has been installed, tested, and recorded according to Leviton certification procedures and practices for Leviton Integrated Networks platforms, systems, and components, and the cable manufacturer's warranty for the installed cable.

Residence Location ___Mesa___

Address ___100 Broad St.___

City,State,Zip ___Mesa, AZ 85233___

Installation Company ___Ace Installation Co.___

Business Address ___100 Main St.___

City,State,Zip ___Anytown, AZ 1000___

Telephone Number ___602-222-1000___

Installer Name ___John Doe___

Signature _____**Date** _____

Figure 16–4
Certification Results can Help Assure Customers that the Infrastructure will Support any Current or Future Application

and configured devices on the network, some of the items to demonstrate to the client may include:

- How to log on to the network
- How to use remote devices
- Basic system configuration issues
- Network security
- Content protection and software-licensing issues

The next section discusses customer orientation and training in the various subsystems.

CUSTOMER ORIENTATION AND TRAINING

When residential systems are automated, the goal is to make the life of residents more pleasant. In many cases, this means that the building will respond to the residents rather than forcing the residents to respond to the building. The purchasing family should be trained in the use and simple maintenance of the system so that they will feel satisfied with its performance.

- Keep the training to a reasonable time. Too much information could be overwhelming to the homeowner.
- Make sure the homeowners have hands-on training in order to correctly operate the systems.
- Adults in a household should receive full training, but it is a good idea to involve children. Parents should ultimately be responsible for training their children.

The homeowner should be given a user notebook with the following:

- User guides
- Troubleshooting checklists
- Warranties
- The HTI's business card and any other contact information

Planning the Training

Installers should begin with a plan of instruction. Will the instruction begin with the most commonly used commands of each system and then move into the less-frequently needed functions? Or will the training cover each sub-system completely before moving on to the next? Either approach can work, but the installer should know which direction will be taken before beginning. Skipping randomly through commands and functions is confusing for the individual being trained, and runs the risk of neglecting training in one or more important areas.

- Cover proper operation of the system. Be sure to also discuss the results of improper operation. In the early days of operating a new system, it is likely that input will be given out of sequence on one or more occasions. The user should know what to expect when this happens and how to correct the situation.
- Look at worst-case scenarios. What happens when the power fails? Do systems automatically reboot, or is user intervention required? What happens if one system becomes "confused"? Can the user force a reset of a malfunctioning system without causing additional malfunctions?
- Provide complete documentation for the system. This documentation includes all commands and functions, all necessary user intervention to correct malfunctions, block diagrams, blueprints, manufacturer-supplied literature on individual components, and the installer's business card.

Presentation of the system documentation is an important milestone in the project, and it is an opportunity for the installer to present a professional, positive impression to the customer. Providing the documentation in a professional binder, slipcase, portfolio, or other appropriate container will help build the user's confidence in the performance of the installer.

Basic Wiring Training

It is helpful for the homeowner to understand the basics of residential wiring infrastructure and its components. Such training can center on giving them basic knowledge and understanding of aspects of the installation such as:

- **Location of the ADO**—Showing the homeowner where the ADO is located allows the homeowner to switch between different services or disconnect service to different rooms of the home for maintenance and troubleshooting.
- **Basic system configuration**—Provides information so that the customer can move equipment around. It also shows the wiring for specific devices and appliances.
- **Which circuit serves which room**—Provides information so that the homeowner can check the circuit breaker to turn appliances on and off, for example, in reaction to electrical faults or lightning surges.

Computer Network Training

Customers who are not yet high-end users become accustomed to the system only over time. Show the customer any specific configuration issues and the best way to use the network to serve their needs. One of the most important aspects of this orientation process is demonstrating how to log on and use high-speed Internet access.

Logging onto the network is a crucial aspect of customer training and orientation because it has to do with keeping out unauthorized users. You have to orient the home network user on the basic procedure for logging on to use Internet access and help them understand how the process works.

Internet Access Training

In a home networking environment, the broadband connection to the Internet is always on. Providers such as telecommunication companies typically supply a logon mechanism for authenticating the user. The process of authenticating a user on a home network involves the establishment of a *session*. The concept of the session is already widely accepted and deployed in the telecommunications business. All Internet dial-up services rely on a protocol called the point-to-point protocol (PPP) to create a session. PPP is the Internet standard for transmission of IP packets over serial lines.

Session

DSL and cable technology providers extend this communications model with a variation of PPP called *Point-to-Point Tunneling Protocol* (PPTP). Microsoft and several vendor companies, known collectively as the PPTP forum, have jointly developed this protocol. With PPTP, home networking users create a logon session every time they access the broadband network.

Point-to-Point Tunneling Protocol (PPTP)

Follow these steps for logging on to a broadband network:

1. After a digital appliance on the home network is powered up, the Dynamic Host Configuration Protocol (DHCP) server on the broadband network allocates a private IP address to the device. Thus, without going though the complicated logon procedure, the device on the home network has limited access to specific resources.

2. When you want to use this device to reach content on the Internet, a type of virtual networking link is established between the home network and the PPTP servers on the broadband network.

3. Next, the home network initiates a second virtual connection based on the PPP protocol. This connection will then prompt you for authentication details (that is, username and password).

4. After the service provider verifies the logon credentials, a public IP address is assigned to the home networking session for the duration of the browsing session.

The advantage of this solution for connecting to a broadband network is that it allows service providers to control access to different types of services. Today, most service providers do not enforce the logon mechanism process for browsing (as described previously); however, they are deploying this mechanism for accessing value-added applications such as online shopping.

Audio/Video Subsystem Training

Although user manuals come with just about all the devices, training gives the homeowner, especially those not very electronics-savvy, a better understanding of how the equipment works.

Basic use and maintenance of the installed devices is important for training and orientation for the homeowner. Perform the following steps:

- Show how the equipment works and be sure that the homeowner can operate it.
- Review all safety procedures, especially the danger to small children. Some equipment with heat-producing components (for example, power amplifiers and some receivers) can become hot enough to cause injury.
- Review issues related to cooling and general climate control around them to protect the users and the equipment.

- Another point of training and orientation relates to basic equipment configuration. This training is useful for temporary disconnections, including equipment testing and troubleshooting.
- The homeowner should be able to provide information for warranty claims regarding equipment or general installations—such as media room locations, equipment locations, distribution device location, and warranty information.

Although the whole-house audio/video system can be easy to understand, more specialized installations such as home theater may at times need professional maintenance and repair of problems that arise. One way to assure your customer of your continued availability for servicing is to offer a service and maintenance contract.

Security Subsystem Training

Training and orientation are critical for the homeowner to properly operate the security system. All individuals who need access to a residence should be trained on the use of the security system after the installation is complete. The training should not be limited to a walkthrough, but should also include basic features programming and password uses. The homeowner should know how to read the "vital signs" of a failure or malfunction. When a false alarm is detected, the homeowner should notify the monitoring company immediately to avoid police notification and possible fines.

The homeowner should know the location of the RJ-31x termination point that interfaces the system with the home phone line. Knowing where to plug or unplug the system from the phone line is important for vital repairs, testing, or troubleshooting the phone line or security system.

Passwords

An important issue for the smooth functioning of a network is that the passwords are documented. Clients need to set up passwords for different types of services and devices (for example, Internet access, security passwords, and so on). A password is a series of letters and numbers, usually four to eight characters in length, that protects a system from unauthorized use. Passwords are a crucial part of programming and configuring the security system.

Passwords often get lost, expire, or are forgotten. There are different options for solving this problem. The HTI can keep a backup of all passwords in the project documentation. Another alternative is for the HTI to set up initial passwords and let clients customize them. Yet another option is to have the HTI keep administering rights for the network and grant the client user rights. In this way, the administrator can always override the user settings and reverse corrupted or malfunctioning settings. Whatever works best depends on what the customer feels comfortable with.

Each authorized user with access to a security system should know the password used to arm and disarm the system. Ideally, authorized users should have their own passwords. Passwords on advanced security systems can be configured with different levels of access. For example, the homeowner would be assigned as the administrator with privileges that provide access to all functions of the system. Other users, such as repair personnel who only need to arm and disarm the system, would be assigned specific access just for that purpose. Passwords can then be changed when the homeowner no longer wants that person to have access to the home.

The most important aspect of passwords is keeping them secure. The following tasks are highly recommended:

- Change passwords regularly.
- Use a combination of letters and numbers in the password.

- Do not use familiar names and dates, such as a middle name or birthday. It is too easy for unauthorized users to figure this out and gain entry.
- Do not use words that can be found in a dictionary. There are programs designed to crack this kind of password.

For more on passwords and security, see Chapter 6.

Teaching the Homeowner to Troubleshoot

Training the homeowner to troubleshoot is required upon completion of the project. Training related to equipment hookup and power issues should be focused on helping the homeowner differentiate between problems that are symptomatic of power supply failures and those linked to possible equipment failure. This training helps the homeowner determine where to direct service calls, such as to the power company instead of the HTI. Service trips to the customer site—only to find that the power wasn't turned on, the equipment was not properly plugged in, or there is a problem with the power supply and not the equipment—can be costly.

Training should also include information to help the homeowner understand when their equipment is not functioning properly due to intermittent power failures or power interruptions such as surges, spikes, and brownouts.

UPGRADES AND MAINTENANCE CONTRACTS

After the home network integration project is complete, you may find that upgrades and maintenance contracts present a unique opportunity for additional sales. You may be able to recommend a new technology or enhancement that benefits the homeowner. You may also recommend network upgrades based on newly emerged technologies that can expand the customer's existing network or make it possible to integrate previously incompatible systems.

Maintenance contracts for the computer network and security systems can provide additional future revenue, as well as keep the customer happy by having a trusted source to turn to if a problem arises.

Upgrade Opportunities

Bring sales materials with you to a service call so as to be prepared to discuss an upgrade opportunity, or to leave them behind for the customer to review at a later time. The biggest upgrade opportunity, however, is probably before the project starts. If the homeowner plans on using more advanced (bandwidth-hungry) applications on the home network in the future, explain the advantages of an advanced wire grade such as Category 6 or fiber-optic cable. The capability to provide cutting-edge services may be a differentiator between companies.

Maintenance Contracts

Even with the best orientation, the bigger problems and failures that occur with your client's system still need professional-level expertise to resolve them. Additionally, the user always wants to be sure that if things seriously go wrong with their installations, they have a trusted source to turn to for help. It is therefore good practice to offer your client some sort of service contract at completion. This also constitutes an additional source of revenue for your business.

Service maintenance contracts can take several formats. It could be service at a large discount to the client if they make a commitment to retain you for future repairs and maintenance work. Another format is a small monthly fee paid by the client in exchange for any repairs any time they occur, not including the cost of parts. Charges for maintenance contracts vary widely, depending on the market and services included. You need to research other companies in your area to find out what is typically charged for a maintenance contract. Similarly, research the types of services typically offered in your area so that you are competitive with other home automation companies. The nature of the service contract that is finally agreed upon always depends on the needs of the client and their comfort level with any sort of signed agreement. The HTI can offer either a comprehensive contract to cover all subsystems or separate maintenance contracts for particular subsystems.

A maintenance contract benefits you by keeping open a potential revenue stream. More importantly, however, such contracts keep customers happy and confident about having a trusted source to turn to when problems arise, particularly those related to integrated home network and security systems.

Computer Network Maintenance   Because many devices, including other residential subsystems, are connected to the computer network in an integrated home, the network is bound to fail from time to time. However, most home network users do not have specific expertise in troubleshooting and fixing problems with computers and network connectivity. It makes good business sense to keep the customers happy and feeling good about the work you do for them. Since you have designed and installed the infrastructure for a home integration project, including the computer network, you possess the most knowledge about what you did and any potential bottlenecks or breakpoints. Also, you hold and understand the terms of warranty-related information for most of the equipment installed.

These are just some of the reasons why it is useful to complete the residential network and systems integration project by offering your customer a maintenance contract. A computer maintenance contract can cover faulty parts, poor wiring, upgrades, performance, and preventive maintenance.

Security System Maintenance   The goal of a security system is to function at all times and according to the way it is programmed to work. Providing a maintenance contract gives the homeowner peace of mind so that any problem that should occur can be corrected quickly. This early correction ensures that any downtime is kept to a minimum.

To keep the security system up and running all the time, a maintenance or standing contract can be signed with the installer, usually for a monthly fee, to carry out periodic maintenance, testing, and repairs. However, this approach is mostly for the off-the-shelf systems that are not monitored by a company. With most of the more advanced systems, the manufacturer installs everything for a specified down payment and then maintains the network and monitoring services for a monthly fee, using their own specialists.

CUSTOMER SUPPORT

Customer support is a service offered by the HTI that fixes problems encountered by the customer. Customer support can do a lot more than simply keep the customer happy. It is the most direct method of communication between your company and the customer. As such, it plays a critical role in the success of your business. As an HTI professional, you should consider the following when thinking about customer support:

- Understand the needs and concerns of your customer. This fundamental consideration should inform all of your customer support efforts.
- Decide whether your customer support should focus on pre-sales activity, post-sales activity, or both.

- Commit to a level of support that is appropriate for both the customer and the support team.

Regarding the last point, the level of support should be sustainable on an ongoing basis. It sometimes might be enough to call a few days after an installation to see whether the customer has any issues or problems using the new home network.

The sections that follow discuss the role of the customer support representative and types of customer support that can be offered, including telephone-based and web-based support. The types of information that should be provided for the client, such as a glossary of terms, contact information, and some common forms, are also discussed. Finally, troubleshooting techniques are covered.

The Customer Support Representative

Customer support representatives should be experienced and well-versed in all the subsystems integrated on the home network, as well as the functionality of the entire network. For this reason, training is crucial for helping customer support representatives feel comfortable and competent in their positions because it will in turn be conveyed to the client who seeks support. Some companies do not want to spend money to train employees. This is a big mistake; customer support should be a top priority. If customers are treated poorly or if their problems are not dealt with in an appropriate manner, they will simply go elsewhere. A small company should either hire experienced customer support personnel or hire a company that specializes in training people. Larger companies with a large customer support department may offer their own training.

All customer support representatives should be aware of the importance of the customer. High standards should be set for customer support professionals. It is important for customer support personnel to be calm and friendly to all customers, regardless of how the customer treats them. It is also important to make sure that the customer support staff has the necessary resources to do their jobs effectively.

Troubleshooting and Maintenance Issues

Providing troubleshooting for customer problems is one facet of customer support. Training is required so that customer support personnel have at least a basic understanding of all the subsystems integrated within the residential network. This makes determining the true nature of the customer's problem easier.

Because providing good customer service is important for every company, the HTI should not cut corners by training people himself or herself. It is advisable for the HTI to either hire people with experience in customer service or hire a customer service training company to train people.

To solve customer problems, support technicians can use the following techniques:

- Interpret and relay the customer's perceived problem or needs.
- Communicate the problem verbally or in writing.
- Recommend a sophisticated new method of identifying problems by using *remote network management* (RNM), which essentially uses technology, instead of a repair technician, to make the service call. To check for signal problems, the RNM software sends out a series of signals to target devices at the client site to check for signal-receiving problems. If a device is equipped with RNM technology, the signal can also be sent from the device to check for signal-sending problems and potential solutions to the problem.
- Coordinate with technicians or subcontractors and oversee the repairs.

Remote network
management
(RNM)

- Test the specific repair work and ensure that the entire network remains functioning properly after repair work is complete.
- Document the problem and the solution that was provided as well as the date, time, and contacts involved.

It is important for the customer support representative to have templates at hand to fill out while dealing with a client's support issue. All problems, questions, and solutions need to be documented meticulously so that the repair technicians can review them and take action, if necessary. There is also value in having a complete collection of trouble reports in an easily accessible form, such as a database. This collection enables management personnel to look for common problems that might be addressed through policy, procedures, or training.

If troubleshooting does not yield an understanding of the problem through telephone, web, or RNM methods, the customer support person should schedule a service call to further investigate the problem.

Telephone-Based Support

Customer support has traditionally been telephone-based; callers were usually redirected to a specific department or to a geographical region. There have been two major call center technologies: *interactive voice response* (*IVR*) and *computer-telephony integration* (*CTI*). IVR systems provide some degree of self-service to customers, who press buttons on touch-tone telephones to arrive at locations at the customer support center where their problems can be resolved or their information needs addressed. By means of CTI, the caller (as well as information about his or her account history) can be directed by computer to the available call center representative. Although telephone-based support is expensive and time-consuming, sometimes the customer is better served by talking to an individual who will take responsibility for solving the problem.

With the popularity of the Internet, many companies have switched most of their customer support to the World Wide Web.

Web-Based Support

The Internet is revolutionizing customer support functions. The technology for the simultaneous transmission of voice, video, and data over the Internet has many different names. These names include multimedia communications, integrated services, web collaboration, convergence, teleweb, web browser sharing, and browser synchronization.

For many customers, the Internet has become the preferred channel to use in order to interact with a business. If a home integration business has not done so already, now is an excellent time to develop a web-enabled call center, which can provide multiple contact points through which clients can access the customer support contacts at the company.

A web-enabled call center may provide a variety of solutions, including the following:

- **Callback:** By pressing a button on their web browser, the customer can request a callback from their contact person in the home integration company. The customer provides information in a predetermined format that includes a phone number at which the customer can be reached, the nature of the problem or request, and a preferred time for the callback. The request is then routed to the appropriate customer support representative based on the user's language, project, and history.
- **Automated e-mail response:** E-mail arriving from the customer can undergo a natural language search for key words and phrases. These are applied to a knowledge base (FAQ, URL, user manual, and so on) to locate content that is relevant to the user's needs. A response is directed back to the user

Interactive voice response (IVR)

Computer-telephony integration (CTI)

automatically by return e-mail. E-mail for which there is no existing solution in the knowledge base is directed to customer support representatives or elsewhere for a response.

- **Interactive chat:** The customer presses a button or link on their browser to request an interactive chat session. The customer's request is then routed to the appropriate customer support representative. Each customer support representative communicates with many users, and each user communicates with a single representative.
- **Voice over Internet Protocol (VoIP):** By pressing a button or a link on their browser and by using headphones and a microphone attached to their computer, the customer can initiate an Internet Protocol (IP) telephone call to their respective call center. Typically, information about the caller is conveyed to the customer support representative. The call occurs over the infrastructure of the Internet, not over the infrastructure of the traditional telephone companies' plain old telephone system (POTS). This saves the caller and the cabling company a little money.
- **Shared web browser sessions:** The customer support agent and customer communicate by VoIP. The customer support representative synchronizes the customer's browser so that the customer support agent can lead the user to specific website pages while communicating with them by voice.

The extent to which a company utilizes one or more of these solutions depends on business strategy, products, and customer needs. These solutions are typically layered, with visitor callback often occurring as the starting point. For example, a customer support function that offers automated e-mail responses would want to offer callback as an option as well. Companies that have resources to support shared web browser sessions also support the remaining four solutions.

The challenges of implementing a web-enabled call center are well worth the investment. Think of the web-enabled call center as a means to differentiate your company from your competitors and as a means to achieve or maintain high customer satisfaction.

Contact Information

Few things are more important than providing reliable contact information. Customers should be able to get in touch with an authoritative representative of the company within a few minutes on most workdays and within a few hours on weekends or at night. Always provide an e-mail address and telephone number on every document (including proposals, contracts, blueprints, business cards, and letters to the customer) and on the website.

Another good reason for making customer support available is to answer questions before they become problems. Specifically, customers often need to talk to product experts before placing an order, or need special assistance from a customer support representative after placing an order.

Finally, at a psychological level, the availability of contact information shows customers the value you place on customer support because you are, in effect, encouraging them to use customer support. Sometimes, customers need contact information just to satisfy their desire for human presence.

Frequently Asked Questions (FAQ)

If many of the customers are asking the same questions, it is a good idea to include answers to these questions on the website or in a document that is given to the customer. This document, usually called Frequently Asked Questions (FAQ), provides answers to common questions so that customers can complete their tasks quickly and easily. A home integration company should also use its expertise in residential systems components to address product-related questions from users in the FAQ. Doing so also saves both the company and the customer time and expense because a customer support representative does not have to keep answering these questions several times each day. Figure 16–5 shows a sample FAQ.

Figure 16–5
Sample FAQ

APPENDIX B – FREQUENTLY ASKED QUESTIONS

Following is a list of general questions that you may have about your Leviton Integrated Network structured wiring system. For questions about your particular option, see the appropriate insert in this guide. If you have any questions that are not addressed in that section call your installer or Leviton customer service. The telephone numbers are listed on the inside cover of the SMC.

How do I know where I can put specific pieces of equipment?

Refer to the appropriate insert in this user guide for information. We have provided you with a page showing the specific jacks/outlets that your particular system contains. Look around your home to determine what types of jacks are placed in each room. Once you have made this determination, you are ready to consider the types of equipment that you can place in a particular location.

Are their minimum standards that my equipment must meet to be compatible with this system?

Yes. Certain pieces of equipment in your home (for example your stereo amplifier) may need to be of a certain grade to be fully compatible with this system. Talk to your certified installer for more information. If you purchase new equipment, be sure to discuss your system with a qualified sales representative. If your sales representative is unsure of compatibility call your installer or Leviton's customer service.

Can I add components to the structured wiring system that I presently have in my home?

Ask your builder, contractor, or certified installer this question. The answer depends on the particular system that has been installed in your home. For example, what preparation for future technology was made in the original design and what locations are prewired?

What, if any, protection does my system have from outside interference such as electrical surges?

Leviton offers whole-house surge protection to protect your electrical equipment. Ask your builder or installer about this option.

Web link: http://www.leviton.com/sections/techsupp/faq.htm

Glossary
Because residential networking has many technical terms, you should provide simple definitions and explanations of important terms that customers will encounter. This book provides a glossary of the terms used and can be used to explain common computer terms and devices used in the computer subsystem. Similarly, give definitions for the types of devices that customers can choose from in the home entertainment subsystem and security subsystem. Some terms can be defined in FAQs or other types of assistance, but a separate glossary, on a website or in printed form, allows the customer to quickly look up a term rather than having to scan the FAQs for a word. Within the glossary, links to related terms should be provided. The glossary should include company information as part of homeowner information. A sample glossary is shown in Figure 16–6.

Figure 16–6
Sample Glossary

APPENDIX A – GLOSSARY

Analog – A continuous signal that varies in voltage and frequency. Sound (e.g. voice, music) is analog because it continually varies in pitch and volume. Analog technology is traditionally used for audio and video recording to capture the complex nature of analog waveforms. Modern electronic equipment converts analog into digital signals (see definition of "Digital") that are easier to preserve and transmit at higher speeds.

Bandwidth – Bandwidth is the total frequency spectrum, in Hertz or cycles per second, that is allocated or available to a channel. It is related to the amount of data that can be carried in bits per second (BPS) by a channel.

Bridging Connection – A parallel connection that draws some of the signal energy from a circuit, often with inconsequential effect on the circuit's operation.

Category 5 or 5e – A category of performance for unshielded twisted pair copper cabling. Used in support of voice and data applications requiring a carrier frequency of up to 100 MHz for a distance of up to 100 meters.

CATV – Community antenna television usually transmitted to your house over coaxial cable from a local Cable TV company.

Certified Installer – An installer trained and/or approved by Leviton to install your structured wiring system. No one except a certified installer should do anything to your system except adding new equipment at the jacks and relocating patch cords within the SMC.

Coaxial Cable – A transmission medium that consists of one or more central wire conductors. Coax carries a much higher bandwidth than a traditional wire pair; however, it is not as easy to work with as the newer category rated "twisted pair". Coaxial cable is usually used for CATV, CCTV, and broadband video.

Composite Video Signal – The complete video signal including the brightness (luminance), the blanking and sync pulses, and the color (chrominance).

Digital – A signal that only varies between two states, represented as a binary code of "one" and "zero". When information, audio and video are turned intro binary digital signals, they can be electronically manipulated, pre served and regenerated at higher speeds.

Digital Data – Any coded information that can be processed or output by a computer or other machine.

Fiber Optic Cable – Network cabling that employs one or more optical fibers, consisting of a central glass or plastic fiber, glass cladding, and a plastic outer sheath. Fiber-optic cable carries information as light instead of electricity, and can carry much more information over greater distances than copper cabling.

Fixed Wiring – Permanent wiring within building walls.

Hub – A group of circuits connected at one point on a network. A hub enables economic traffic concentration.

Integral Bridging – A technique that combines different media and topologies. It requires the installation of additional network interface cards in a server.

Jacks – The low voltage receptacles mounted within a wallplate into which your telephone, audio, and video equipment is plugged. The shape of the opening determines the type of equipment that can be installed at the particular jack location.

Jumpers – Cable that has connectors installed on both ends.

Lead – The short length of a conductor that hangs free in a box or service panel (i.e. a wire end).

Module – An identifiable piece of hardware or software.

Mode – The method of operation of a system – the method of processing data.

Modem – A device that converts digital signals for transmission over analog carrier lines. Modems provide computer access to the Internet through cable or basic phone lines.

Electronic Forms

Another reason to have a website is that it can provide information about the company as well as forms for requesting information. Customers or potential customers can fill out and submit these forms. The requests can be processed and responded to in a timely manner after the request is received. Electronic forms can also be saved to a database.

Also, the company website may function as an update center for software installed on the home network. Some of the subsystems actually come with proprietary, customized software solutions, so it is likely that these applications will require more than one update during their lifetime. The HTI can offer these software updates on the company website for the homeowner to download.

SUMMARY

This chapter discussed the importance of customer service and support.

- The "final walkthrough" is when the HTI shows the customer the finished project.
- After the customer is satisfied with the work, obtain the customer's signature. This is called a "sign-off." You can then submit an invoice to be paid.
- The customer should receive a packet with all relevant documentation.
- The customer should be trained on how to use the home automation system and subsystems.

- Additional sales opportunities can be made by suggesting upgrades to equipment when visiting the customer's home during a service call.
- Maintenance contracts are another opportunity for additional sales.
- A customer support department provides the customer with answers to questions and solutions to problems. A web-based customer support department can save time, man-hours, and money.

GLOSSARY

Computer-telephony integration (CTI)   Technology used by customer support centers to direct calls. Information about the caller is used to direct the call.

Final walkthrough   Process through which the HTI walks the customer through the entire project and verifies that each item was completed to the customer's satisfaction.

Interactive voice response (IVR)   Technology used by customer support centers to direct calls. Callers must choose from lists of options.

Point-to-Point Tunneling Protocol (PPTP)   This protocol (set of rules) allows a LAN to be extended beyond the physical boundaries of the office through private lines over the Internet. This kind of interconnection is used for a virtual private network (VPN).

Punch list   A document generated during the final walkthrough that lists problems to be corrected.

Remote network management (RNM)   Software that talks to the devices at the customer's home to determine the problem.

Session   A session is a series of interactions between two endpoints that occur during the span of a single connection. The session begins when the connection is established at both ends and terminates when the connection is ended.

Sign-off   This document states that the customer has inspected the installation and that it is complete, minus any discrepancies noted on the punch list.

CHECK YOUR UNDERSTANDING

1. An HTI should keep a copy of the source code to protect the customer against the worst effects of:
   a. a brownout.
   b. the programmer going out of business.
   c. software bugs.
   d. a power surge.

2. Which software is used to replace service calls?
   a. Remote network management
   b. Remote service management
   c. Remote interactive response
   d. Computer-telephony integration

3. RNM works by sending signals from the service center to the customer's devices. The customer's devices cannot send signals to the service center.
   a. True
   b. False

4. Which telephone-based customer support technology requires the user to select from option lists?
   a. Interactive call response
   b. Interactive voice response
   c. Remote voice response
   d. Computer-telephony integration

5. What is the term for the document that contains problems that were identified during the final walkthrough?
   a. Sign-off
   b. Punch list
   c. Faults list
   d. Checklist

6. Where should you place your company's contact information?
   a. On the contract only
   b. On the proposal and contract only
   c. On your website only
   d. On every document and your website

7. Which of the following describes the scenario of a customer placing a call over the Internet using a web browser?
   a. RNM call
   b. VoIP call
   c. POTS call
   d. PSTN call

8. Which of the following tasks should a homeowner leave to the HTI?
   a. Turning on equipment
   b. Checking cable connections
   c. Configuring IP addresses
   d. Changing passwords

9. If there is no power to equipment, which of the following is a possible cause?
   a. Equipment not turned on
   b. Equipment not plugged in
   c. Power failure
   d. All of the above

10. Which of the following is not an example of intermittent power failure?
   a. Surges
   b. Spikes
   c. Brownouts
   d. Faulty power supply

11. Training on the resident security system should include:
   a. a walkthrough.
   b. basic feature programming.
   c. setting passwords.
   d. All of the above

12. A password should include:
   a. only numerical characters.
   b. only alphabetical characters.
   c. a combination of alphabetical and numerical characters.
   d. any word found in the dictionary.

13. Passwords should be set once and never changed.
   a. True
   b. False

14. RNM stands for:
   a. Remote Network Management
   b. Radical Network Management
   c. Repair Network Manager
   d. Required Network Management

15. RNM is used to:
   a. test stereo components.
   b. to signal devices at one location from another location to test for functionality.
   c. to allow users to communicate from different rooms in a residence.
   d. to detect a security breach.

16. What does CTI stand for?

17. What does a web-enabled call center provide?

18. Which document provided to the customer includes the HTI contact information?

19. Which document provides answers to the most common questions?

20. Which document explains important terms?

Answers To Check Your Understanding

Chapter 1
1. b
3. a
5. d
7. c
9. b
11. b
13. c
15. c

Chapter 2
1. a
3. c
5. b
7. a
9. c
11. d

Chapter 3
1. a
3. b
5. c
7. a
9. d
11. c
13. b
15. b

Chapter 4
1. b
3. a
5. a
7. a
9. c
11. a
13. a
15. a
17. a

Chapter 5
1. a
3. b
5. b
7. b

9. a
11. a
13. d
15. a

Chapter 6
1. c
3. c
5. b
7. b
9. d
11. b
13. a
15. a

Chapter 7
1. b
3. d
5. d
7. c
9. b
11. d
13. a
15. d

Chapter 8
1. b
3. c
5. b
7. a
9. b
11. b
13. a
15. d

Chapter 9
1. b
3. a
5. d
7. c
9. d
11. a
13. a
15. c

Chapter 10
1. b
3. b
5. a
7. a
9. a
11. d
13. d
15. c

Chapter 11
1. c
3. c
5. a
7. b
9. b
11. c
13. b
15. a

Chapter 12
1. a
3. b
5. a
7. d
9. a
11. c
13. b
15. d

Chapter 13
1. d
3. a
5. d
7. a
9. d
11. d
13. c
15. b

Chapter 14
1. b
3. a
5. a

Chapter 10
7. c
9. b
11. c
13. b
15. c

Chapter 15
1. b
3. d
5. c
7. b
9. f
11. c
13. b
15. b

Chapter 16
1. b
3. b
5. b
7. b
9. d
11. d
13. b
15. b

Index

AC (alternating current), 245, 247
  communications outlets and, 280–281
  DC and, 18, 234, 242–243, 285
  interference from, 60
  low-voltage media *v.*, 191
  telephone ring and, 92, 108
AC cable (armored cable), 246–247
Access, 3, 21, 380. *See also* Internet access;
  MAC (Media Access Control)
  addresses; Remote access; XCA
  (Extended Conditional Access)
  automation of, 163, 188
  controls for, 180–181
Access Providers (APs), 25, 30, 210
Adapters, 60, 312. *See also* Banjos
  for backward-compatibility, 82–83
  for connectors, 360
  for linking TV/computers, 118
  for power, 238
  for testing, 349
Add-ons, 255
Addresses, 47, 48, 76, 86. *See also* IP
  (Internet protocol) addresses; MAC
  (Media Access Control) addresses;
  Network Address Translation (NAT)
  with lighting automation, 164, 174
  TCP/IP protocol and, 50, 85–86, 267
ADO. *See* Auxiliary Disconnect Outlet
  (ADO)
ADSL (form of DSL), 71, 72
Agilent Technologies, 362, 388
Air Conditioning, 2–3. *See also* HVAC (heat-
  ing, ventilation and air-conditioning)
Air flow, 175–176, 178, 179
Alarm panel, 151
Alarm systems, 2, 144–150. *See also* False
  alarms; Fire alarms; RJ-31x
  cabling/wiring for, 27, 209
  components of, 150–156, 161
  monitoring of, 146, 148, 150
  supervision for, 145–146, 161
  terminations for, 327
  tools for, 15
American Wire Gauge (AWG) system, 198,
  234–235, 246, 247
Ammeters, 366
Ampere, 233–234, 247
Amplifiers, 27, 140, 220–221
  in audio/video distribution, 112–113, 139
  receivers and, 124–125, 170, 325
  televisions and, 125, 139
Analysts, 5
Anixter (cable distributor), 32, 200
ANSI (American National Standards
  Institute), 26, 232

ANSI/NFPA-70: National Electrical Code
  (NEC), 34
ANSI/NFPA-71: Installation, Maintenance,
  and use of Signaling Systems for the
  Central Station, 34
ANSI/NFPA-72: National Fire Alarm
  Code, 34
ANSI/NFPA-75: Protection of Electronic
  Computer/Data Processing
  Equipment, 34
ANSI/TIA/EIA-232, 188, 205
  automation and, 165, 168, 176, 180, 183
  control processor and, 260
ANSI/TIA/EIA-422, 166–167, 188, 205–206
ANSI/TIA/EIA-423, 205
ANSI/TIA/EIA-485, 188, 206
  automation and, 166–167, 176, 183
  control processor and, 259–260
ANSI/TIA/EIA-492AAAA-A/
  AAAC/CAAA, 376
ANSI/TIA/EIA-606, Administration
  Standard for the Telecommunications
  Infrastructure of Commercial
  Buildings, 1993, 26
ANSI/TIA/EIA-607, Commercial Building
  Grounding and Bonding Requirements
  for Telecommunications, 1994, 26, 233
ANSI/TIA/EIA-568-A, Commercial
  Building Telecommunications
  Pathways and Spaces, 1998, 26, 32, 41,
  103–104, 316
ANSI/TIA/EIA-568-A-3, 127–128, 140
ANSI/TIA/EIA-568-B, Commercial Building
  Telecommunications Standard, 2001, 26,
  80, 103–104, 108, 199
ANSI/TIA/EIA-570-A, Residential
  Telecommunications Cabling
  Standards, 41
  on cabling, 199–200, 229, 314
  on jacks, 319
  on outlets, 222–223
  standards and, 26–31, 40, 97, 100, 104,
  108, 128, 139, 355
  on testing, 134
  on twisted-pair cabling, 59
ANSI/TIA/EIA-570-B, Residential
  Telecommunications Cabling
  Standards, 97
Antennas, 133, 327. *See also* Community
  antenna TV
  in A/V distribution, 113, 127, 139
  balun in, 282
  cables/wires and, 196, 197, 258
  for television, 117
Apartments, 27, 40, 377

Appliances, 2, 4, 9, 10, 21
  amperage/wattage of, 235
  CEBus and automation of, 165
  noise and, 196
  PLC systems and, 60
Application layer, 46–50, 88
Applications, 8–9, 50, 263–265, 281–283.
  See also CAL (Common Application
  Language)
Architects, 5, 21
As-built drawings, 225, 306, 308
Asia, 73
Assisted living facilities, 27
Associations, 5, 7–8, 23–24, 40. *See also*
  Organizations; *specific associations*
AT & T, 357
Atheros Communications, 65
@home e-mail service, 53
Attenuation, 194–195, 229, 345
  amplifiers and, 220
  testing for, 20, 134, 226, 354, 365
Attenuators, 139, 170, 195
Audio, 116, 118–119. *See also* Sound
  systems; Stereo systems
  cables for, 27, 35, 227, 321–322
  connections for, 130–131
  network hookups for, 113–114
  testing of, 353
Audio/video (A/V), 2, 21
  components for, 325–327, 339
  customer training on, 391–392
  distribution devices for, 121–122
  fiber-optic cables for, 283
  floor plans for, 111–112, 113, 135–136
  installation/troubleshooting of,
  130–137
  outlets for, 280
  PCs, television and, 10–11
  PLC systems and, 60
  in SMC, 284
  source devices for, 120–121, 130, 139
  wall plates for, 346
  whole-house distribution of, 111–114, 124,
  136, 139
Australia, 73, 80, 167
Authentication, 68, 391. *See also* Passwords
Authority Having Jurisdiction (AHJ), 25, 40
Automation, 2, 6, 27, 163, 188
Automation systems, 163–167, 179,
  183–188, 270
  components and, 333, 340
  control processor for, 252
  malfunctions of, 188, 390
  software for, 264–265
  testing/documentation of, 268–270, 366